RANDOM PROCESSES:

A MATHEMATICAL APPROACH FOR ENGINEERS

RANDOM PROCESSES:

A MATHEMATICAL APPROACH FOR ENGINEERS

ROBERT M. GRAY

Electrical Engineering Department
Stanford University
Stanford, Calif.

LEE D. DAVISSON

Electrical Engineering Department
University of Maryland
College Park, Md.

Prentice-Hall, Inc., Englewood Cliffs, New Jersey 07632

Library of Congress Cataloging in Publication Data

Gray, Robert M.
    Random processes.

    Bibliography: p.
    1. Stochastic processes. I. Davisson, Lee D.
II. Title.
QA274-G73 1986   519.2   84-26459
ISBN 0-13-752882-**5**

Editorial/production supervision: *Mary Carnis*
Cover design: *20/20 Services, Inc.*
Manufacturing buyer: *Anthony Caruso and Rhett Conklin*

Printed in the United States of America

10 9 8 7 6 5 4 3 2 1

ISBN 0-13-752882-5 01

Prentice-Hall International (UK) Limited, *London*
Prentice-Hall of Australia Pty. Limited, *Sydney*
Prentice-Hall Canada Inc., *Toronto*
Prentice-Hall Hispanoamericana, S.A., *Mexico*
Prentice-Hall of India Private Limited, *New Delhi*
Prentice-Hall of Japan, Inc., *Tokyo*
Prentice-Hall of Southeast Asia Pte. Ltd., *Singapore*
Editora Prentice-Hall do Brasil, Ltda., *Rio de Janeiro*
Whitehall Books Limited, *Wellington, New Zealand*

to our families

CONTENTS

PREFACE

Nothing in nature is random... A thing appears random only through the incompleteness of our knowledge. —Spinoza, *Ethics I*

I do not believe that God rolls dice. —Einstein

One can argue that given complete knowledge of the physics of an experiment, the outcome must always be predictable, at least with the aid of a sufficiently powerful computer. This metaphysical argument must be tempered with several facts. The relevant parameters may not be measurable with sufficient precision due to mechanical or theoretical limits. For example, the uncertainty principle prevents the simultaneous accurate knowledge of both position and momentum. The deterministic functions may be too complex to compute in finite time. The computer itself may make errors due to power failures, lightning, or the general perfidy of inanimate objects. The experiment could take place in a remote location with the parameters unknown to the observer; for example, in a communication link, the transmitted message is unknown *a priori*, for if it were not, there would be no need for communication. The results of the experiment could be reported by an unreliable witness—either incompetent or dishonest. For these and other reasons, it is useful to have a theory for the analysis and synthesis of processes that behave in a random or unpredictable manner. The goal is to construct mathematical models that lead to reasonably accurate prediction of the long-term average behavior of random processes. The theory should produce good estimates of the average behavior of real processes and thereby connect theoretical derivations with measurable results.

In this book we attempt a development of the basic theory and applications of random processes that uses the language and viewpoint of rigorous mathematical treatments of the subject but which requires only a typical bachelor's degree level of electrical engineering education including elementary linear systems theory, elementary probability, and transform theory and applications. Detailed proofs are presented only when within the scope of this background. These simple proofs, however, often provide the groundwork for "handwaving" justifications of more general and complicated results that are semi-rigorous in that they can be made rigorous by the appropriate delta-epsilontics of real analysis or measure theory. A primary goal of this approach is thus to use intuitive arguments that accurately reflect the underlying mathematics and which will hold up under future scrutiny if the student continues to more advanced courses. Another goal is to enable the student who might not continue to more advanced courses to be able to read and generally follow the modern literature on applications of random processes to information and communication theory, estimation and detection, control, signal processing, and stochastic systems theory.

ACKNOWLEDGMENTS

Both of us would like to thank our universities, Stanford University and the University of Maryland, for the financial support of their sabbatical leaves, during which much of this book was written. The first author would also like to thank the John Simon Guggenheim Memorial Foundation for its support and encouragement of related research and of the writing of this book. He would also like to thank the Industrial Affiliates Program of the Information Systems Laboratory of Stanford University for providing computer facilities, both mini and micro, which greatly facilitated the writing of this book. We gratefully acknowledge the extensive comments and corrections of the students who suffered through the earlier drafts of this book and of the anonymous referees who reviewed it. Thanks also are due to Dr. Richard Blahut for his careful reading and numerous helpful suggestions.

Special thanks are owed to our wives and the rest of our families for their support and perseverance through the many years that this project has taken.

Robert M. Gray
Lee D. Davisson

RANDOM PROCESSES:

A MATHEMATICAL APPROACH
FOR ENGINEERS

1

INTRODUCTION

A random or stochastic process is a mathematical model for a phenomenon that evolves in time in an unpredictable manner from the viewpoint of the observer. The phenomenon may be a sequence of real-valued measurements of voltage or temperature, a binary data stream from a computer, a modulated binary data stream from a modem, a sequence of coin tosses, the daily Dow-Jones average, radiometer data or photographs from deep space probes, a sequence of images from a cable television, and so on. It may be unpredictable due to such effects as interference or noise in a communication link or storage medium, or it may be an information-bearing signal—deterministic from the viewpoint of an observer at the transmitter but random to an observer at the receiver.

The theory of random processes quantifies the above notions so that one can construct mathematical models of real phenomena that are both tractable and meaningful. Tractability is required in order for the engineer (or anyone else) to be able to perform analyses and syntheses of random processes, perhaps with the aid of computers. The "meaningful" requirement is that the models provide a reasonably good approximation of the actual phenomena. An oversimplified model may provide results and conclusions that do not apply to the real phenomenon being modeled. Perhaps the most distinguishing characteristic between an average engineer and an outstanding engineer is the ability to derive effective models.

Courses and texts on the theory and application of random processes usually fall into either of two general and distinct categories. One category is the common engineering approach, which involves fairly elementary probability theory, standard undergraduate Riemann calculus, and a large dose of "cookbook" formulas—often with little attention paid to conditions under which the formulas are valid. The results are often justified by nonrigorous and occasionally mathematically inaccurate handwaving or intuitive plausibility arguments that may not reflect the actual underlying mathematical structure and may not be supportable by a precise proof. While intuitive arguments can be extremely valuable in providing insight into deep theoretical results, they can be a handicap if they do not capture the essence of a rigorous proof.

A development of random processes that is insufficiently mathematical leaves the student ill prepared to generalize the techniques and results when faced with a real-world example not covered in the text. For example, if one is faced with the problem of designing signal processing equipment for predicting or communicating measurements being made for the first time by a space probe, how does one construct a mathematical model for the physical process that will be useful for analysis?

An additional problem with an insufficiently mathematical development is that it does not leave the student adequately prepared to read modern research literature. The more advanced mathematical language of recent work is increasingly used even in simple cases because it is precise and universal and focuses on the structure common to all random processes. Even if an engineer is not directly involved in research, knowledge of the current literature can often provide useful ideas and techniques for tackling specific problems. Engineers unfamiliar with basic terms such as *sigma field* and *conditional expectation* will find many potentially valuable references shrouded in mystery.

The other category of courses and texts on random processes is the typical mathematical approach, which requires an advanced mathematical background of real analysis, measure theory, and integration theory; it involves precise and careful theorem statements and proofs, and it is far more careful to specify precisely the conditions required for a result to hold. Most engineers do not, however, have the required mathematical background, and the extra care required in a completely rigorous development severely limits the number of topics that can be covered in a typical course—in particular, the applications that are so important to engineers tend to be neglected. In addition, too much time can be spent with the formal details, obscuring the often simple and elegant ideas behind a proof. Little, if any, physical motivation for the topics is usually given.

This book attempts a compromise between the two approaches by giving the basic, elementary theory and a profusion of examples in the language and notation of the more advanced mathematical approaches. The intent is to make the crucial concepts clear in the traditional elementary cases, such as coin flipping, and thereby to emphasize the mathematical structure of all random processes in the simplest possible context. The structure is then further developed by numerous increasingly complex examples of random processes that have proved useful in stochastic systems analysis. Careful proofs are constructed only in elementary cases. For example, the fundamental theorem of expectation is proved only for discrete random variables where it is proved simply by a change of variables in a sum. The continuous analog is subsequently given without a careful proof but with the explanation that it is simply the integral analog of the summation formula and hence can be viewed as a limiting form of the discrete result.

By this means we strive to capture the spirit of important proofs without undue tedium and to make plausible the required assumptions and constraints. This, in turn, should aid the student in determining when certain tools do or do not apply and what additional tools might be necessary when new generalizations are required.

A distinct aspect of the mathematical viewpoint is the "grand experiment" view of random processes as being a probability measure on sequences (for discrete time) or waveforms (for continuous time) rather than being an infinity of smaller experiments representing individual outcomes (called random variables) that are somehow glued together. From this point of view random variables are merely special cases of random processes. In fact, the grand experiment viewpoint was popular in the early days of applications of random processes to systems and was called the "ensemble" viewpoint (e.g., in the work of Norbert Wiener and his students). By viewing the random process as a whole instead of as a collection of pieces, many basic ideas, such as stationarity and ergodicity, that characterize the dependence on time of probabilistic descriptions and the relation between time averages and probabilistic averages are much easier to define and study. This also permits a more complete discussion of processes that violate such probabilistic regularity requirements yet still have useful relations between time and probabilistic averages.

Even though a student completing this book will not be able to follow the details in the literature of many proofs of results involving random processes, the basic results and their development and implications should be fairly accessible, and the most common examples of random processes and classes of random processes should be fairly familiar. In particular, the student should be well equipped to follow the gist of most arguments in the various *IEEE Transactions* dealing with random processes.

It also should be mentioned that the authors are electrical engineers and, as such, have written this text with an electrical engineering flavor. However, the required knowledge of classical electrical engineering is slight, and engineers in other fields should be able to follow the material presented.

This book is intended to provide a one-quarter or one-semester course that develops the basic ideas and language of the theory of random processes and provides a rich collection of examples of commonly encountered processes, properties, and calculations. Although in some cases these examples may seem somewhat artificial, they are chosen to illustrate the way engineers should think about random processes and for simplicity and conceptual content rather than to present the method of solution to some particular application. *Sections that can be skimmed or omitted for the shorter one-quarter curriculum are marked with an asterisk.* Discrete time processes are given more emphasis than in many texts because they are simpler to handle and because they are of increasing practical importance in sampled data and computer systems. For example, linear filter input/output relations are carefully developed for discrete time and then the continuous time analogs are obtained by replacing sums with integrals.

Most examples are developed by beginning with simple processes and then filtering or modulating them to obtain more complicated processes. This provides many examples of typical probabilistic computations and introduces several examples of modeling complicated processes as the output of operations on simple processes. The examples are drawn from the areas of communications, signal processing, systems theory, and estimation. Extra tools are introduced as needed to develop properties of the examples.

The prerequisites for this book are elementary set theory, elementary probability, and some familiarity with linear systems theory (Fourier analysis, convolution, linear filters, and transfer functions).

It is possible for more capable students to follow the material presented without prior knowledge of probability theory since the fundamental ideas of probability theory are embedded within the framework of random processes. Prior experience with elementary probability, however, can be a valuable aid for building intuition of the more advanced (and likely more rapid) development of this book than would be encountered in an elementary probability course.

ORGANIZATION OF THE BOOK

Chapter 2 sketches several prerequisite definitions and concepts from elementary set theory and linear systems theory using examples to be encountered later in the book. The first subject is crucial at an early stage and should be reviewed before proceeding to chapter 3. The second subject is not required until chapter 8, but it serves as a reminder of material that the student should already be familiar with. Elementary probability is not reviewed, as our basic development includes elementary probability.

This review of prerequisite material serves to collect together some notations and many definitions that will be used throughout the book. It is, however, only a brief review and cannot serve as an adequate substitute for a complete course on the material. This chapter can be given as a first reading assignment and either skipped or briefly skimmed in class; lectures can proceed from an introduction, perhaps incorporating some preliminary material, directly to chapter 3.

Chapter 3 provides a careful development of the fundamental concept of probability theory—a probability space or experiment. The notions of sample space, event space, and probability measure are introduced, and several examples are toured.

Chapter 4 treats the theory of measurements made on experiments: random variables, which are scalar-valued measurements; random vectors, which are a vector or finite collection of measurements; and random processes, which can be viewed as sequences or waveforms of measurements. These measurements or random objects are shown to inherit their own probabilistic description and structure from the underlying experiment and the form of the measurement. As a result, many of the basic properties of random variables, vectors, and processes follow from those of probability spaces. Probability distributions are introduced and used to consider elementary conditional probability and independence. Many simple and complicated examples of random variables and vectors are introduced, but only simple random processes are considered in this chapter.

Chapter 5 develops the idea of a specification or description of a random process through its joint probability distributions and thereby permits the introduction of many more advanced examples of random processes, including independent identically distributed (*i.i.d.*) random processes and the ubiquitous Gaussian random process. Included in the development are the fundamentals of elementary conditional probability and independence. There are usually many ways to describe a given random process. Specification provides a canonical description that is often, but not always, the most useful description for computation.

Chapter 6 begins with a discussion of the basic limit theorems of random processes. These results are the foundation results of random processes since they relate the probabilistic computations of probability theory to the long-term behavior that one can expect from a random process in the real world. A somewhat cavalier development of the possible behavior of long-term sample or time averages of measurements motivates the introduction of the idea of the expectation or probabilistic average of a measurement. After some discussion of basic properties and examples of expectation, the remainder of the chapter and the following chapter are devoted to results relating the time-average behavior of a random process to the expectation of the process, results known as laws of large numbers or ergodic theorems. In these two chapters two particular examples of expectation—the mean and covariance of a random process—are seen to play a fundamental role in determining the long-term average behavior of a random process. Chapter 8 is devoted to a more detailed study of these examples, called the second-order moments of a random process.

Chapter 8 concentrates on the computation of second-order moments—the mean and covariance—of a variety of random processes. The primary example is a form of derived distribution problem: Suppose that a given random process with known second-order moments is put into a linear system. What are the second-order moments of the resulting output random process? This problem is treated for linear systems represented by convolutions and for linear modulation systems. Transform techniques are shown to often provide a simplification in the computations, much like their ordinary role in elementary linear systems theory. The chapter closes with a development of several simple results from the theory of linear least-squares estimation. This provides an example of both the computation and the application of second-order moments.

Chapter 9 takes a step toward a more complete probabilistic description of the output of a linear system driven by a known random process; that is, it develops more properties than just the second-order moments. To accomplish this, however, one must make additional assumptions on the input process and on the form of the linear filter. The general model of a linear filter driven by a memoryless process is used to develop several popular models of discrete time random processes. Analogous continuous time random process models are then developed by direct description of their behavior. The basic class of random processes considered is the class of independent increment processes, but other processes with similar definitions but quite different properties are also introduced. Among the models considered are autoregressive processes, moving-average processes, ARMA (autoregressive-moving average) processes, counting processes, random walks, independent increment processes, Markov processes, Poisson and Gaussian processes, and the random telegraph wave. The development provides a setting for the discussion of characteristic functions—transforms of probability distribution functions—and nonelementary conditional probability. We also briefly consider an example of a nonlinear system where the output random processes can at least be partially described—the exponential function of a Gaussian or Poisson process which models phase or frequency modulation.

Chapter 10 expands on the ideas of chapter 9 and considers the class of compound random processes, a type of "doubly stochastic" random process formed by taking linear combinations of a random number of samples of an input random process. This type of process provides a motivation and application for the concept of conditional expectation.

Chapter 11 provides a sketch of the development of the Gaussian and Poisson distributions from underlying physical assumptions. These ideas are combined with a bit of physics to sketch the development of the mathematical model for thermal noise, one of the most important noise sources in modern communication theory. The chapter is out of the mainstream of this book, but it is useful to show the origins of some of the random process models considered.

The book concludes with a section on supplementary reading and occasional historical notes for each chapter. We assemble in that section references on additional background material as well as on books that pursue the various topics in more depth or on a more advanced level. We feel that these comments and references are supplementary to the development and that less clutter results by putting them in a single appendix rather than strewing them throughout the text. The section is intended as a guide for further study, not as an exhaustive description of the relevant literature, the latter goal being beyond the authors' interests and stamina.

2

PRELIMINARIES:
SET THEORY, MAPPINGS,
AND LINEAR SYSTEMS

The theory of random processes is constructed on a large number of abstractions. These abstractions are necessary to achieve generality with precision while keeping the notation used manageably brief. The student will probably find learning facilitated if, with each abstraction, he keeps in mind (or on paper) a concrete picture or example of a special case of the abstraction. From this the general situation should rapidly become clear. Concrete examples and exercises are introduced throughout the book to help with this process.

SET THEORY

In this section the basic set theoretic ideas that are used throughout the book are introduced. The starting point is an *abstract space*, or simply a *space*, consisting of *elements* or *points*, the smallest quantities with which we shall deal. This space, often denoted by Ω, is sometimes referred to as the universal set. To describe a space we may use braces notation with either a list or a description contained within the braces { }. Examples are:

[2.1]

The abstract space with only the two elements *zero* and *one* to denote the possible receptions of a radio receiver of binary data at one particular signaling time instant. Equivalently, we could give different names to the elements and have a space {0,1}, the binary numbers, or a space with the elements *heads* and *tails*. Clearly the structure of all of these spaces is the same; only the names have been changed.   Notationally   we   describe   these   spaces   as   {*zero,one*},   {0,1},   and {*heads,tails*}, respectively.

[2.2]

Given a fixed positive integer k, the abstract space consisting of all possible binary k-tuples, that is, all 2^k k-dimensional binary vectors. This space could model the possible sequences of k flips of the same coin or a single flip of k coins. Note that example [2.1] is a special case of example [2.2].

[2.3]

The abstract space with elements consisting of all infinite sequences of *ones* and *zeros* or 1's and 0's denoting the sequence of possible receptions of a radio receiver of binary data over all signaling times. The sequences could be single-sided (or one-sided or unilateral) in the sense of beginning at time zero and continuing forever, or they could be double-sided (or doubly infinite or bilateral) in the sense of beginning in the infinitely remote past (time $-\infty$) and continuing into the infinitely remote future.

[2.4]

The abstract space consisting of all ASCII (American Standard Code for Information Interchange) codes for characters (letters, numerals, and control characters such as line feed, rub out, etc.). These might be in decimal, hexadecimal, or binary form. In general, we can consider this space as just a space $\{a_i, i = 1,...,n\}$ containing a finite number of elements (which here might well be called symbols, letters, or characters).

[2.5]

Given a fixed positive integer k, the space of all k-dimensional vectors with coordinates in the space of example [2.4]. This could model all possible contents of an ASCII buffer used to drive a serial printer.

[2.6]

The abstract space of all infinite (single-sided or double-sided) sequences of ASCII character codes.

[2.7]

The abstract space with elements consisting of all possible voltages measured at the output of a radio receiver at one instant of time. Since all physical equipment has limits to the values of voltage (called "dynamic range") that it can support, one model for this space is a subset of the real line such as the closed interval $[-V,V]$ $= \{r: -V \leq r \leq V\}$, i.e., the set of all real numbers r such that $-V \leq r \leq +V$. If, however, the dynamic range is not precisely known or if we wish to use a single space as a model for several measurements with different dynamic ranges, then we might wish to use the entire real line $\mathbf{R} = (-\infty,\infty) = \{r: -\infty < r < \infty\}$. The fact that the space includes "impossible" as well as "possible" values is acceptable in a model.

[2.8]

Given a positive integer k, the abstract space of all k-dimensional vectors with coordinates in the space of example [2.7]. If the real line is chosen as the coordinate space, then this is k-dimensional Euclidean space.

[2.9]

The abstract space with elements being all infinite sequences of members of the space of example [2.7], e.g., all single-sided real-valued sequences of the form $\{x_n, n = 0,1,2,...\}$, where $x_n \in \mathbf{R}$ for all $n = 1,2,....$

[2.10]

Instead of constructing a new space as sequences of elements from another space, we might wish to consider a new space consisting of all waveforms whose instantaneous values are elements in another space, e.g., the space of all waveforms $\{x(t); t \in (-\infty,\infty)\}$, where $x(t) \in \mathbf{R}$, all t. This would model, for instance, the space of all possible voltage-time waveforms at the output of a radio receiver. Examples of members of this space are $x(t) = \cos \omega t$, $x(t) = e^{st}$, $x(t) = 1$, $x(t) = t$, and so on. As with sequences, the waveforms may begin in the remote past or they might be defined for t running from 0 to ∞.

The preceding examples focus on three related themes that will be considered throughout the book: Examples [2.1], [2.4], and [2.7] present models for the possible values of a single measurement. The mathematical model for such a measurement with an unknown outcome is called a *random variable*. Such simple spaces describe the possible values that a random variable can assume. Examples [2.2], [2.5], and [2.8] treat vectors (finite collections or finite sequences) of individual measurements. The mathematical model for such a vector-valued measurement is called a *random vector*. Since a vector is made up of a finite collection of scalars, we can also view this random object as a collection (or family) of random variables. These two viewpoints—a single random vector-valued measurement and a collection of random scalar-valued measurements—will both prove useful. Examples [2.3], [2.6], and [2.9] consider infinite sequences of values drawn from a common alphabet and hence the possible values of an infinite sequence of individual measurements. The mathematical model for this is called a *random process* (or a *random sequence* or a *random time series*). Example [2.10] considers a waveform taking values in a given coordinate space. The mathematical model for this is also called a *random process*. When it is desired to distinguish between random sequences and random waveforms, the first is called a *discrete time random process* and the second is called a *continuous time random process*.

In chapter 4 we shall define precisely what is meant by a random variable, a random vector, and a random process. For now, random variables, random vectors, and random processes can be viewed simply as abstract spaces such as in the preceding examples for scalars, vectors, and sequences or waveforms together with a probabilistic description of the possible outcomes; that is, a means of quantifying how likely certain outcomes are. It is a crucial observation at this point that the three notions are intimately connected: Random vectors and processes can be viewed as collections or families of random variables. Conversely, we can obtain the scalar random variables by observing the coordinates of a random vector or random process. That is, if we "sample" a random process once, we get a random variable. Thus we shall often be interested in several different, but related, abstract spaces. For example, the individual scalar outputs may be drawn from one space, say A, which could be any of the spaces in examples [2.1], [2.4], or [2.7]. We then may also wish to look at all possible k-dimensional vectors with coordinates in A, a space that is often denoted by A^k, or at spaces of infinite sequences or waveforms of A. These latter spaces are called *product spaces* and will play an important role in modeling random phenomena.

Usually one will have the option of choosing any of a number of spaces as a model for the outputs of a given random variable. For example, in flipping a coin one could use the binary space {*head*,*tail*}, the binary space {0,1} (obtained by assigning 0 to *head* and 1 to *tail*), or the entire real line **R**. Obviously the last space is much larger than needed, but it still captures all of the possible outcomes (along with many "impossible" ones). Which view and which abstract space is the "best" will depend on the problem at hand, and the choice will usually be made for reasons of convenience.

Given an abstract space, we shall consider groupings or collections of the elements that may be (but are not necessarily) smaller than the whole space and larger than single points. Such groupings are called *sets*. Examples (corresponding respectively to the previous abstract space examples) are:

[2.11]

The set consisting of the single element *one*.

[2.12]

The set of all k-dimensional binary vectors with exactly one *zero* coordinate.

[2.13]

The set of all infinite sequences of *one*s and *zero*s with exactly 50% of the symbols being *one* (as defined by an appropriate mathematical limit).

[2.14]

The set of all ASCII characters for capital letters.

[2.15]

The set of all four-letter English words.

[2.16]

The set of all infinite sequences of ASCII characters excluding those representing control characters.

[2.17]

Intervals such as the set of all voltages lying between 1 volt and 20 volts are useful subsets of the real line. These come in several forms, depending on whether or not the end points are included. Given $b > a$, define the "open" interval $(a,b) = \{r: a < r < b\}$, and given $b \geq a$, define the "closed" interval $[a,b] = \{r: a \leq r \leq b\}$. That is, we use a bracket if the end point is included and a parenthesis if it is not. We will also consider "half-open" or "half-closed" intervals of the form $(a,b] = \{r: a < r \leq b\}$ and $[a,b) = \{r: a \leq r < b\}$.

[2.18]

The set of all vectors of k voltages such that the largest value is less than 1 volt.

[2.19]

The set of all sequences of voltages which are all nonnegative.

[2.20]

The set of all voltage-time waveforms that lie between 1 and 20 volts for all time.

Given a set F of points in an abstract space Ω, we shall write $\omega \epsilon F$ for "the point ω is contained in the set F" and $\omega \notin F$ for "the point ω is not contained in the set F." The symbol ϵ is referred to as the *element inclusion symbol*. We shall often describe a set using this notation in the form $F = \{\omega: \omega$ has some property$\}$. Thus $F = \{\omega: \omega \epsilon F\}$. For example, a set in the abstract space $\Omega = \{\omega: -\infty < \omega < \infty\}$ (the real line **R**) is $\{\omega: -2 \leq \omega < 4.6\}$. The abstract space itself is a grouping of elements and hence is also called a set. Thus $\Omega = \{\omega: \omega \epsilon \Omega\}$.

If a set, say F, is included within another set, say G, that is, if every point in F is also contained in G, then F is said to be a *subset* of G, and we write $F \subset G$. The symbol $\subset$ is called the *set inclusion symbol*. Since a set is included within itself, every set is a subset of itself.

We shall denote by $\varnothing$ the empty set, that is, the set of no points at all. The empty set is considered to be a subset of all other sets by mathematical convention.

An individual element or point ω_0 in F can be considered both as a point or element in the space and as a one-point set or single-element or singleton set $\{\omega_0\} = \{\omega: \omega = \omega_0\}$. Note, however, that the braces notation is more precise when we are considering the one-point set and that $\omega_0 \epsilon \Omega$ while $\{\omega_0\} \subset \Omega$.

The three basic operations on sets are complementation, intersection, and union. The definitions are given next. Refer also to Figure 2.1 as an aid in visualizing the definitions. In Figure 2.1 Ω is pictured as the outside box and the sets F and G are pictured as arbitrary blobs within the box. Such diagrams are called *Venn diagrams*.

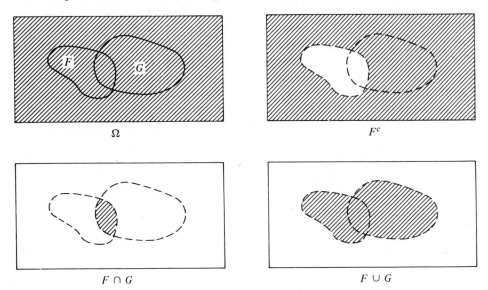

Ω F^c

$F \cap G$ $F \cup G$

Figure 2.1 Basic set operations.

Given a set F, the *complement* of F is denoted by F^c, which is defined by

$$F^c = \{\omega: \omega \notin F\},$$

that is, the complement of F contains all of the points of Ω that are not in F.

Given two sets F and G, the *intersection* of F and G is denoted by $F \cap G$, which is defined by

$$F \cap G = \{\omega: \omega \epsilon F \text{ and } \omega \epsilon G\},$$

that is, the intersection of two sets F and G contains the points which are in both sets.

If F and G have no points in common, then $F \cap G = \varnothing$, the null set, and F and G are said to be *disjoint* or *mutually exclusive*.

Given two sets F and G, the *union* of F and G is denoted by $F \cup G$, which is defined by

$$F \cup G = \{\omega: \omega \epsilon F \text{ or } \omega \epsilon G\},$$

that is, the union of two sets F and G contains the points that are either in one set or the other, or both.

Observe that the intersection of two sets is *always* a subset of each of them, e.g., $F \cap G \subset F$. The union of two sets, however, is not a subset of either of them (unless one set is a subset of the other). Both of the original sets are subsets of their union, e.g., $F \subset F \cup G$.

In addition to the three basic set operations, there are two others that will come in handy. Both can be defined in terms of the three basic operations. Refer to Figure 2.2 as a visual aid in understanding the definitions.

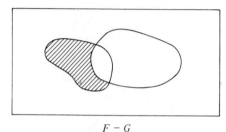
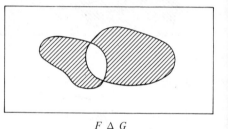

$$F - G \qquad\qquad\qquad\qquad\qquad F \triangle G$$

Figure 2.2 Set differences.

Given two sets F and G, the *set difference* of F and G is denoted by $F - G$, which is defined as

$$F - G = \{\omega: \omega \epsilon F \text{ and } \omega \notin G\} = F \cap G^c$$

that is, the difference of F and G contains all of the points in F that are not also in G. Note that this operation is not completely analogous to the "minus" of ordinary arithmetic because there is no such thing as a "negative set."

Given two sets F and G, their *symmetric difference* is denoted by $F \triangle G$, which is defined as

$$F \triangle G = \{\omega: \omega \epsilon F \text{ or } \omega \epsilon G \text{ but not both }\}$$

$$= (F - G) \cup (G - F) = (F \cap G^c) \cup (F^c \cap G)$$

$$= (F \cup G) - (F \cap G),$$

that is, the symmetric difference between two sets is the set of points that are in one of the two sets but are not common to both sets. If both sets are the same, the symmetric difference consists of no points, that is, it is the empty set. If $F \subset G$, then obviously $F \triangle G = G - F$.

Observe that two sets F and G will be equal if and only if $F \subset G$ and $G \subset F$. This observation is often useful as a means of proving that two sets are identical: First prove that each point in one set is in the other and hence the first set is a subset of the second. Then prove the opposite inclusion. Surprisingly, this technique is frequently much easier than a direct proof that two sets are identical by a pointwise argument of commonality.

We will often wish to combine sets in a series of operations and to reduce the expression for the resulting set to its simplest and most compact form. Although the most compact form frequently can be seen quickly with the aid of a Venn diagram, as in Figures 2.1 and 2.2, to be completely rigorous, the use of set theory or set algebra to manipulate the basic operations is required. Table 2.1 collects the most important such identities. The first seven relations can be taken as axioms in an algebra of sets and used to derive all other relations, including the remaining relations in the table. Some examples of such derivations follow the table. Readers who are familiar with Boolean algebra will find a one-for-one analogy between the algebra of sets and Boolean algebra.

DeMorgan's "laws" (2.6) and (2.10) are useful when complementing unions or intersections. Relation (2.16) is useful for writing the union of overlapping sets as a union of disjoint sets. A set and its complement are always disjoint by relation (2.5).

Examples of Proofs

Relation (2.8). From the definition of intersection and Figure 2.1 we verify the truth of (2.8). Algebraically, we show the same thing from the basic seven axioms: From (2.4) and (2.6) we have that

$$A \cap B = ((A \cap B)^c)^c = (A^c \cup B^c)^c,$$

and using (2.1), (2.4), and (2.6), this becomes

$$(B^c \cup A^c)^c = (B^c)^c \cap (A^c)^c$$

as desired.

Relation (2.18). Set $F = \Omega$ in (2.5) to obtain $\Omega \cap \Omega^c = \varnothing$, which with (2.7) and (2.8) yields (2.18).

Relation (2.11). Complement (2.5), $(F^c \cap F)^c = \varnothing^c$, and hence, using (2.6), $F^c \cup F = \varnothing^c$, and finally, using (2.4) and (2.18), $F^c \cup F = \Omega$.

Relation (2.12). Using F^c in (2.7): $F^c \cap \Omega = F^c$. Complementing the result: $(F^c \cap \Omega)^c = (F^c)^c = F$ (by (2.4)). Using (2.6): $(F^c \cap \Omega)^c = F \cup \Omega^c = F$. From (2.18) $\Omega^c = \varnothing$, yielding (2.12).

TABLE 2.1 Useful Set Theoretic Relations

$F \cup G = G \cup F$ (*commutative law for union*)   (2.1)

$F \cup (G \cup H) = (F \cup G) \cup H$ (*associative law* for *union*)   (2.2)

$F \cap (G \cup H) = (F \cap G) \cup (F \cap H)$ (2.3)

(*distributive law — intersection over union*)

$(F^c)^c = F$ (2.4)

$F \cap F^c = \varnothing$ (2.5)

$(F \cap G)^c = F^c \cup G^c$ (*DeMorgan's "law"*)   (2.6)

$F \cap \Omega = F$ (2.7)

$F \cap G = G \cap F$ (*commutative law for intersection*)   (2.8)

$F \cap (G \cap H) = (F \cap G) \cap H$ (*associative law for intersection*)   (2.9)

$(F \cup G)^c = F^c \cap G^c$ (*DeMorgan's other "law"*)   (2.10)

$F \cup F^c = \Omega$ (2.11)

$F \cup \varnothing = F$ (2.12)

$F \cup (F \cap G) = F = F \cap (F \cup G)$ (2.13)

$F \cup \Omega = \Omega$ (2.14)

$F \cap \varnothing = \varnothing$ (2.15)

$F \cup G = F \cup (F^c \cap G) = F \cup (G - F)$ (2.16)

$F \cup (G \cap H) = (F \cup G) \cap (F \cup H)$ (2.17)

(*distributive law — union over intersection*)

$\Omega^c = \varnothing$ (2.18)

$F \cup F = F$ (2.19)

$F \cap F = F$ (2.20)

Relation (2.20). Set $G = F$ and $H = F^c$ in (2.3) to obtain $F \cap (F \cup F^c) = (F \cap F) \cup (F \cap F^c) = F \cap F$ using (2.5) and (2.12). Applying (2.11) and (2.7) to the left-hand side of this relation yields $F \cap \Omega = F = F \cap F$.

Relation (2.19). Complement (2.20) using (2.6) and replace F^c by F.

We will have occasion to deal with more general unions and intersections, that is, unions or intersections of more than two or three sets. As long as the number of unions and intersections is finite, the generalizations are obvious. However, we can define these operations for quite general infinite collections of sets as well (where the generalizations are not so obvious). Say that we have an indexed collection of sets $\{A_i;\ i \in \mathbf{I}\}$, sometimes denoted $\{A_i\}_{i \in \mathbf{I}}$, for some index set $\mathbf{I}$. In other words, this collection is a set whose elements are sets—one set A_i for each possible value of an index i drawn from $\mathbf{I}$. We call such a collection a *family* or *class* of sets. (To avoid confusion we never say a "set of sets.") The index set $\mathbf{I}$ can be thought of as numbering the sets. Typical index sets are the set $\mathbf{Z}_+$ of all nonnegative integers, $\mathbf{Z}_+ = \{0,1,2,...\}$, the set $\mathbf{Z}$ of all integers, $\mathbf{Z} = \{...,-1,0,1,...\}$, or the real line $\mathbf{R}$. The index set may be finite in that it has only a finite number of entries, say $\mathbf{I} = \{0,1,...,k-1\}$. The index set is said to be *countably infinite* if its elements can be counted, that is, can be put into a one-to-one correspondence with a subset of the nonnegative integers $\mathbf{Z}_+$; e.g., $\mathbf{Z}_+$ or $\mathbf{Z}$ itself. If an index set has an infinity of elements, but the elements cannot be counted, then it is said to be uncountably infinite, for example $\mathbf{R}$ or the unit interval $[0,1]$ (see exercise 2.11).

The family of sets is said to be finite, countable, or uncountable if the respective index set is finite, countable, or uncountable. As an example, the family of sets $\{[0,1/r)\ ;\ r \in \mathbf{I}\}$ is countable if $\mathbf{I} = \mathbf{Z}$ and uncountable if $\mathbf{I} = \mathbf{R}$.

The obvious extensions of the pairwise definitions of union and intersection will now be given. Given an indexed family of sets $\{A_i; i \in \mathbf{I}\}$ define the union by

$$\bigcup_{i \in \mathbf{I}} A_i = \{\omega: \omega \in A_i \text{ for at least one } i \in \mathbf{I}\}$$

and define the intersection by

$$\bigcap_{i \in \mathbf{I}} A_i = \{\omega: \omega \in A_i \text{ for all } i \in \mathbf{I}\}.$$

In certain special cases we shall make the notation more specific for particular index sets. For example, if $\mathbf{I} = \{0,...,n-1\}$, then we write the union and intersection as

$$\bigcup_{i=0}^{n-1} A_i \text{ and } \bigcap_{i=0}^{n-1} A_i ,$$

respectively.

A collection of sets $\{F_i; i \in \boldsymbol{I}\}$ is said to be *disjoint* (or *pairwise disjoint* or *mutually exclusive*) if

$$F_i \cup F_j = \varnothing; \text{ all } i, j \in \boldsymbol{I}, i \neq j ,$$

that is, if no sets in the collection contain points contained by other sets in the collection.

The class of sets is said to be *collectively exhaustive* or to *exhaust* the space if

$$\bigcup_{i \in \boldsymbol{I}} F_i = \Omega ,$$

that is, together the F_i contain all the points of the space.

A collection of sets $\{F_i; i \in \boldsymbol{I}\}$ is called a *partition* of the space Ω if the collection is both disjoint and collectively exhaustive. A collection of sets $\{F_i; i \in \boldsymbol{I}\}$ is said to partition a set G if the collection is disjoint and the union of all of its members is identical to G.

MAPPINGS AND FUNCTIONS

We shall make much use of mappings or functions from one space to another. This is of importance in a number of applications. For example, the waveforms and sequences that we considered as members of an abstract space describing the outputs of a random process are just func-tions of time, e.g., for each value of time t in some continuous or discrete collection of possible times we assigned some output value to the function. As a more complicated example, consider a binary digit that is transmitted to a receiver at some destination by sending either plus or minus V volts through a noisy environment called a "channel." At the receiver a decision is made whether $+V$ or $-V$ was sent. The receiver puts out a 1 or a 0, depending on the decision. In this example three mappings are involved: The transmitter maps a binary symbol in $\{0,1\}$ into either $+V$ or $-V$. During transmission, the channel has an input either $+V$ or $-V$ and pro-duces a real number, not usually equal to $0,1$, $+V$, or $-V$. At the receiver, a real number is viewed and a binary number produced.

We will encounter a variety of functions or mappings, from simple arithmetic operations to general filtering operations. We now introduce some common terminology and notation for handling such functions. Given two abstract spaces Ω and A, an A-valued *function* or *mapping* f or, in more detail, $f: \omega \to A$ is an assignment of a unique point in A to each point in Ω; that is, given any point $\omega \in \Omega$, $f(\omega)$ is some value in A. Ω is called the *domain* of the function f, A is called the *range* of f. Given any sets $F \subset \Omega$ and $G \subset A$, define the *image* of F (under f) as the set

$$f(F) = \{a: a = f(\omega) \text{ for some } \omega \in F\}$$

and the *inverse image* of G (under f) as the set

$$f^{-1}(G) = \{\omega : f(\omega) \in G\} .$$

Thus $f(F)$ is the set of all points in A obtained by mapping points in F, and $f^{-1}(G)$ is the set of all points in Ω that map into G.

For example, let $\Omega = [-1,1]$ and $A = [-10,10]$. Given the function $f(\omega) = \omega^2$ with domain Ω and range A, define the sets $F = (-1/2,1/2) \subset \Omega$ and $G = (-1/4,1) \subset A$. Then $f(F) = (0,1/4)$ and $f^{-1}(G) = [-1,1]$. As you can see from this example, not all points in G have to correspond to points in F. In fact, the inverse image can be empty; e.g., continuing the same example, $f^{-1}((-1/4,0)) = \varnothing$.

The image of the entire space Ω is called the *range space* of f, and it need not equal the range; e.g., the function f could map the whole input space into a single point in A. For example, $f: \mathbf{R} \to \mathbf{R}$ defined by $f(r) = 1$, all r, has a range space of a single point. If the range space equals the range, the mapping is said to be *onto*. (Is the mapping f of the preceding example onto? What is the range space? Is the range unique?)

A set with exactly one point is called a *singleton set* (or a one-point set). A mapping is called *one − to − one* if the inverse image of all singleton sets are also singleton sets, and hence the inverse image is also a mapping of points; that is, $f^{-1}(a)$ consists of a single point for every point a in the range space.

LINEAR SYSTEM FUNDAMENTALS

In general, a *system* L is a mapping of an input time function, $x = \{x(t); t \in \mathbf{I}\}$ into an output time function, $L(x) = y = \{y(t); t \in \mathbf{I}\}$. Usually the functions take on real or complex values for each value of time t in $\mathbf{I}$. The system is called a discrete time system if $\mathbf{I}$ is discrete; e.g., $\mathbf{Z}$ or $\mathbf{Z}_+$, and it is called a continuous time system if $\mathbf{I}$ is continuous; e.g., $\mathbf{R}$ or $[0,\infty)$. If only nonnegative times are allowed, e.g., $\mathbf{I}$ is $\mathbf{Z}_+$ or $[0,\infty)$, the system is called a one-sided or single-sided system. If time can go on infinitely in both directions, then it is said to be a two-sided or double-sided system.

A system L is said to be *linear* if the mapping is linear, that is, for all complex (or real) constants a and b and all input functions x_1 and x_2

$$L(ax_1 + bx_2) = aL(x_1) + bL(x_2) . \qquad (2.21)$$

There are many ways to define or describe a particular linear system: One can provide a constructive rule for determining the output from the input; e.g., the output may be a weighted sum or integral of values of the input. Alternatively, one may provide a set of equations whose solution determines the output from the input, e.g., differential or difference equations involving the input and output at various times. Our emphasis will be on the former constructive technique, but we shall occasionally consider examples of other techniques.

The most common and the most useful class of linear systems comprises systems that can be represented by a convolution, that is, where the output is described by a weighted integral or sum of input values. We first consider continuous time systems and then turn to discrete time systems.

For $t \in I \subset \mathbf{R}$, let $x(t)$ be a continuous time input to a system with output $y(t)$ defined by the convolution integral

$$y(t) = \int_{s: t-s \in I} x(t-s)h_t(s)ds \ . \tag{2.22}$$

The function $h_t(t)$ is called the *impulse response* of the system since it can be considered the output of the system at time t which results from an input of a unit impulse or Dirac delta function $x(t) = \delta(t)$ at time 0. The index set is usually either $(-\infty, \infty)$ or $[0, \infty)$ for continuous time systems. The linearity of integration implies that the system defined by (2.22) is a linear system. A system of this type is called a *linear filter*. If the impulse response does not depend on time t, then the filter is said to be *time-invariant* and the convolution integral can be written as

$$y(t) = \int_{s: t-s \in I} x(t-s)h(s)ds = \int_{s \in I} x(s)h(t-s)ds \ . \tag{2.23}$$

We shall deal almost exclusively with time-invariant filters. Such a linear time-invariant system is often depicted using a block diagram as in Figure 2.3.

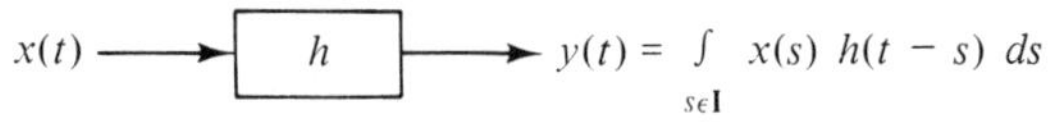

$$x(t) \longrightarrow \boxed{\ h\ } \longrightarrow y(t) = \int_{s \in I} x(s)\ h(t-s)\ ds$$

Figure 2.3   Linear filter.

If $x(t)$ and $h(t)$ are absolutely integrable, i.e.,

$$\int_I |x(t)|\, dt \ , \ \int_I |h(t)|\, dt \ < \ \infty \ , \tag{2.24}$$

then their Fourier transforms exist:

$$X(f) = \int_I x(t)e^{-j2\pi ft}dt \ , \ H(f) = \int_I h(t)e^{-j2\pi ft}dt \ . \tag{2.25}$$

Continuous time filters satisfying (2.24) are said to be *stable*. $H(f)$ is called the filter transfer function or the system function. We point out that (2.24) is a sufficient but not necessary condition for the existence of the transform. We shall not usually be concerned with the fine points of the existence of such transforms and their inverses. The inverse transforms that we require will be accomplished either by inspection or by reference to a table.

A basic property of Fourier transforms is that convolution in the time domain corresponds to multiplication in the frequency domain, and hence the output transform is given by

$$Y(f) = H(f)X(f) . \qquad (2.26)$$

Even if a particular system has an input that does not have a Fourier transform, (2.26) can be used to find the transfer function of the system by using some other input that does have a Fourier transform.

As an example, consider Figure 2.4, where two linear filters are concatenated or cascaded: $x(t)$ is input to the first filter, and the output $y(t)$ is input to the second filter, with final output $z(t)$. If both filters are stable and $x(t)$ is absolutely integrable, the Fourier transforms satisfy

$$Y(f) = H_1(f)X(f) , \quad Z(f) = H_2(f)Y(f) , \qquad (2.27)$$

or

$$Z(f) = H_2(f)H_1(f)X(f) .$$

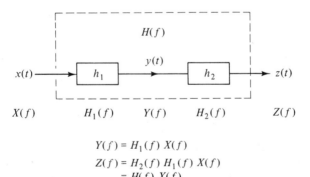

$$Y(f) = H_1(f) X(f)$$
$$Z(f) = H_2(f) H_1(f) X(f)$$
$$= H(f) X(f)$$

Figure 2.4 Cascade filter.

Obviously the overall filter transfer function is $H(f) = H_2(f)H_1(f)$. The overall impulse response is then the inverse transform of $H(f)$.

Frequently (but not necessarily) the output of a linear filter can also be represented by a finite order differential equation in terms of the differential operator, $D = d/dt$:

$$\sum_{k=0}^{n} a_k D^k y(t) = \sum_{i=0}^{m} b_i D^i x(t) \;. \tag{2.28}$$

The output is completely specified by the input, the differential equation, and appropriate initial conditions. Under suitable conditions on the differential equation, the linear filter is stable, and the transfer function can be obtained by transforming both sides of (2.28). However, we shall not pursue this approach further.

Turn now to Figure 2.5. Here we show an idealized sampled data system to demonstrate the relationship between discrete and continuous time filters. The input function $x(t)$ is input to a mixer, which forms the product of $x(t)$ with a pulse train, $p(t) = \sum_{k \in I} \delta(t - k)$, of Dirac delta functions spaced one second apart in time. I is a suitable subset of $\mathbf{Z}$. If we denote the sampled values $x(k)$ by x_k, the product is $x(t)p(t) = \sum_k x_k \delta(t - k)$. The product is the input to a linear filter with impulse response $h(t)$. Applying the convolution integral of equation (2.23) and sampling the output with a switch at one-second intervals, we have as an output function at time n

$$y_n = \sum_{k\,:\,k \in I} x_k h_{n-k} = \sum_{k\,:\,n-k \in I} x_{n-k} h_k \;. \tag{2.29}$$

$$x(t) \xrightarrow{\hspace{1cm}} \boxed{\times} \xrightarrow{\; x(t)p(t) \;} \boxed{h} \xrightarrow{\hspace{1cm}} y(t)$$

$$p(t) = \sum_{k \in I} \delta(t - k)$$

$$x(t)p(t) = \sum_{k \in I} x(k)\, \delta(t - k)$$

$$y(t) = \int x(\tau)\, p(\tau)\, h(t - \tau)\, d\tau$$
$$= \sum_{k \in I} x(k)\, h(n - k)$$

Figure 2.5 Sampled data system.

Thus, macroscopically the filter is a discrete time linear filter with a discrete convolution sum in place of an integral. $\{h_k\}$ is called the *unit pulse response* or *Kronecker δ response* of the discrete time filter. Its name is derived from the fact that h_k is the output of the linear filter at time k when a Kronecker delta function is input at time zero. If only a finite number of the h_k are nonzero, then the filter is sometimes referred to as an FIR (finite impulse response) filter. If a filter is not an FIR filter, then it is an IIR (infinite impulse response) filter.

If $\{h_k\}$ and $\{x_k\}$ are both absolutely summable,

$$\sum_k |h_k| < \infty \ , \ \sum_k |x_k| < \infty \ , \tag{2.30}$$

then their discrete Fourier transforms exist:

$$H(f) = \sum_k h_k e^{-j2\pi kf} \ , \ X(f) = \sum_k x_k e^{-j2\pi kf} \ . \tag{2.31}$$

Discrete time filters satisfying (2.30) are said to be *stable*. $H(f)$ is called the filter transfer function. The output transform is given by

$$Y(f) = H(f)X(f) \ . \tag{2.32}$$

The example of Figure 2.4 applies for discrete time as well as continuous time.

For convenience and brevity, we shall occasionally use a general notation $\boldsymbol{F}$ to denote both the discrete and continuous Fourier transform; that is,

$$\boldsymbol{F}(x) = \begin{cases} \int_{\boldsymbol{I}} x(t) e^{-j2\pi ft} dt, & \boldsymbol{I} \text{ continuous }, \\[2ex] \sum_{k \in \boldsymbol{I}} x_k e^{-j2\pi fk}, & \boldsymbol{I} \text{ discrete } . \end{cases} \tag{2.33}$$

A more general discrete time linear system is described by a difference equation of the form

$$\sum_k a_k y_{n-k} = \sum_i b_i x_{n-i} \ . \tag{2.34}$$

Observe that the convolution of (2.29) is a special case of the above where only one of the a_k is not zero. Observe also that the difference equation (2.34) is a discrete time analog of the differential equation (2.28). As in that case, to describe an output completely one has to specify initial conditions.

A continuous time or discrete time filter is said to be *causal* if the pulse response or impulse response is zero for negative time; that is, if a discrete time pulse response h_k satisfies $h_k = 0$ for $k < 0$ or a continuous time impulse response $h(t)$ satisfies $h(t) = 0$ for $t < 0$.

EXERCISES

2.1 Use the first seven relations to prove relations (2.10), (2.13), and (2.16).

2.2 Use relation (2.16) to obtain a partition $\{G_i; i = 1,2,...,k\}$ of Ω from an arbitrary finite class of collectively exhaustive sets $\{F_i; i = 1,2,...,k\}$ with the property that $G_i \subset F_i$ for all i and

$$\bigcup_{j=0}^{i} G_j = \bigcup_{j=0}^{i} F_j; \text{ all } i .$$

Repeat for a countable collection of sets $\{F_i\}$. (You must prove that the given collection of sets is indeed a partition.)

2.3 If $\{F_i\}$ partitions Ω, show that $\{G \cap F_i\}$ partitions G.

2.4 Show that $F \subset G$ implies that $F \cap G = F$, $F \cup G = G$, and $G^c \subset F^c$.

2.5 Show that if F and G are disjoint, then $F \subset G^c$.

2.6 Show that $F \cap G = (F \cup G) - (F \Delta G)$.

2.7 Let $F_r = [0, 1/r)$, $r \in (0,1]$. Find $\bigcup_{r \in (0,1]} F_r$ and $\bigcap_{r \in (0,1]} F_r$.

2.8 Prove the countably infinite version of DeMorgan's "laws." For example, given a sequence of sets $F_i; i = 1,2,...$, then

$$\bigcap_{i=1}^{\infty} F_i = \left[\bigcup_{i=1}^{\infty} F_i^c \right]^c .$$

2.9 Define the subsets of the real line

$$F_n = \{r : |r| > \frac{1}{n}\} ,$$

and

$$F = \{0\} .$$

Show that

$$F^c = \bigcup_{n=1}^{\infty} F_n .$$

2.10 Let F_i, $i = 1,2,...$ be a countable sequence of "nested" closed intervals whose length is not zero, but tends to zero; i.e., for every i, $F_i = [a_i, b_i] \subset F_{i-1} \subset F_{i-2} \cdots$ and $b_i - a_i \to 0$ as $i \to \infty$. What are the points in $\bigcap_{i=1}^{\infty} F_i$?

2.11 Prove that the interval $[0,1]$ cannot be put into one-to-one correspondence with the set of integers as follows: Suppose that there is such a correspondence so that $x_1, x_2, x_3, ...$ is a listing of all numbers in $[0,1]$. Use exercise 2.10 to construct a set that consists of a point not in this listing. This contradiction proves the statement.

2.12 Show that inverse images preserve set theoretic operations, that is, given $f\Omega \to A$ and sets F and G in A, then

$$f^{-1}(F^c) = (f^{-1}(F))^c \ ,$$

$$f^{-1}(F \cup G) = f^{-1}(F) \cup f^{-1}(G) \ ,$$

and

$$f^{-1}(F \cap G) = f^{-1}(F) \cap f^{-1}(G) \ .$$

If $\{F_i, i \in I\}$ is an indexed family of subsets of A that partitions A, show that $\{f^{-1}(F_i); i \in I\}$ is a partition of Ω. Do images preserve set theoretic operations in general? (Prove that they do or provide a counterexample.)

2.13 An experiment consists of rolling two four-sided dice (each having faces labeled 1, 2, 3, 4) on a glass table. Depict the space Ω of possible outcomes. Define two functions on Ω: $X_1(\omega) = $ the sum of the two down faces and $X_2(\omega) = $ the product of the two down faces. Let A_1 denote the range space of X_1, A_2 the range space of X_2, and A_{12} the range space of the vector-valued function $\mathbf{X} = (X_1, X_2)$, that is, $\mathbf{X}(\omega) = (X_1(\omega), X_2(\omega))$. Draw in both Ω and A_{12} the set $\{\omega: X_1(\omega) < X_2(\omega)\}$. The cartesian product $\prod_{i=1}^{2} A_i$ of two sets is defined as the collection of all pairs of elements, one from each set, that is,

$$\prod_{i=1}^{2} A_i = \{ \text{all } a,b: a \in A_1, b \in A_2 \} \ .$$

Is it true above that $A_{12} = \prod_{i=1}^{2} A_i$?

2.14 Let $\Omega = [0,1]$ and A be the set of all infinite binary vectors. Find a one-to-one mapping from Ω to A, being careful to note that some rational numbers have two infinite binary representations (e.g., $1/2 = .1000... = .0111...$ in binary).

2.15 Find a one-to-one mapping from:
  (a) $[0,1]$ to $[0,2)$.
  (b) $[0,1]$ to the unit square in two-dimensional Euclidean space.
  (c) $\mathbf{Z}$ to $\mathbf{Z}_+$. When is it possible to find a one-to-one mapping from one space to another?

2.16 Suppose that a voltage is measured that takes values in $\Omega = [0,15]$. The voltage is mapped into the finite space $A = \{0,1,...,15\}$ for transmission over a digital channel. A mapping of this type is called a *quantizer*. What is the best mapping in the sense that the maximum error is minimized?

2.17 Let A be as in exercise 2.16, i.e., the space of 16 messages which is mapped into the space of 16 waveforms, $B = \{\cos nt, n = 0,1,....,15; t \in [0,2\pi]\}$. The selected waveform from B is transmitted on a *waveform channel,* which adds noise; i.e., B is mapped into C = {set of all possible waveforms {$y(t)$ = cos nt + $noise(t)$; $t \in [0,2\pi]$}}. (This is a random mapping in a sense that will be described in subsequent chapters.) Find a good mapping from C into D = A. D is the *decision space* and the mapping is called a *decision rule*. (In other words, how would you perform this mapping knowing little of probability theory. Your mapping should at least give the correct decision if the noise is absent or small.)

2.18 Given a continuous time linear filter with impulse response $h(t)$ given by e^{-at} for $t \geq 0$ and 0 for $t < 0$, where a is a positive constant, find the transfer function $H(f)$ of the filter. Is the filter stable? What happens if a = 0?

2.19 Given a discrete time linear filter with pulse response h_k given by r^k for $k \geq 0$ and 0 for $k < 0$, where r has magnitude strictly less than 1, find the transfer function $H(f)$. (Hint: Use the geometric series formula.) Is the filter stable? What happens if r = 1? Assume that $|r| < 1$. Suppose that the input $x_k = 1$ for all nonnegative k and x_k = 0 for all negative k is put into the filter. Find a simple expression for the output as a function of time. Does the transform of the output exist?

2.20 A continuous time system is described by the following relation: Given an input x = {$x(t)$; $t \in \mathbf{R}$}, the output y = {$y(t)$; $t \in \mathbf{R}$} is defined for each t by

$$y(t) = (a_0 + a_1 x(t))\cos(2\pi f_0 t + \theta) ,$$

where a_0, a_1, f_0, and θ are fixed parameters. (This system is called an amplitude modulation (AM) system.) Under what conditions on the parameters is this system linear? Is it time-invariant?

3

PROBABILITY SPACES

The theory of random processes is a branch of probability theory and probability theory is a special case of the branch of mathematics known as measure theory. Probability theory and measure theory both concentrate on functions that assign real numbers to certain sets in an abstract space according to certain rules. These set functions can be viewed as measures of the size or weights of the sets. For example, the precise notion of area in two-dimensional Euclidean space and volume in three-dimensional Euclidean space are both examples of measures on sets. Alternative measures of sets in three-dimensional space are mass and weight. Observe that from elementary calculus we can find volume by integrating a constant over the set. From physics we can find mass by integrating a mass density or summing point masses over a set. In both cases the set is a region of three-dimensional space. In a similar manner, probabilities will be computed by integrals of densities of probability or sums of "point masses" of probability.

Both probability theory and measure theory consider only nonnegative real-valued set functions. The value assigned by the function to a set is called the probability or the measure of the set, respectively. The basic difference between probability theory and measure theory is that the former considers only set functions that are normalized in the sense of assigning the value of 1 to the entire abstract space, corresponding to the intuition that the abstract space contains every possible outcome of an experiment and hence should happen with certainty or probability 1.

Subsets of the space have some uncertainty and hence have probability less than 1.

Probability theory begins with the concept of a *probability space,* which is a collection of three items: (1) an *abstract space,* such as those encountered in chapter 2, called a *sample space,* which contains all distinguishable *elementary outcomes* or results of an experiment; (2) an *event space* consisting of a collection of subsets of the abstract space which we wish to consider as possible events and to which we wish to assign a probability; and (3) a *probability measure*—an assignment of a number between 0 and 1 to every event, that is, to every set in the event space. We require that the event space have an algebraic structure in the following sense: Any sequence of set-theoretic operations (union, intersection, complementation, difference, symmetric difference) on events must produce other events, even countably infinite sequences of operations. Uncountably infinite sequences of operations on events need not, however, produce other events. Before making these ideas precise, several comments are in order.

First of all, we emphasize that a probability space is composed of three parts; an abstract space is only one part. Do not let the terminology confuse you: "Space" has more than one usage. Having an abstract space model all possible distinguishable outcomes of an experiment should be an intuitive idea since it is simply giving a precise mathematical name to an imprecise English description. Since subsets of the abstract space corre- spond to collections of elementary outcomes, it should also be possible to assign probabilities to such sets. It is a little harder to see, but we can also argue that we should focus on the sets and not on the individual points when assigning probabilities since in many cases a probability assignment known only for points will not be very useful. For example, if we spin a fair wheel and the outcome is known to be equally likely to be any number between 0 and 1, then the probability that any particular point such as .3781984637 or exactly $1/\pi$ occurs is 0 because there are an uncountable infinity of possible points, none more likely than the others. Hence knowing only that the probability of each and every point is zero, we would be hard pressed to make any meaningful inferences about the probabilities of other events such as the outcome being between 1/2 and 3/4.

The difficulty inherent in this example leads to a less natural aspect of the probability space triumvirate—the fact that we must specify an event space or collection of subsets of our abstract space to which we wish to assign probabilities. In the example it is clear that taking the individual points and their *countable* combinations is not enough (see also exercise 3.1). On the other hand, why not just make the event space the class of *all* subsets of the abstract space? Why require the specification of which subsets are to be deemed sufficiently important to be blessed with the name "event"? In fact, this is one of the principal differences between elementary probability theory and advanced probability theory (and the point at which the student's intuition frequently runs into trouble). When the abstract space is finite or even countably infinite, one can consider all possible subsets of the space to be events, and one can build a useful theory. When the abstract space is uncountably infinite, however, as in the case of the space consisting of the real line, one cannot build a useful theory without confining the allowable subsets to which one will assign a probability. Roughly speaking, this is because probabilities of sets in uncountable spaces are found by integrating over sets, and some sets are simply too nasty to be integrated over. Although it is difficult to show, for such spaces there does not exist a reasonable and consistent means of assigning probabilities to all subsets without contradiction or without violating desirable properties. In fact, it is so difficult to show that such "non-probability-measurable" subsets of the real line exist that we will not attempt to do so in this book. The reader should only be aware of the problem so that the need for specifying an event space is understood.

Thus a probability space must make explicit not just the elementary outcomes or "finest-grain" outcomes that constitute our abstract space; it must also specify the collections of sets of these points to which we intend to assign probabilities. Subsets of the abstract space that do not belong to the event space will simply not have probabilities defined. The algebraic structure that we have postulated for the event space will ensure that if we take (countable) unions of events (corresponding to a logical "or") or intersections of events (corresponding to a logical "and"), then the resulting sets are also events and hence will have probabilities. In fact, this is one of the main functions of probability theory: Given a probabilistic description of a collection of events, find the probability of some new event formed by set-theoretic operations on the given events.

The probability space will serve as the foundation for the definition of random variables, random vectors, and random processes, which will inherit their basic properties from the underlying probability space. These properties will in turn yield new probability spaces. Much of the theory of random processes consists of developing the implications of certain operations on probability spaces: Beginning with some probability space we form new ones by operations called variously mappings, filtering, sampling, coding, communicating, estimating, detecting, averaging, measuring, or other names denoting linear or nonlinear operations. Stochastic systems theory is the combination of systems theory with probability theory. The essence of stochastic systems theory is the connection of a system to a probability space with a description of the output. Thus a precise formulation and a good understanding of probability spaces are prerequisites to a precise formulation and correct development of examples of random processes and stochastic systems.

We now provide the abstract mathematical definition of a probability space and then make several comments on the intuition and goals behind the definitions and on some implications and alternative characterizations of the definitions. Examples are provided to illustrate the basic concepts.

A *sample space* Ω is an abstract space, a nonempty collection of points or members or elements called *sample points* (or *elementary events* or *elementary outcomes*).

An *event space* (or *sigma field* or *sigma algebra*) $\boldsymbol{F}$ of a sample space Ω is a nonempty collection of subsets of Ω called *events* with the following properties:

$$\text{If } F \in \boldsymbol{F}, \text{ then also } F^c \in \boldsymbol{F}, \tag{3.1}$$

that is, if a given set is an event, then its complement must also be an event.

If for some finite n, $F_i \in \boldsymbol{F}$, $i = 1,2,...,n$, then also

$$\bigcup_{i=1}^{n} F_i \in \boldsymbol{F}, \tag{3.2}$$

that is, a finite union of events must also be an event.

If $F_i \in \boldsymbol{F}$, $i = 1,2,...$, then also

$$\bigcup_{i=1}^{\infty} F_i \in \boldsymbol{F}, \tag{3.3a}$$

that is, a countable union of events must also be an event.

The latter equation is numbered (3.3a) with malice aforethought: We shall later see alternative but equivalent forms which will be numbered (3.3b) and (3.3c). The abbreviated number (3.3) refers to any (or all) of these forms. Observe that (3.2) is a special case of (3.3a), but it is convenient to consider the finite case separately. If a collection of sets satisfies only (3.1) and (3.2), then it is called a *field* or *algebra* of sets. For this reason, in elementary probability theory one often refers to the "algebra of events." Both (3.1) and (3.2) can be considered as "closure" properties; that is, an event space must be closed under complementation and unions in the sense that performing a sequence of complementations or unions of events must yield a set that is also in the collection, that is, a set that is also an event. Observe also that (3.1), (3.2), and (2.11) imply that

$$\Omega \in \boldsymbol{F}, \tag{3.4}$$

that is, the whole sample space considered as a set must be in $\boldsymbol{F}$; that is, it must be an event. Intuitively, Ω is the "certain event," the event that "something happens."

Note that if F is a subset of Ω, then we write $F \subset \Omega$, and if the subset F is also in the event space, then we write $F \in \boldsymbol{F}$; that is, we use set inclusion when considering F as a subset of an abstract space and element inclusion when considering F as a member of the event space and hence as an event. Alternatively, the elements of Ω are points, and a collection of these points is a subset of Ω; but the elements of $\mathbf{F}$ are sets—subsets of Ω—and not points. A student should ponder the different natures of abstract spaces of points and event spaces consisting of sets until the reasons for set inclusion in the former and element inclusion in the latter space are clear. Consider especially the difference between an element of Ω and a subset of Ω that consists of a single point. The latter *may* or *may not* be an element of $\mathbf{F}$, the former is *never* an element of $\mathbf{F}$.

A *measurable space* $(\Omega, \mathbf{F})$ is a pair consisting of a sample space Ω and an event space or sigma field $\mathbf{F}$ of subsets of Ω. The strange name "measurable space" reflects the fact that we can assign a nonunique measure, viz., a probability measure, to such a space and thereby form a probability space or probability measure space.

A *probability measure* P on a measurable space $(\Omega, \mathbf{F})$ consisting of an event space or sigma field $\mathbf{F}$ of subsets of a sample space Ω is an assignment of a real number $P(F)$ to every member F of the sigma field (that is, to every event) such that P obeys the following rules, which we refer to as the axioms of probability.

Axiom 1.

$$P(F) \geq 0 \text{ for all } F \in \mathbf{F} \tag{3.5}$$

i.e., no event has negative probability.

Axiom 2.

$$P(\Omega) = 1 \tag{3.6}$$

i.e., the probability of "everything" is one.

Axiom 3.

If F_i, $i = 1,2,\ldots,n$ are disjoint, then

$$P\left(\bigcup_{i=1}^{n} F_i \right) = \sum_{i=1}^{n} P(F_i) . \tag{3.7}$$

Axiom 4a.

If F_i, $i = 1,2,\ldots$ are disjoint, then

$$P\left(\bigcup_{i=1}^{\infty} F_i \right) = \sum_{i=1}^{\infty} P(F_i) . \tag{3.8a}$$

(As with property (3.3a) of an event space, the fourth axiom of probability is qualified as 4a because we shall later see other equivalent versions.)

As with the defining properties of an event space, for the purposes of discussion, we have listed separately the finite special case (3.7) of the general condition (3.8a). To emphasize an important point: A function P which assigns numbers to elements of an event space of a sample space is a probability measure *if and only if* it satisfies all of the above properties!

A *probability space* or *experiment* is a triple $(\Omega, \mathbf{F}, P)$ consisting of a sample space Ω, an event space or sigma field $\mathbf{F}$ of subsets of Ω, and a probability measure P defined for all members of $\mathbf{F}$.

Before developing each idea in more detail and providing several examples of each piece of a probability space, we pause to consider two simple examples of the complete construction. The first example is the simplest possible probability space and is commonly referred to as the trivial probability space. We depart from our customary indexing to emphasize just how useless this example is. It does serve a purpose, however, by showing that a well-defined model need not be interesting.

[3.0]

Let Ω be any abstract space and let $\mathbf{F} = \{\Omega, \varnothing\}$; that is, $\mathbf{F}$ consists of exactly two sets—the sample space (everything) and the empty set (nothing). This is called the trivial sigma field. This is a model of an experiment where only two events are possible: "Something happens" or "nothing happens"—not a very interesting description. There is only one possible probability measure for this measurable space: $P(\Omega) = 1$ and $P(\varnothing) = 0$. This probability measure meets the required rules that define a probability measure; they can be directly verified since there are only two possible events. Equations (3.5) and (3.6) are obvious. Equations (3.7) and (3.8a) follow since the only possible values for F_i are Ω and $\varnothing$. At most one of the F_i can be Ω since they are disjoint. If one of the F_i is indeed Ω, then both sides of the equality are 1. Otherwise, both sides are 0.

[3.1]

Let $\Omega = \{0, 1\}$ as in example [2.1]. Let $\mathbf{F} = \{ \{0\}, \{1\}, \Omega = \{0, 1\}, \varnothing \}$. Since $\mathbf{F}$ contains *all* of the subsets of Ω, the properties (3.1) through (3.3a) are satisfied, and hence it is an event space. (There is one other possible event space that could be defined for Ω in this example. What is it?) Define the set function P by

$$
P(F) = \begin{cases}
1/4 & \text{if } F = \{0\} \\
3/4 & \text{if } F = \{1\} \\
0 & \text{if } F = \varnothing \\
1 & \text{if } F = \Omega .
\end{cases}
$$

It is easily verified that P satisfies the axioms of probability and hence is a probability measure. Therefore $(\Omega, \mathbf{F}, P)$ is a probability space. Note that we had to give the value of $P(F)$ for *all* events F. Note also that the choice of $P(F)$ is not unique for the given measurable space $(\Omega, \mathbf{F})$; we could have chosen any value in $[0,1]$ for $P(\{1\})$ and used the axioms to complete the definition.

The preceding example is the simplest nontrivial example of a probability space and provides a rigorous mathematical model for applications such as the binary transmission of a single bit or for the flipping of a single biased coin once.

We now develop in more detail properties and examples of the three components of probability spaces: sample spaces, event spaces, and probability measures.

SAMPLE SPACES

Intuitively, a sample space is a listing of all conceivable finest-grain, distinguishable outcomes of an experiment to be modeled by a probability space. Mathematically it is just an abstract space. We encountered in chapter 2 several examples of such spaces that are useful for modeling the outputs of stochastic systems. We repeat some of these examples here for reference.

Examples of Sample Spaces

[3.2]

A finite space $\Omega = \{a_k; k = 1,2,...,K\}$. Specific examples are the binary space $\{0,1\}$ and the finite space of integers $\{0,1,2,...,k-1\}$, a space we will denote by $\mathbf{Z}_k$.

[3.3]

A countably infinite space $\Omega = \{a_k; k = 1,2,...\}$, for some sequence $\{a_k\}$. Specific examples are the space of all nonnegative integers $\{0,1,2,...\}$, which we denote by $\mathbf{Z}_+$, and the space of all integers $\{...,-2,-1,0,1,2,...\}$, which we denote by $\mathbf{Z}$. Other examples are the space of all rational numbers, the space of all even integers, and the space of all periodic sequences of integers.

Both examples [3.2] and [3.3] are called *discrete* spaces. Spaces with finite or countably infinite numbers of elements are called discrete spaces.

[3.4]

An interval of the real line $\mathbf{R}$, for example, $\Omega = (a,b)$. We might consider an open interval, a closed interval, a half-open interval, or even the entire real line itself.

Spaces such as example [3.4] that are not discrete are said to be *continuous*.

[3.5]

A space consisting of k-dimensional vectors with coordinates taking values in one of the previously described spaces. A useful notation for such vector spaces is a *product space.* Let A denote one of the abstract spaces previously considered. Define the cartesian product A^k by

$$A^k = \{\text{all vectors } \mathbf{a} = (a_0, a_1, \ldots, a_{k-1}) \text{ with } a_i \in A\} \ .$$

Thus, for example, $\mathbf{R}^k$ is k-dimensional Euclidean space. $\{0,1\}^k$ is the space of all binary k-tuples, that is, the space of all k-dimensional binary vectors. $[0,1]^2$ is the unit square in the plane. $[0,1]^3$ is the unit cube in three-dimensional Euclidean space.

Alternative notations for a cartesian product space are

$$\prod_{i \in \mathbf{Z}_k} A_i = \prod_{i=0}^{k-1} A_i = A^k \ ,$$

where the A_i are all replicas or copies of A, that is, where $A_i = A$, all i. Other notations for such a finite-dimensional cartesian product are

$$\underset{i \in \mathbf{Z}_k}{\times} A_i = \underset{i=0}{\overset{k-1}{\times}} A_i = A^k \ .$$

This and other product spaces will prove to be a useful means of describing abstract spaces modeling sequences of elements from another abstract space.

Observe that a finite-dimensional vector space constructed from a discrete space is also discrete since if one can count the number of possible values one coordinate can assume, then one can count the number of possible values that a finite number of coordinates can assume.

[3.6]

A space consisting of infinite sequences drawn from one of the examples [3.2] through [3.4]. This is also a product space. Let A be a sample space and let A_i be replicas or copies of A. We will consider both singly and doubly infinite products to model sequences with and without a finite origin, respectively. Define the doubly infinite or two-sided space

$$\prod_{i \in \mathbf{Z}} A_i = \{\text{all sequences } \{a_i; \ i = \ldots, -1, 0, 1, \ldots\}; \ a_i \in A_i\} \ ,$$

and the singly infinite or one-sided space

$$\prod_{i \in \mathbf{Z}_+} A_i = \{\text{all sequences } \{a_i; \ i = 0, 1, \ldots\}; \ a_i \in A_i\} \ .$$

These two spaces are also denoted by $\prod_{i=-\infty}^{\infty} A_i$ or $\underset{i=-\infty}{\overset{\infty}{\times}} A_i$ and $\prod_{i=0}^{\infty} A_i$ or $\underset{i=0}{\overset{\infty}{\times}} A_i$, respectively. This notation should be used with caution because, except for the index used (the convention is that usually i, j, k, l, m, n imply integers), it is not clear whether it is a space of sequences or a space of waveforms, as in the next example.

The two spaces under discussion are often called *sequence spaces*. Even if the original space A is discrete, the sequence space constructed from A will be continuous. For example, there is an uncountable infinity of single-sided binary sequences. To see this, recall that there is an uncountable infinity of real numbers in the unit interval [0,1] and that we can express all of these numbers in the binary number system as sequences to the right of the "decimal" point (exercise 2.11).

[3.7]

Let A be one of the sample spaces of examples [3.2] through [3.4]. Form a new abstract space consisting of all waveforms or functions of time with values in A, for example, all real-valued time functions. This space is also modeled as a product space. For example, the infinite two-sided space for a given A is

$$\prod_{t \in \mathbf{R}} A_t = \{\text{all waveforms } \{x(t); \, t \in (-\infty, \infty)\}; \, x(t) \in A, \text{all } t\}$$

with a similar definition for one-sided spaces and for time functions on a finite time interval. This space is also denoted by $\displaystyle\prod_{t=-\infty}^{\infty} A_t$ or $\displaystyle\mathop{\times}_{t=-\infty}^{\infty} A_t$ but, as previously noted, this notation can be ambiguous, as it is not clear whether the space contains sequences or waveforms, and the use of the integer index for sequences and t ("time") for waveforms is a bit sloppy mathematically; that is, it is bad practice to have the meaning of the product depend on the dummy variable or index. Nonetheless, this abuse of notation is frequently employed for simplicity.

Note that we indexed sequences using subscripts, as in x_n, and we indexed waveforms using parentheses, as in $x(t)$. In fact, the notations are interchangeable; we could denote waveforms as $\{x(t); \, t \in \mathbf{R}\}$ or as $\{x_t; \, t \in \mathbf{R}\}$. The notation using subscripts for sequences and parentheses for waveforms is the most common, and we will usually stick to it. It is worth remembering, however, that vectors, sequences, and waveforms are all just indexed collections of numbers; the only difference is the index set: finite for vectors, countably infinite for sequences, and continuous for waveforms.

All of the product spaces we have described can be viewed as special cases of the general product space defined next.

Let $\mathbf{I}$ be an index set such as a finite set of integers $\mathbf{Z}_k$, the set of all integers $\mathbf{Z}$, the set of all nonnegative integers $\mathbf{Z}_+$, the real line $\mathbf{R}$, or the nonnegative reals $[0, \infty)$. Given a family of spaces $\{A_t; \, t \in \mathbf{I}\}$, define the product space

$$\prod_{t \in \mathbf{I}} A_t = \{\text{all } \{a_t; \, t \in \mathbf{I}\}; \, a_t \in A_t, \text{ all } t\} \, .$$

The notation $\mathop{\times}_{t \in \mathbf{I}} A_t$ is also used for the same thing. Thus product spaces model spaces of vectors, sequences, and waveforms whose coordinate values are drawn from some fixed space. We shall often denote the general product space by the abbreviated form:

$$A^I = \prod_{t \in I} A_t .$$

This leads to two notations for the space of all k-dimensional vectors with coordinates in A: A^k and A^{Z_k}. We will usually use the shorter and simpler notation when convenient.

All of the sample spaces in this chapter were described verbally in the examples of chapter 2. The examples describe the most important sample spaces for our applications.

EVENT SPACES

Intuitively, an event space is a collection of subsets of the sample space or groupings of elementary events which we shall consider as physical events and to which we wish to assign probabilities. Mathematically, an event space is a collection of subsets that is closed under certain set-theoretic operations; that is, performing certain operations on events or members of the event space must give other events. Thus, for example, if in the example of a single voltage measurement example we have $\Omega = \mathbf{R}$ and we are told that the set of all voltages greater than 5 volts $= \{\omega: \omega \geq 5\}$ is an event, that is, is a member of a sigma field $\mathbf{F}$ of subsets of $\mathbf{R}$, then necessarily its complement $\{\omega: \omega < 5\}$ must also be an event, that is, in the sigma field $\mathbf{F}$. If the latter set is not in $\mathbf{F}$, then $\mathbf{F}$ cannot be an event space! Observe that no problem arises if the complement physically cannot happen—events that "cannot occur" can be included in $\mathbf{F}$ and then assigned probability zero when choosing the probability measure P. For example, even if you know that the voltage does not exceed 5 volts, if you have chosen the real line $\mathbf{R}$ as your sample space, then you must include the set $\{r: r > 5\}$ in the event space if the set $\{r: r \leq 5\}$ is an event. The impossibility of a voltage greater than 5 is then expressed by assigning $P(\{r: r > 5\}) = 0$. (Besides the given subset and its complement, what else must be added so that you have an event space? Would this normally be a useful event space?)

While the definition of a sigma field requires only that the class be closed under complementation and countable unions, these requirements immediately yield additional closure properties. The countably infinite version of DeMorgan's "laws" of elementary set theory require that if F_i, $i = 1, 2, \ldots$ are all members of a sigma field, then so is

$$\bigcap_{i=1}^{\infty} F_i = (\bigcup_{i=1}^{\infty} F_i^c)^c .$$

It follows by similar set-theoretic arguments that any countable sequence of any of the set-theoretic operations (union, intersection, complementation, difference, symmetric difference) performed on events must yield other events. Observe, however, that there is no guarantee that *uncountable* operations on events will produce new events; they may or may not. For example, if we are told that $\{F_r;\ r \in [0,1]\}$ is a family of events, then it is not necessarily true that $\bigcup_{r \in [0,1]} F_r$ is an event (see exercise 3.1 for an example).

The requirement that a finite sequence of set-theoretic operations on events yields other events is an intuitive necessity and is easy to verify for a given collection of subsets of an abstract space: It is intuitively necessary that logical combinations (*and* and *or* and *not*) of events corresponding to physical phenomena should also be events to which probability can be assigned. If you know the probability of a voltage being greater than zero and you know the probability that a voltage is greater than 5 volts, then you should also be able to know the probability that the voltage is greater than zero but not greater than 5 volts. It is easy to verify that finite sequences of set-theoretic combinations yield events because the finiteness of elementary set theory usually yields simple proofs.

A natural question arises in regard to (3.1) and (3.2): Why not try to construct a useful probability theory on the more general notion of a field rather than a sigma field? The response is that it unfortunately does not work: Probability theory requires many results involving limits, and such asymptotic results require the infinite relations of (3.3a) and (3.8a) to work. In some special cases, such as single coin flipping or single die rolling, the simpler finite results suffice because there are only a finite number of possible outcomes, and hence limiting results become trivial—any field is automatically a sigma field. If, however, one flips a coin forever, then there is an uncountable infinity of possible outcomes, and the asymptotic relations become necessary. Let Ω be the space of all one-sided binary sequences. Suppose that you consider the smallest field formed by all finite set-theoretic operations on the individual one-sided binary sequences, that is, singleton sets in the sequence space. Then many countably infinite sets of binary sequences (say the set of all periodic sequences) are not events since they cannot be expressed as finite sequences of set-theoretic operations on the singleton sets. Obviously, the sigma field formed by including countable set-theoretic operations does not have this defect. This is why sigma fields must be used rather than fields.

The condition (3.3a) can be related to a condition on limits by defining the notion of a limit of a sequence of sets. This notion will also be useful shortly in considering the axioms of probability. Consider a sequence of nested sets F_n, $n = 1,2,\dots$ with the property that each set contains its predecessor, that is, that $F_{n-1} \subset F_n$ for all n. Such a nested sequence of sets is said to be *increasing*. For example, the sequence $F_n =$ $[1,2-1/n)$ of subsets of the real line is increasing. The sequence $(-n,a)$ is also increasing. Intuitively, the first example increases to a limit of $[1,2)$ in the sense that every point in the set $[1,2)$ is eventually included in one of the F_n. Similarly, the sequence in the second example increases to $(-\infty,a)$. Formally, the limit of an increasing sequence of sets can be defined as the union of all of the sets in the sequence since the union contains all of the points in all of the sets in the sequence and does not contain any points not contained in at least one set (and hence an infinite number of sets) in the sequence:

$$\lim_{n \to \infty} F_n = \bigcup_{n=1}^{\infty} F_n \, .$$

Figure 3.1a illustrates such a sequence in a Venn diagram.

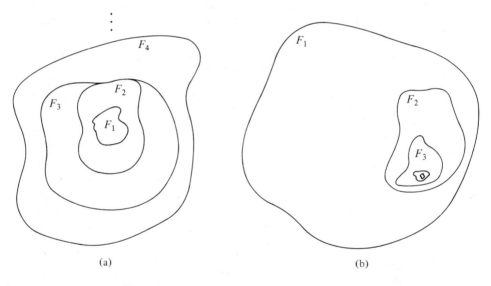

Figure 3.1   (a) Increasing Sets. (b) Decreasing Sets.

Thus the limit of the sequence of sets $[1,2-1/n)$ is indeed the set $[1,2)$, as desired, and the limit of $(-n,a)$ is $(-\infty,a)$.

Similarly, suppose that F_n; $n = 1,2,...$ is a *decreasing* sequence of nested sets in the sense that $F_n \subset F_{n-1}$ for all n as illustrated by the Venn diagram in Figure 3.1b. For example, the sequences of sets $[1,1+1/n)$ and $(1-1/n,1+1/n)$ are decreasing. Again we have a natural notion of the limit of this sequence: Both these sequences of sets collapse to the point or singleton set $\{1\}$—the point in common to all the sets. This suggests a formal definition based on the countably infinite intersection of the sets.

Given a decreasing sequence of sets F_n; $n = 1,2,...$, we define the limit of the sequence by

$$\lim_{n \to \infty} F_n = \bigcap_{n=1}^{\infty} F_n \, ,$$

that is, a point is in the limit of a decreasing sequence of sets *if and only if* it is contained in all the sets of the sequence.

Thus, given a sequence of increasing or decreasing sets, the limit of the sequence can be defined in a natural way: the union of the sets of the sequence or the intersection of the sets of the sequence, respectively.

Say that we have a sigma field **F** and an increasing sequence of sets F_n; $n = 1,2,...$ of sets in the sigma field. Since the limit of the sequence is defined as a union and since the union of a countable number of events must be an event, then the limit must be an event. For example, if we are told that the sets $[1,2-1/n)$ are all events, then the limit $[1,2)$ must also be an event. If we are told that all finite intervals of the form (a,b), where a and b are finite, are events, then the semi-infinite interval $(-\infty,b)$ must also be an event, since it is the limit of the sequence of sets $(-n,b)$ as $n \to \infty$.

By a similar argument, if we are told that each set in a decreasing sequence F_n is an event, then the limit must be an event, since it is an intersection of a countable number of events. Thus, for example, if we are told that all finite intervals of the form (a,b) are events, then the points or singleton sets must also be events, since a point $\{a\}$ is the limit of the decreasing sequence of sets $(a-1/n, a+1/n)$.

If a class of sets is only a field rather than a sigma field, that is, if it satisfies only (3.1) and (3.2), then there is no guarantee that the class will contain *all* limits of sets. Hence, for example, knowing that a class of sets contains all half-open intervals of the form $(a,b]$ for a and b finite does not ensure that it will also contain points or singleton sets! In fact, it is straightforward to show that the collection of all such half-open intervals together with the complements of such sets and all finite unions of the intervals and complements forms a field. The singleton sets, however, are not in the field! (See exercise 3.4.)

Thus if we tried to construct a probability theory based on only a field, we might have probabilities defined for events such as (a,b) meaning "the output voltage of a measurement is between a and b" yet not have probabilities defined for a singleton set $\{a\}$ meaning "the output voltage is exactly a." By requiring that the event space be a sigma field instead of only a field, we are assured that all such limits are indeed events.

It is a straightforward exercise to show that given (3.1) and (3.2), property (3.3a) is equivalent to either of the following:

If $F_n \in \mathbf{F}$; $n = 1,2,...$, is a decreasing sequence, then

$$\lim_{n \to \infty} F_n \in \boldsymbol{F} \ . \tag{3.3b}$$

If $F_n \in \boldsymbol{F}$; $n = 1,2,...$, is an increasing sequence, then

$$\lim_{n \to \infty} F_n \in \boldsymbol{F} \ . \tag{3.3c}$$

We have already seen that (3.3a) implies (3.3b) and (3.3c). For example, if (3.3c) is true and G_n is an arbitrary sequence of events, then define the increasing sequence

$$F_n = \bigcup_{i=1}^{n} G_i \ .$$

Obviously $F_{n-1} \subset F_n$, and then (3.3c) implies (3.3a), since

$$\bigcup_{i=1}^{\infty} G_i = \bigcup_{n=1}^{\infty} F_n = \lim_{n \to \infty} F_n \in \boldsymbol{F} \ .$$

Examples of Event Spaces

As we have noted, for a given sample space the selection of an event space is not unique; it depends on the events to which it is desired to assign probabilities and also on analytical limitations on the ability to assign probabilities. We begin with two examples that represent the extremes of event spaces—one possessing the minimum quantity of sets and the other possessing the maximum. We then study event spaces useful for the sample space examples of the preceding section.

[3.8]

Given a sample space Ω, then the collection $\{\Omega, \varnothing\}$ is a sigma field. This is just the trivial event space already treated in example [3.0]. Observe again that this is the smallest possible event space for any given sample space because no other event space can have fewer elements.

[3.9]

Given a sample space Ω, then the collection *all subsets of* Ω is a sigma field. This is true since any countable sequence of set-theoretic operations on subsets of Ω must yield another subset of Ω and hence must be in the collection of *all* possible subsets. The collection of all subsets of a space is called the *power set* of the space. Observe that this is the largest possible event space for the given sample space, because it contains every possible subset of the sample space.

This sigma field is a useful event space for the sample spaces of examples [3.2] and [3.3]; that is, for sample spaces that are discrete. Hence we shall virtually always take our event space as the power set when dealing with a discrete sample space. A discrete sample space with n

elements has a power set with at most 2^n elements (exercise 3.3). For example, the power set of the binary sample space $\Omega = \{0,1\}$ is the collection $\{\{0\},\{1\},\Omega = \{0,1\},\varnothing\}$, a list of all possible subsets of the space.

Unfortunately, the power set is too large to be useful for continuous spaces. To treat the reasons for this is beyond the scope of a book at this level, but we can say that it is not possible in general to construct interesting probability measures on the power set of a continuous space. There are special cases where we can construct particular probability measures on the power set of a continuous space by mimicking the construction for a discrete space (see, e.g., exercises 3.5 and 3.8). Truly continuous experiments cannot, however, be rigorously defined for such a large event space because integrals cannot be defined over all events in such spaces.

While both of the preceding examples can be used to provide event spaces for the special case of $\Omega = \mathbf{R}$, the real line, neither leads to a useful probability theory in that case. In the next example we consider another event space for the real line that is more useful (and, in fact, is used almost always for $\mathbf{R}$). First, we note that any useful event space for the real line should usually include as members all intervals of the form (a,b) since we certainly will wish to consider events of the form "the output voltage is between 3 and 5 volts." Second, we obviously require that the event space satisfy the defining properties for an event space, that is, that we have a collection of subsets of Ω that satisfy properties (3.1) through (3.3a). A means of accomplishing both of these goals in a relatively simple fashion is to define our event space as the *smallest* sigma field that contains the desired subsets, to wit, the intervals and all of their countable set-theoretic combinations (bewildering as it may seem, this is not the same as all subsets of $\mathbf{R}$). Of course, although a sigma field that is based on the intervals is most useful, it is also possible to consider other starting points. These considerations motivate the following general definition.

Given a sample space Ω (such as the real line $\mathbf{R}$) and an arbitrary class $\boldsymbol{G}$ of subsets of Ω—usually the class of all open intervals of the form (a,b) when $\Omega = \mathbf{R}$—define $\sigma(\boldsymbol{G})$, the sigma field *generated* by the class $\boldsymbol{G}$, to be the smallest sigma field containing all of the sets in $\boldsymbol{G}$, where by "smallest" we mean that if $\boldsymbol{F}$ is any sigma field and it contains $\boldsymbol{G}$, then it also contains $\sigma(\boldsymbol{G})$.

It is straightforward but irrelevant to show that this is a valid definition, that is, that the sigma field generated by a class of subsets of a sample space is well-defined. For our purposes it suffices simply to understand the definition, that is, to obtain an event space by requiring (1) that it contain a specified collection of physically important events and (2) that it be a sigma field in the sense of satisfying (3.1) through (3.3). (Recall that when we refer to equation (3.3), we mean any or all of the equations (3.1a) through (3.1c), which are equivalent given that (3.1) and (3.2) hold.)

For example, as noted before, we might require that a sigma field of the real line contain all intervals; then it would also have to contain at least all complements of intervals and all countable unions and intersections of intervals (and all countable complements, unions, and intersections of these results, ad infinitum). This technique will be used several times to specify useful event spaces in complicated situations such as continuous sample spaces, sequence spaces, and function spaces. We are now ready to provide the proper, most useful event space for the real line.

[3.10]

Given the real line **R**, the *Borel field* (or, more accurately, the *Borel sigma field*) is defined as the sigma field generated by all the open intervals of the form (a,b). The members of the Borel field are called *Borel sets*. We shall denote the Borel field by $\boldsymbol{B}(\mathbf{R})$, and hence

$$\boldsymbol{B}(\mathbf{R}) = \sigma(\text{all open intervals}) .$$

Since $\boldsymbol{B}(\mathbf{R})$ is a sigma field and since it contains all of the open intervals, it must also contain limit sets of the form

$$(-\infty,b) = \lim_{n \to \infty} (-n,b) ,$$

$$(a,\infty) = \lim_{n \to \infty} (a,n) ,$$

and

$$\{a\} = \lim_{n \to \infty} (a-1/n, a+1/n) ,$$

that is, the Borel field must include semi-infinite open intervals and the singleton sets or individual points. Furthermore, since the Borel field is a sigma field it must contain differences. Hence it must contain semi-infinite half-open sets of the form

$$(-\infty,b] = (-\infty,\infty) - (b,\infty) ,$$

and since it must contain unions of its members, it must contain half-open intervals of the form

$$(a,b] = (a,b)\cup\{b\} \text{ and } [a,b) = (a,b)\cup\{a\} .$$

In addition, it must contain all closed intervals and all finite or countable unions and complements of intervals of any of the preceding forms. Roughly speaking, the Borel field contains all subsets of the real line that can be obtained as an approximation of countable combinations of intervals. It is a deep and difficult result of measure theory that the Borel field of the real line is in fact different from the power set of the real line; that is, there exist subsets of the real line that are not in the Borel field. While we will not describe such a subset, we can guarantee that these

"unmeasurable" sets have no physical importance, that they are very hard
to construct, and that an engineer will never encounter such a subset in
practice. It may, however, be necessary to demonstrate that some weird
subset is in fact an event in this sigma field. This is typically accomplished
by showing that it is the limit of simple Borel sets.

In some cases we wish to deal not with a sample space that is the
entire real line, but one that is some subset of the real line. In this case
we define the Borel field as the Borel field of the real line "cut down" to
the smaller space.

Given that the sample space, Ω, is a Borel subset of the real line **R**,
the *Borel field* of Ω, denoted $\boldsymbol{B}(\Omega)$, is defined as the collection of all sets
of the form $F \cap \Omega$, for $F \in \boldsymbol{B}(\mathbf{R})$; that is, the intersection of Ω with all of
the Borel sets of **R** forms the class of Borel sets of Ω.

It can be shown (exercise 3.2) that, given a discrete subset A of the
real line, the Borel field $\boldsymbol{B}(A)$ is identical to the power set of A. Thus,
for the first three examples of sample spaces, the Borel field serves as a
useful event space since it reduces to the intuitively appealing class of all
subsets of the sample space.

The remaining examples of sample spaces are all product spaces.
The construction of event spaces for such product spaces—that is, spaces
of vectors, sequences, or waveforms—is more complicated and less intuitive than the constructions for the preceding event spaces. In fact, there
are several possible techniques of construction, which in some cases lead to
different event spaces. We wish to convey an understanding of the structure of such event spaces, but we do not wish to dwell on the technical
difficulties that can be encountered. Hence we shall study only one of the
possible constructions—the simplest possible definition of a product sigma
field by making a direct analogy to a product sample space. This
definition will suffice for most systems studied herein, but it has shortcom-
ings. At this time we mention one particular weakness: The event space
that we shall define may not be big enough when studying the theory of
continuous time random processes.

[3.11]

Given an abstract space A, a sigma field $\boldsymbol{F}$ of subsets of A, an index set $\boldsymbol{I}$, and a
product sample space of the form

$$A^{\boldsymbol{I}} = \prod_{t \in \boldsymbol{I}} A_t ,$$

where the A_t are all replicas of A, the product sigma field

$$\boldsymbol{F}^{\boldsymbol{I}} = \prod_{t \in \boldsymbol{I}} \boldsymbol{F}_t$$

is defined as the sigma field generated by all "one-dimensional" sets of the form

$$\{\{a_t; t \in \boldsymbol{I}\}: a_t \in F \text{ for } t = s \text{ and } a_t \in A_t \text{ for } t \neq s\}$$

for some $s\in I$ and some $F\in F$; that is, the product sigma field is the sigma field generated by all "one-dimensional" events formed by collecting all of the vectors or sequences or waveforms with one coordinate constrained to lie in a one-dimensional event and with the other coordinates unrestricted. The product sigma field must contain *all* such events; that is, for all possible indices s and all possible events F.

Thus, for example, given the one-dimensional abstract space **R**, the real line, along with its Borel field, Figure 3.2 depicts three examples of one-dimensional sets in $\mathbf{R}^2$, the two-dimensional Euclidean plane. Note, for example, that the unit sphere $\{(x,y): x^2+y^2\le 1\}$ is *not* a one-dimensional set since it requires simultaneous constraints on two coordinates.

More generally, for a fixed finite k the product sigma field $\boldsymbol{B}(\mathbf{R})^{Z_k}$ (or simply $\boldsymbol{B}(\mathbf{R})^k$) of k-dimensional Euclidean space $\mathbf{R}^k$ is the smallest sigma field containing all one-dimensional events of the form $\{\mathbf{x} = (x_0,x_1,...,x_{k-1}): x_i\in F\}$ for some $i = 0,1,...,k-1$ and some Borel set F of **R**. The two-dimensional example (Figure 3.2a) has this form with $k = 2$, $i = 0$, and $F = (1,3)$. This one-dimensional set consists of all values in the infinite rectangle between 1 and 3 in the x_1 direction and all values between $-\infty$ and ∞ in the x_0 direction.

To summarize, we have defined a space A with event space $\boldsymbol{F}$, and an index set $\boldsymbol{I}$ such as $\mathbf{Z}_+$, $\mathbf{Z}$, $\mathbf{R}$, or $[0,1)$, and we have formed the product space $A^{\boldsymbol{I}}$ and the associated product event space $\boldsymbol{F}^{\boldsymbol{I}}$. We know that this event space contains all one-dimensional events by construction. We next consider what other events must be in $\boldsymbol{F}^{\boldsymbol{I}}$ by virtue of its being an event space.

After the one-dimensional events that pin down the value of a single coordinate of the vector or sequence or waveform, the next most general kinds of events are finite-dimensional sets that separately pin down the values of a finite number of coordinates. Let **K** be a finite subset of $\boldsymbol{I}$, that is, a finite collection of members of $\boldsymbol{I}$ and hence $\mathbf{K}\subset\boldsymbol{I}$. Say that **K** has K members, which we shall denote as $\{k_i; i = 0,1,...,K-1\}$. These K numbers can be thought of as a collection of sample times such as $\{1, 4, 8, 256, 1027\}$ for a sequence or $\{1.5, 9.07, 40.0, 41.2, 41.3\}$ for a waveform. We assume for convenience that the sample times are ordered in increasing fashion. Let $F_{k_i}; i = 0,1,...,K-1$ be a collection of members of $\boldsymbol{F}$. Then a set of the form

$$\{\{x_t; t\in\boldsymbol{I}\}: x_{k_i}\in F_{k_i}; i = 0,1,...,K-1\}$$

is an example of a finite-dimensional set. Note that it collects all sequences or waveforms such that a finite number of coordinates are constrained to lie in one-dimensional events. Two examples of two-dimensional sets of this form in two-dimensional space are illustrated in

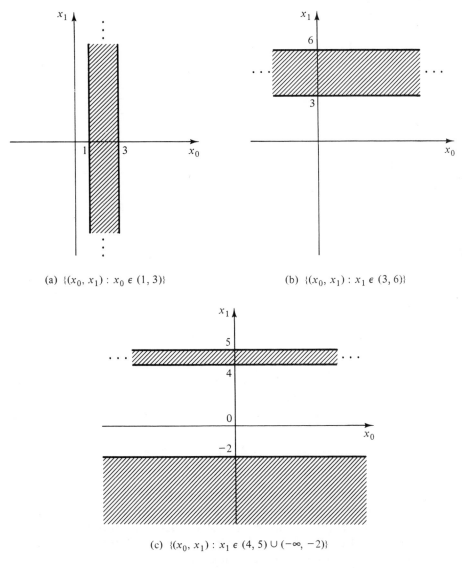

(a) $\{(x_0, x_1) : x_0 \in (1, 3)\}$

(b) $\{(x_0, x_1) : x_1 \in (3, 6)\}$

(c) $\{(x_0, x_1) : x_1 \in (4, 5) \cup (-\infty, -2)\}$

Figure 3.2   One-dimensional events in two-dimensional space.

Figure 3.3. Observe there that when the one-dimensional sets constraining the coordinates are intervals, then the two-dimensional sets are rectangles. Analogous to the two-dimensional example, finite-dimensional events having separate constraints on each coordinate are called *rectangles*. Observe, for example, that a circle or sphere in Euclidean space is not a rectangle because it cannot be defined using separate constraints on the coordinates; the constraints on each coordinate depend on the values of the others—e.g., in two dimensions we require that $x_0^2 \leq 1 - x_1^2$.

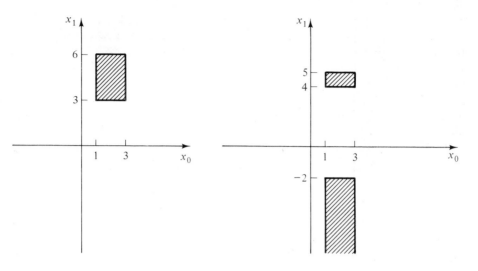

(a) $\{(x_0, x_1) : x_0 \in (1, 3), \quad x_1 \in (3, 6)\}$ (b) $\{(x_0, x_1) : x_0 \in (1, 3), \quad x_1 \in (4, 5) \cup (-\infty, -2)\}$

Figure 3.3   Two-dimensional events.

Note that example (a) of Figure 3.3 is just the intersection of examples (a) and (b) of Figure 3.2 and that example (b) of Figure 3.3 is the intersection of examples (a) and (c) of Figure 3.2. In fact, in general we can express finite-dimensional rectangles as intersections of one-dimensional events as follows:

$$\{\{x_t; \ t \in \boldsymbol{I}\}: x_{k_i} \in F_{k_i}; \ i = 0, 1, \dots, K-1 \} = \bigcap_{i=0}^{K-1} \{\{x_t; \ t \in \boldsymbol{I}\}: x_{k_i} \in F_i\} ,$$

that is, a set constraining a finite number of coordinates to each lie in one-dimensional events or sets in $\boldsymbol{F}$ is the intersection of a collection of one-dimensional events. Since $\boldsymbol{F}^{\boldsymbol{I}}$ is a sigma field and since it contains the one-dimensional events, it must contain such finite intersections, and hence it must contain such finite-dimensional events.

By concentrating on events that can be represented as the finite intersection of one-dimensional events we do not mean to imply that all events in the product event space can be represented in this fashion—the event space will also contain all possible limits of finite unions of such rectangles, complements of such sets, and so on. For example, the unit circle in two dimensions is not a rectangle, but it can be considered as a limit of unions of rectangles and hence is in the event space generated by the rectangles. (See exercise 3.22.)

The moral of this discussion is that the product sigma field for spaces of sequences and waveforms must contain (but not consist exclusively of) all sets that are described by requiring that the outputs or coordinates for a finite number of events lie in sets in the one-dimensional event space $\boldsymbol{F}$.

We shall further explore such product event spaces when considering random processes, but for the moment the key points are (1) a product event space is a sigma field, and (2) it contains all "one-dimensional events" consisting of subsets of the product sample space formed by grouping together all vectors or sequences or waveforms having a single fixed coordinate lying in a one-dimensional event. In addition, it contains all rectangles or finite-dimensional events consisting of all vectors or sequences or waveforms having a finite number of coordinates constrained to lie in one-dimensional events.

PROBABILITY MEASURES

The defining axioms of a probability measure as given in equations (3.5) through (3.8a) also correspond to intuitive notions, at least for the first three properties. The first property requires that a probability be a nonnegative number. In a purely mathematical sense, this is an arbitrary restriction, but it is in accord with the long history of intuitive and combinatorial developments of probability. Probability measures share this property with other measures such as area, volume, weight, and mass.

The second defining property corresponds to the notion that the probability that *something* will happen or that an experiment will produce one of its possible outcomes is one. This, too, is mathematically arbitrary but is a convenient and historical assumption. (From childhood we learn about things that are "100% certain"; obviously we could as easily take 100 or π (but *not* infinity—why?) to represent certainty.)

The third property is the key one and is called "additivity" or "finite additivity." In English it reads that the probability of occurrence of a finite collection of events having no points in common must be the sum of the probabilities of the separate events. More generally, the basic assumption of measure theory is that *any* measure—probabilistic or not—such as weight, volume, mass, and area should be additive: The mass of a group of disjoint regions of matter should be the sum of the separate masses; the weight of a group of objects should be the sum of the individual weights. Equation (3.7) only pins down this property for finite collections of events. The additional restriction of (3.8a), called *countable additivity,* is a limiting or asymptotic or infinite version, analogous to (3.3a) for set algebra. This again leads to the rhetorical questions of why the more complicated, more restrictive, and less intuitive infinite version is required. In fact, it was the addition of this limiting property that provided the fundamental idea for Kolmogorov's development of modern probability theory in the 1930s.

The response to the rhetorical question is essentially the same as that for the asymptotic set algebra property: Countably infinite properties are required to handle asymptotic and limiting results. Such results are crucial because we often need to evaluate the probabilities of complicated events that can only be represented as a limit of simple events. We need, in turn, for the probability of the complicated event to be the limit of the sum of the probabilities of the simple events. (This is analogous to the way that integrals are obtained as limits of finite sums.)

Note that it is *countable* additivity that is required. Uncountable additivity cannot be defined sensibly. This is easily seen in terms of the fair wheel mentioned at the beginning of the chapter. If the wheel is spun, any particular number has probability zero. On the other hand, the probability of the event made up of all of the *uncountable* numbers between 0 and 1 is obviously one. If you consider defining the probability of all the numbers between 0 and 1 to be the uncountable sum of the individual probabilities, you see immediately the essential contradiction that results.

Before exploring countable additivity further, we develop some simple but important elementary properties of probability measures that follow from the axioms of probability, (3.5) through (3.8a).

Elementary Properties of a Probability Space (Ω, F, P).   (All sets are assumed to be events, that is, in F.)

(a) $P(F^c) = 1 - P(F)$.

(b) $P(F) \leq 1$.

(c) $P(\emptyset) = 0$.

(d)   If $\{ F_i \}$ is a finite or countably infinite partition of Ω, then $P(G) = \sum_i P(G \cap F_i)$ for any event G.

Proof

(a) $F \cup F^c = \Omega$ implies $P(F \cup F^c) = 1$ (axiom 2). $F \cap F^c = \emptyset$ implies $1 = P(F \cup F^c) = P(F) + P(F^c)$ (axiom 3), which implies (a).

(b) $P(F) = 1 - P(F^c) \leq 1$ (axiom 1 and (a) above).

(c)   Axiom 2 and (a) above.

(d) $P(G) = P(G \cap \Omega) = P(G \cap (\bigcup_i F_i)) = P(\bigcup_i (G \cap F_i)) = \sum_i P(G \cap F_i)$.

At times we are interested in finding the probability of the limit of a sequence of events. To relate the countable additivity property of (3.8a) to limiting properties, recall the discussion of the limiting properties of events given earlier in this chapter in terms of increasing and decreasing sequences of events. Say we have an increasing sequence of events F_n; $n = 0,1,2,...$, $F_{n-1} \subset F_n$, and let F denote the limit set, that is, the union of all of the F_n. We have already argued that the limit set F is itself an event. Intuitively, since the F_n converge to F, the probabilities of the F_n should converge to the probability of F. Such convergence is called a *continuity* property of probability and is very useful for evaluating the probabilities of complicated events as the limit of a sequence of probabilities of simpler events. We shall show that countable additivity implies such continuity. To accomplish this, define the sequence of sets $G_0 = F_0$ and $G_n = F_n - F_{n-1}$ for $n = 1,2,....$ The G_n are disjoint and have the same union as do the F_n (see Figure 3.1a as a visual aid). Thus we have from countable additivity that

$$P(\lim_{n \to \infty} F_n) = P(\bigcup_{k=0}^{\infty} F_k) = P(\bigcup_{k=0}^{\infty} G_k) = \sum_{k=0}^{\infty} P(G_k) .$$

Since the probability on the left is bounded above by one, so is the sum on the right. Rewriting the sum to group the first n terms together and using finite additivity we have that $F_n = \bigcup_{k=0}^{n} G_k$, so that

$$P(F) = P(F_n \cup \bigcup_{k=n+1}^{\infty} G_k) = P(F_n) + P(\bigcup_{k=n+1}^{\infty} G_k)$$

$$= P(F_n) + \sum_{k=n+1}^{\infty} P(G_k) .$$

Taking the limit of the right-hand side above as $n \to \infty$, the rightmost sum must converge to zero as $n \to \infty$ because the final sum of the previous equation was finite. Thus we have proved the following:

If F_n is a sequence of increasing events, then

$$P(\lim_{n \to \infty} F_n) = \lim_{n \to \infty} P(F_n) , \tag{3.8b}$$

that is, the probability of the limit of a sequence of increasing events is the limit of the probabilities. Note that the sequence of probabilities on the right-hand side of (3.8b) is increasing with increasing n.

A similar argument can be used to show that one can also interchange the limit with the probability measure given a sequence of decreasing events; that is,

If F_n is a sequence of decreasing events, then

$$P(\lim_{n \to \infty} F_n) = \lim_{n \to \infty} P(F_n) . \tag{3.8c}$$

Note that the sequence of probabilities on the right-hand side of (3.8c) is decreasing with increasing n. It can be shown (see exercise 3.17) that, given (3.5) through (3.7), the three conditions (3.8a), (3.8b), and (3.8c) are equivalent; that is, any of the three could serve as the fourth axiom of probability.

Property (3.8b) is called *continuity from below,* and (3.8c) is called *continuity from above.* The designations "from below" and "from above" relate to the direction from which the respective sequences of probabilities approach their limit. These continuity results are the basis for using integral calculus to compute probabilities, since integrals can be expressed as limits of sums.

Examples of Probability Spaces

We now provide several examples of probability measures on our examples of sample spaces and sigma fields and thereby give some examples of complete probability spaces.

[3.12]

Let Ω be a finite set and let $\pmb{F}$ be the power set of Ω. Suppose that we have a function $p(\omega)$ that assigns a real number to each sample point ω in such a way that

$$p(\omega) \geq 0, \text{ all } \omega \epsilon \Omega \tag{3.9}$$

and

$$\sum_{\omega \epsilon \Omega} p(\omega) = 1 . \tag{3.10}$$

Define the set function P by

$$P(F) = \sum_{\omega \epsilon F} p(\omega) , \text{ all } F \epsilon \pmb{F} . \tag{3.11}$$

P is easily verified to be a probability measure: It obviously satisfies axioms 1 and 2. It is finitely and countably additive from the properties of sums. In particular, given a sequence of disjoint events, only a finite number can be distinct (since the power set of a finite space has only a finite number of members). To be disjoint, the balance of the sequence must equal $\varnothing$. The probability of the union of these sets will be the finite sum of the $p(\omega)$ over the points in the union which equals the sum of the probabilities of the sets in the sequence.

Example [3.1] is a special case of example [3.12].

A function $p(\omega)$ satisfying (3.9) and (3.10) is called a *probability mass function* or *pmf*. It is important to observe that the probability mass function is defined only for *points* in the sample space, while a probability measure is defined for a collection of *sets* (which can include, of course, the sets having single points). Intuitively, the probability of a set is given by the sum of the probabilities of the points as given by the pmf. Obviously it is much easier to describe the probability function than the probability measure since it is only specified for points. The axioms of probability then guarantee that the probability function can be used to compute the probability measure. Note that given one, we can always determine the other. In particular, given the pmf p, we can construct P using (3.11). Given P, we can find the corresponding pmf p from the formula

$$p(\omega) = P(\{\omega\}) .$$

While the details will differ, this idea will prove to hold in general for finite or countable spaces with $\boldsymbol{F}$ the power set. A generalization of this idea will later be seen to hold for quite general situations: An abstract probability measure, a set function for which the basic theory is developed, is equivalent to a probability function, a point function that can be used to compute probabilities. Here summations are used to find probabilities from probability functions.

We list below several of the most common examples of pmf's. The reader should verify that they are all indeed valid pmf's, that is, that they satisfy (3.9) and (3.10).

The binary pmf. $\Omega = \{0,1\}$; $p(0) = 1-p$, $p(1) = p$, where p is a parameter in $(0,1)$.

The uniform pmf. $\Omega = \boldsymbol{Z}_n = \{0,1,...,n-1\}$ and $p(k) = 1/n$; $k \in \boldsymbol{Z}_n$.

The binomial pmf. $\Omega = \boldsymbol{Z}_{n+1} = \{0,1,...,n\}$ and

$$p(k) = \binom{n}{k} p^k (1-p)^{n-k}; \ k \in \boldsymbol{Z}_{n+1},$$

where

$$\binom{n}{k} = \frac{n!}{k!(n-k)!}$$

is the binomial coefficient.

These pmf's have many important applications. In particular, the binary pmf is a probability model for coin flipping with a biased coin or for a single sample of a binary data stream. The uniform pmf on $\boldsymbol{Z}_6$ can model the roll of a fair die. Observe that it would not be a good model for ASCII data since, for example, the letters t and e and the symbol for space have a higher probability than other letters. The binomial pmf is a probability model for the number of heads in n successive biased coin flips.

The same construction provides a probability measure on countably infinite spaces such as $\mathbf{Z}$ and $\mathbf{Z}_+$. It is no longer as simple to prove countable additivity, but it should be fairly obvious that it holds and, at any rate, it follows from standard results in elementary analysis for convergent series. Hence we shall only state the following example without proving countable additivity, but bear in mind that it follows from the properties of infinite summations.

[3.13]

Let Ω be a space with a countably infinite number of elements and let $\boldsymbol{F}$ be the power set of Ω. Then if $p(\omega)$; $\omega\in\Omega$ satisfies (3.9) and (3.10), the set function P defined by (3.11) is a probability measure.

Two common examples of pmf's on countably infinite sample spaces follow. The reader should test their validity.

The geometric pmf. $\Omega = \{1,2,3,...\}$ and $p(k) = (1-p)^{k-1}p$; $k = 1, 2,...$, where $p\in(0,1)$ is a parameter.

The Poisson pmf. $\Omega = \mathbf{Z}_+ = \{0,1,2,...\}$ and $p(k) = \dfrac{\lambda^k e^{-\lambda}}{k!}$, where λ is a parameter in $(0, \infty)$.

We will later see the origins of several of these pmf's and their applications, but for the moment they should be considered as common important examples. Various properties of these pmf's and a variety of calculations involving them are explored in the exercises at the end of the chapter.

While the foregoing ideas were developed for scalar sample spaces such as $\mathbf{Z}_+$, they also apply to vector sample spaces. For example, if A is a discrete space, then so is the vector space $A^k = \{$all vectors $\mathbf{x} = (x_0,...x_{k-1})$ with $x_i\in A$, $i = 0,1,...,k-1\}$.

We now turn to continuous spaces.

[3.14]

Let $(\Omega,\boldsymbol{F}) = (\mathbf{R},\boldsymbol{B}(\mathbf{R}))$, the real line together with its Borel field. Suppose that we have a real-valued function f on the real line that satisfies the following properties:

$$f(r) \geq 0, \text{ all } r\in\Omega. \tag{3.12}$$

$$\int_\Omega f(r)dr = 1, \tag{3.13}$$

that is, the function $f(r)$ has a well-defined integral over the real line. Define the set function P by

$$P(F) = \int_F f(r)dr, \ F\in\boldsymbol{B}(\mathbf{R}). \tag{3.14}$$

Alternatively, we can write the definition of P without constraining the limits of integration by defining the *indicator function* of a set F by

$$1_F(r) = \begin{cases} 1 \text{ if } r \in F \\ 0 \text{ otherwise} \end{cases}$$

and writing

$$P(F) = \int f(r) 1_F(r) dr, \ F \in \boldsymbol{B}(\mathbf{R}) . \tag{3.15}$$

Is this set function actually a probability measure? It certainly should be, since (3.12) to (3.14) are the integral analogs of the summations of (3.9) to (3.11). We take the time to consider this issue since it leads us to the reasons behind the requirements for sigma fields and Borel fields.

The first issue is fundamental: Do the integrals make sense; i.e., are they well-defined for all events of interest? Suppose first that we take the common engineering approach and use Riemann integration—the form of integration used in elementary calculus. Then the above integrals are defined at least for events F that are intervals. This implies from the linearity properties of Riemann integration that the integrals are also well-defined for events F that are finite unions of intervals. It is not difficult, however, to construct sets F for which the indicator function 1_F is so nasty that the function $f(r) 1_F(r)$ does not have a Riemann integral. For example, suppose that $f(r)$ is 1 for $r \in [0,1]$ and 0 otherwise. Then the Riemann integral $\int 1_F dr$ is not defined for the set F of all irrational numbers, yet intuition should suggest that the set has probability 1. Thus the definition of (3.15) has a basic problem: The integral in the formula giving the probability measure of a set may not be well-defined.

A natural approach to escaping this dilemma would be to use the Riemann integral when possible, i.e., to define the probabilities of events that are finite unions of intervals, and then to obtain the probabilities of more complicated events by expressing them as a limit of finite unions of intervals, if the limit makes sense. This would hopefully give us a reasonable definition of a probability measure on a class of events much larger than the class of all finite unions of intervals. Intuitively, it should give us a probability measure for all sets that can be expressed as increasing or decreasing limits of finite unions of intervals.

This larger class is, in fact, the Borel field, but the Riemann integral has the unfortunate property that in general we cannot interchange limits and integration; that is, the limit of a sequence of integrals of converging functions may not be itself an integral of a limiting function.

This problem is so important to the development of a rigorous probability theory that it merits additional emphasis: Even though the familiar Riemann integrals of elementary calculus suffice for most engineering and computational purposes, they are too weak for building a useful theory, proving theorems, and evaluating the probabilities of some events which can be most easily expressed as limits of simple events. The problems are that the Riemann integral does not exist for sufficiently general functions and that limits and integration cannot be interchanged in general.

The solution is to use a different definition of integration—the Lebesgue integral. Here we need only concern ourselves with a few simple properties of the Lebesgue integral. The first property is unimportant for practice but may aid intuition. The Riemann integral of a function $f(r)$ "carves up" or partitions the domain of the argument r and effectively considers weighted sums of the values of the function $f(r)$ as the partition becomes ever finer. Conversely, the Lebesgue integral "carves up" the values of the function itself and effectively defines an integral as a limit of simple integrals of quantized versions of the function. This simple change of definition results in two fundamentally important properties of Lebesgue integrals that are not possessed by Riemann integrals:

1. The integral is defined for all Borel sets.
2. Subject to suitable technical conditions (such as integrands with bounded absolute value), one can interchange the order of limits and integration; e.g., if $F_n \uparrow F$, then

$$P(F) = \int 1_F(r)f(r)dr = \int \lim_{n \to \infty} 1_{F_n}(r)f(r)dr$$

$$= \lim_{n \to \infty} \int 1_{F_n}(r)f(r)dr = \lim_{n \to \infty} P(F_n) ,$$

that is, (3.8b) holds, and hence the set function is continuous from below.

The first property means that the definition of (3.15) is well-defined. From the linearity property of integration (of either kind), the resulting set function is additive: if F and G are disjoint, then

$$\int_{F \cup G} f(r)dr = \int_F f(r)dr + \int_G f(r)dr .$$

It also follows from the basic properties of integration (any kind) that $P(F) \geq 0$ for any Borel set F. Thus the mere existence of the integral in (3.15) is sufficient to imply that the first three axioms of probability are satisfied. The second property of Lebesgue integration implies that P satisfies the continuity property of (3.8b) and hence, since it is additive, it is also countably additive. Thus if we interpret the integral in a Lebesgue sense, (3.15) gives a well-defined probability measure. We observe in passing that even if we confined interest to events for which the Riemann integral made sense, it would not follow that the resulting probability measure would be countably additive: As with continuity, these asymptotic properties hold for Lebesgue integration but not for Riemann integration.

How do we reconcile the use of a Lebesgue integral given the assumed prerequisite of traditional engineering calculus courses based on the Riemann integral? Here a standard result of real analysis comes to our aid: If the Riemann integral exists, then so does the Lebesgue integral, and the two are the same. If the Riemann integral does not exist, then we can try to find the probability as a limit of probabilities of simple events for which the Riemann integrals do exist, e.g., as the limit of probabilities of finite unions of intervals. In other words, Riemann calculus will usually suffice for computation (at least if $f(r)$ is Riemann integrable) provided we realize that we may have to take limits of Riemann integrals for complicated events. Observe, for example, that in the case mentioned where $f(r)$ is 1 on [0,1], the probability of a single point 1/2 can now be found easily as a limit of Riemann integrals:

$$P(\{\frac{1}{2}\}) = \lim_{\epsilon \to 0} \int_{(\frac{1}{2}-\epsilon, \frac{1}{2}+\epsilon)} dr = \lim_{\epsilon \to 0} 2\epsilon = 0 ,$$

as expected.

In summary, our engineering compromise is this: We must realize that for the theory to be valid and for (3.15) indeed to give a probability measure on subsets of the real line, the integral must be interpreted as a Lebesgue integral and Riemann integrals may not exist. For computation, however, one will almost always be able to find probabilities by either Riemann integration or by taking limits of Riemann integrals over simple events. This distinction between Riemann integrals for computation and Lebesgue integrals for theory is analogous to the distinction between rational numbers and real numbers. Computational and engineering tasks use only arithmetic of finite precision in practice. However, in developing the theory irrational numbers such as $\sqrt{2}$ and π are essential. Imagine how hard it would be to develop a theory without using irrational numbers, and how unwise it would be to do so just because the eventual computations do not use them. So it is with Lebesgue integrals.

The function f used in (3.12) to (3.15) is called a *probability density function* or *pdf* since it is a nonnegative function that is integrated to find a total mass of probability, just as a mass density function in physics is integrated to find a total mass. Like a pmf, a pdf is defined only for *points* in Ω and not for sets. Unlike a pmf, a pdf is not in itself the probability of anything; for example, a pdf can take on values greater than one, while a pmf cannot. Under a pdf, points frequently have probability zero even though the pdf is nonzero. Both probability functions, the pmf and the pdf, can be used to define and compute a probability measure:

The pmf is summed over all points in the event, and the pdf is integrated over all points in the event.

Some of the most common pdf's are listed below. The reader should verify that they are indeed valid pdf's, that is, that they satisfy (3.12) and (3.13). The pdf's are assumed to be 0 outside of the specified domain. b, a, $\lambda > 0$, m, and $\sigma > 0$ are parameters in **R**.

The uniform pdf. Given $b > a$, $f(r) = 1/(b-a)$ for $r \in [a,b]$.

The exponential pdf. $f(r) = \lambda e^{-\lambda r}$; $r \geq 0$.

The doubly exponential pdf. $f(r) = \dfrac{\lambda}{2} e^{-\lambda |r|}$; $r \in$ **R**.

The Gaussian pdf. $f(r) = (2\pi\sigma^2)^{-1/2} e^{-(r-m)^2/2\sigma^2}$; $r \in$ **R**.

Just as we used a pdf to construct a probability measure on the space $(\mathbf{R}, \boldsymbol{B}(\mathbf{R}))$, we can also use it to define a probability measure on any smaller space $(A, \boldsymbol{B}(A))$, where A is a subset of **R**. As a technical detail we note that to ensure that the integrals all behave as expected we must also require that A itself be a Borel set of **R** so that it is precluded from being too nasty a set. Such probability spaces can be considered to have a sample space of either **R** or A, as convenient. In the former case events outside of A will have zero probability.

By considering multidimensional integrals we can also extend this construction to finite-dimensional product spaces, e.g., $\mathbf{R}^k$.

[3.15]

Given the measurable space $(\mathbf{R}^k, \boldsymbol{B}(\mathbf{R})^k)$, say we have a real-valued function f on R^k with the properties that

$$f(\mathbf{x}) \geq 0 \ ; \ \text{all } \mathbf{x} = (x_0, x_1, \ldots, x_{k-1}) \in \mathbf{R}^k \ , \tag{3.16}$$

$$\int_{\mathbf{R}^k} f(\mathbf{x}) d\mathbf{x} = 1 \ . \tag{3.17}$$

Then define a set function P by

$$P(F) = \int_F f(\mathbf{x}) d\mathbf{x}; \ \text{all } F \in \boldsymbol{B}(\mathbf{R})^k, \tag{3.18}$$

where the vector integral is shorthand for the k-dimensional integral, that is,

$$P(F) = \int_{(x_0, x_1, \ldots, x_{k-1}) \in F} f(x_0, x_1, \ldots, x_{k-1}) dx_0 dx_1 \ldots dx_{k-1} \ .$$

Note that (3.16) to (3.18) are exact vector equivalents of (3.12) to (3.14). Is P defined by (3.18) a probability measure? The answer is a qualified yes with exactly the same qualifications as in the one-dimensional case.

As in the one-dimensional sample space, a function f with the above properties is called a probability density function or pdf. To be more concise we will occasionally refer to a pdf on k-dimensional space as a k-dimensional pdf.

There are two common and important examples of k-dimensional pdf's. These are defined next. In both examples the dimension k of the sample space is fixed and the pdf's induce a probability measure on $(\mathbf{R}^k, \boldsymbol{B}(\mathbf{R})^k)$ by (3.18).

[3.16] *The product pdf*

Let f_i; $i = 0,1,...,k-1$, be a collection of one-dimensional pdf's; that is, $f_i(r)$; $r \in \mathbf{R}$ satisfies (3.12) and (3.13) for each $i = 0,1,...,k-1$. Define the product k-dimensional pdf f by

$$f(\mathbf{x}) = f(x_0, x_1,...,x_{k-1}) = \prod_{i=0}^{k-1} f_i(x_i) \, .$$

The product pdf in k-dimensional space is simply the product of k pdf's on one-dimensional space. The one-dimensional pdf's are called the *marginal* pdf's, and the multidimensional pdf is sometimes called a *joint* pdf. It is easy to verify that the product pdf integrates to 1.

The case of greatest importance is when all of the marginal pdf's are identical, that is, when $f_i(r) = f_0(r)$ for all i. Note that any of the previously defined pdf's on $\mathbf{R}$ yield a corresponding multidimensional pdf by this construction. In a similar manner we can construct pmf's on discrete product spaces as a product of marginal pmf's.

[3.17] *The multidimensional Gaussian pdf*

Let $\mathbf{m} = (m_0, m_1,...,m_{k-1})^t$ denote a column vector (the superscript t stands for "transpose"). Let Λ denote a k by k square matrix with entries $\{\lambda(i,j); i = 0,1,...,k-1; j = 0,1,...,k-1\}$. Assume that Λ is symmetric; that is, that $\Lambda^t = \Lambda$ or, equivalently, that $\lambda(i,j) = \lambda(j,i)$, all i,j. Assume also that Λ is *positive definite*; that is, for any nonzero vector $\mathbf{y} \in \mathbf{R}^k$ the quadratic form $\mathbf{y}^t \Lambda \mathbf{y}$ is positive, that is,

$$\mathbf{y}^t \Lambda \mathbf{y} = \sum_{i=0}^{k-1} \sum_{j=0}^{k-1} y_i \lambda(i,j) y_j > 0 \, .$$

A multidimensional pdf is said to be Gaussian if it has the following form for some vector $\mathbf{m}$ and matrix Λ satisfying the above conditions:

$$f(\mathbf{x}) = (2\pi)^{-k/2} (\det\Lambda)^{-1/2} e^{-1/2(\mathbf{x}-\mathbf{m})^t \Lambda^{-1}(\mathbf{x}-\mathbf{m})}; \quad \mathbf{x} \in \mathbf{R}^k.$$

where $\det\Lambda$ is the determinant of the matrix Λ.

Since the matrix Λ is positive definite, the inverse of Λ exists and hence the pdf is well defined. It is also necessary for Λ to be positive definite if the integral of the pdf is to be finite. The Gaussian pdf may appear complicated, but it will later be seen to be one of the simplest to deal with. We shall later develop the significance of the vector **m** and matrix Λ. Note that if Λ is a diagonal matrix, example [3.17] reduces to a special case of example [3.16].

The reader must either accept on faith that the Gaussian pdf also integrates to 1 or seek out a derivation.

[3.18] *Mixtures*

Suppose that P_i, $i = 1, 2, 3,...$ is a collection of probability measures on a common measurable space $(\Omega, \mathbf{F})$, and let a_i, $i = 1, 2,...$ be nonnegative numbers that sum to 1. Then the set function P defined by

$$P(F) = \sum_i a_i P_i(F)$$

is also a probability measure on $(\Omega, \mathbf{F})$. This relation is usually abbreviated to

$$P = \sum_i a_i P_i .$$

The first two axioms are obviously satisfied by P, and countable additivity follows from the properties of sums. (Finite additivity is easily demonstrated for the case of a finite number of P_i.) A probability measure formed in this way is called a *mixture*. Observe that this construction can be used to form a probability measure with both discrete and continuous aspects. For example, let Ω be the real line and $\mathbf{F}$ the Borel field; suppose that f is a pdf and p is a pmf; then for any $\lambda \in (0,1)$ the measure P defined by

$$P(F) = \lambda \sum_{x \in F} p(x) + (1-\lambda) \int_{x \in F} f(x) dx$$

combines a discrete portion described by p and a continuous portion described by f. Note that this construction works for both scalar and vector spaces. This combination of discrete and continuous attributes is one of the main applications of mixtures. Another is in modeling a random process where there is some uncertainty about the parameters of the experiment. For example, consider a probability space for the following experiment: First a fair coin is flipped and a 0 or 1 (tail or head) observed. If the coin toss results in a 1, then a fair die described by a uniform pmf p_1 is rolled, and the outcome is the result of the experiment. If the coin toss results in a 0, then a biased die described by a nonuniform pmf p_2 is rolled, and the outcome is the result of the experiment. The pmf of the overall experiment is then the mixture $p_1/2 + p_2/2$. The mixture model captures our ignorance of which die we will be rolling.

EXERCISES

3.1 Describe the sigma field of subsets of R generated by the points or singleton sets. Does this sigma field contain intervals of the form (a,b) for $b > a$?

3.2 Given a finite subset A of the real line R, prove that the power set of A and $B(A)$ are the same. Repeat for a countably infinite subset of R.

3.3 Given that the discrete sample space Ω has n elements, show that the power set of Ω consists of 2^n elements.

3.4 Let $\Omega = R$, the real line, and consider the collection F of subsets of R defined as all sets of the form

$$\bigcup_{i=0}^{k} (a_i,b_i] \cup \bigcup_{j=0}^{m} (c_i,d_i]^c$$

for all possible choices of nonnegative integers k and m and all possible choices of real numbers $a_i < b_i$, $c_i < d_i$. If k or m is 0, then the respective unions are defined to be empty so that the empty set itself has the form given. In other words, F contains all possible *finite* unions of half-open intervals of this form and complements of such half-open intervals. Every set of this form is in F and every set in F has this form. Prove that F is a field of subsets of Ω. Does F contain the points? For example, is the singleton set $\{0\}$ in F ? Is F a sigma field?

3.5 Let $\Omega = [0,\infty)$ be a sample space and let F be the sigma field of subsets of Ω generated by all sets of the form $(n,n+1)$ for $n = 1,2,\ldots$

(a) Are the following subsets of Ω in F? (i) $[0,\infty)$, (ii) $Z_+ = \{0,1,2,\ldots\}$, (iii) $[0,k]\cup[k+1,\infty)$ for any positive integer k, (iv) $\{k\}$ for any positive integer k, (v) $[0,k]$ for any positive integer k, (vi) $(1/3,2)$.

(b) Define the following set function on subsets of Ω:

$$P(F) = c \sum_{i \in Z_+ : i+1/2 \in F} 3^{-i}$$

(If there is no i for which $i+1/2 \in F$, then the sum is taken as zero.) Is P a probability measure on (Ω,F) for an appropriate choice of c? If so, what is c?

(c) Repeat part (b) with B, the Borel field, replacing F as the event space.

(d) Repeat part (b) with the power set of $[0,\infty)$ replacing F as the event space.

(e)   Find $P(F)$ for the sets F considered in part (a).

3.6 Show that an equivalent axiom to axiom 3 of probability is the following:

If F and G are disjoint, then $P(F \cup G) = P(F) + P(G)$,

that is, we really need only specify finite additivity for the special case of $n = 2$.

3.7 Consider the measurable space $([0,1], \boldsymbol{B}([0,1]))$. Define a set function P on this space as follows:

$$P(F) = \begin{cases} 1/2 & \text{if } 0 \in F \text{ or } 1 \in F \text{ but not both} \\ 1 & \text{if } 0 \in F \text{ and } 1 \in F \\ 0 & \text{otherwise .} \end{cases}$$

Is P a probability measure?

3.8 Let S be a sphere in $\mathbf{R}^3$: $S = \{ (x,y,z): x^2+y^2+z^2 \le r^2 \}$, where r is a fixed radius. In the sphere are fixed N molecules of gas, each molecule being considered as an infinitesimal volume (that is, it occupies only a point in space). Define for *any* subset of S the function

$$n(F) = \{ \text{ the number of molecules in } F \} .$$

Show that $P(F) = n(F)/N$ is a probability measure on the measurable space consisting of S and its power set.

In exercises 3.9 to 3.16 let $(\Omega, \boldsymbol{F}, P)$ be a probability space and assume that all given sets are events.

3.9 If $G \subset F$, prove that $P(F-G) = P(F) - P(G)$. Use this fact to prove that if $G \subset F$, then $P(G) \le P(F)$.

3.10 Show that for arbitrary (not necessarily disjoint) events,

$$P(F \cup G) = P(F) + P(G) - P(F \cap G) .$$

3.11 Let $\{F_i\}$ be a countable partition of a set G. Prove that for any event H,

$$\sum_i P(H \cap F_i) = P(H \cap G) .$$

3.12 If $\{ F_i; i = 1,2,... \}$ forms a partition of Ω and $\{ G_i; i = 1,2,... \}$ forms a partition of Ω, prove that for any H,

$$P(H) = \sum_{i=1}^{\infty} \sum_{j=1}^{\infty} P(H \cap F_i \cap G_j) .$$

3.13 Prove that $|P(F) - P(G)| \leq P(F \Delta G)$.

3.14 Prove that for any events F, G, and H,

$$P(F \Delta G) \leq P(F \Delta H) + P(H \Delta G) .$$

The astute observer may recognize this as a form of the triangle inequality; one can consider $P(F \Delta G)$ as a distance or metric on events.

3.15 Prove that if $P(F) \geq 1 - \delta$ and $P(G) \geq 1 - \delta$, then also $P(F \cap G) \geq 1 - 2\delta$. In other words, if two events have probability nearly one, then their intersection has probability nearly one.

3.16 Prove that for any sequence (i.e., countable collection) of events F_i,

$$P(\bigcup_{i=1}^{\infty} F_i) \leq \sum_{i=1}^{\infty} P(F_i) .$$

This inequality is called the *union bound* . (*Hint*: Use exercise 2.2 or 3.10.)

3.17 Show that given (3.5) through (3.7), (3.8b) or (3.8c) implies (3.8a). Thus (3.8a),(3.8b), and (3.8c) provide equivalent candidates for the fourth axiom of probability.

3.18 Prove that all of the named pmf's introduced in this chapter are indeed pmf's, that is, that they satisfy the required properties.

3.19 Prove that all of the named (one-dimensional) pdf's introduced in this chapter are indeed pdf's, that is, that they satisfy the required properties.

3.20 A probability space consists of a sample space Ω = all pairs of positive integers (that is, $\Omega = \{1,2,3,...\}^2$) and a probability measure P described by the pmf p defined by

$$p(k,m) = p^2(1-p)^{k+m-2} .$$

   (a)   Find $P(\{(k,m):k \geq m\})$.

   (b)   Find the probability $P(\{(k,m):k+m = r\})$ as a function of r for $r = 2,3,....$ Show that the result is a pmf.

   (c)   Find the probability $P(\{(k,m):k$ is an odd number$\})$.

3.21 Given the uniform pdf on $[0,1]$, $f(x) = 1$; $x \in [0,1]$, find an expression for $P((a,b))$ for all real $b > a$. Find a formula for the probability of the event $\{x:x \leq r\}$ as a function of $r \in \mathbf{R}$. (This is called a *cumulative distribution function*.) As an example where Riemann integrals may not work in the definition of a probability measure from a pdf but continuity of probability measures yields a result: Find the probability of the set $\{x: x$ is a rational number$\}$ and the set

$\{x : x$ is an irrational number$\}$. Find the probability of the event

$$F = \{\omega: \omega\in[\frac{1}{2^k},\frac{1}{2^k}+\frac{1}{2^{k+1}}) \text{ for some even } k\}$$

$$= \bigcup_{k \text{ even}} [\frac{1}{2^k},\frac{1}{2^k}+\frac{1}{2^{k+1}}) .$$

3.22 Let Ω be a unit square $\{(x,y): (x,y)\in\mathbf{R}^2, -1/2\leq x\leq 1/2, -1/2\leq y\leq 1/2\}$ and let F be the corresponding product Borel field. Is the circle $\{ (x,y): (x^2+y^2)^{1/2}\leq 1/2\}$ in F? (Give a plausibility argument.) If so, find the probability of this event if one assumes a uniform density function on the unit square.

3.23 Let $\Omega = \mathbf{R}^2$ and suppose we have a pdf $f(x,y)$ such that

$$f(x,y) = \begin{cases} C \text{ if } x\geq 0, y\geq 0, x+y\leq 1 \\ 0 \text{ otherwise .} \end{cases}$$

Find the probability $P(\{ (x,y): 2x>y\})$. Find the probability $P(\{ (x,y): x\leq\alpha\})$ for all real α. Is f a product pdf?

3.24 Prove that the product k-dimensional pdf integrates to 1 over $\mathbf{R}$.

3.25 Given the one-dimensional exponential pdf, find $P(\{ x: x>r\})$ and the cumulative distribution function $P(\{ x: x\leq r\})$ for $r\in\mathbf{R}$.

3.26 Given the k-dimensional product doubly exponential pdf, find the probabilities of the following events in $\mathbf{R}^k$: $\{\mathbf{x}: x_0\geq 0\}$, $\{\mathbf{x}: x_i>0$, all $i = 0,1,...,k-1\}$, $\{ \mathbf{x}: x_0>x_1\}$.

3.27 Let $(\Omega,F) = (\mathbf{R},\boldsymbol{B}(\mathbf{R}))$. Let P_1 be the probability measure on this space induced by a geometric pmf with parameter p and let P_2 be the probability measure induced on this space by an exponential pdf with parameter λ. Form the mixture measure $P = P_1/2+P_2/2$. Find $P(\{ \omega: \omega>r\})$ for all $r\in[0,\infty)$.

3.28 Let $\Omega = \mathbf{R}^2$ and suppose we have a pdf $f(x,y)$ such that

$$f(x,y) = Ce^{-\frac{1}{2\sigma^2}x^2}e^{-\lambda y}; x\in(-\infty,\infty), y\in[0,\infty) .$$

Find the constant C. Is f a product pdf? Find the probability $\Pr(\{(x,y): \sqrt{|x|}\leq\alpha\})$ for all possible values of a parameter α. Find the probability $\Pr(\{(x,y): x^2\leq y\})$.

3.29 Define $g(x)$ by

$$g(x) = \begin{cases} \lambda e^{-\lambda x} & x\in[0,\infty) \\ 0 & \text{otherwise .} \end{cases}$$

Let $\Omega = \mathbf{R}^2$ and suppose we have a pdf $f(x,y)$ such that

$$f(x,y) = Cg(x)g(y-x) .$$

Find the constant C. Find an expression for the probability $P(\{(x,y): y \le \alpha\})$ as a function of the parameter α. Is f a product pdf?

3.30 Let $\Omega = \mathbf{R}^2$ and suppose we have a pdf such that

$$f(x,y) = \begin{cases} C|x| & -1 \le x \le 1; \; -1 \le y \le x \\ 0 & \text{otherwise} . \end{cases}$$

Find the constant C. Is f a product pdf?

4

RANDOM VARIABLES,
VECTORS, AND PROCESSES

This chapter provides the theoretical foundations of random variables, vectors, and processes; that is, it gives the basic definitions and properties together with some simple examples. All three concepts are variations on a single theme and may be included in the general term of *random object*. We will deal specifically with random variables first because they are the simplest conceptually—they can be considered to be special cases of the other two concepts.

The name *random variable* suggests a variable that takes on values randomly. In a loose, intuitive way this is the right interpretation—e.g., an observer who is measuring the amount of noise on a communication link sees a random variable in this sense. However, we require a more precise mathematical definition for analytical purposes. Mathematically a random variable is neither random nor a variable—it is just a function mapping some sample space into another space. The first space is the sample space portion of a probability space, and the second space is a subset of the real line (some authors would call this a "real-valued" random variable; we adopt the majority view and consider a random variable to be real-valued only and thus we require no qualifications).

A random variable is perhaps best thought of as a measurement on a probability space; that is, for each sample point ω the random variable produces some value, denoted functionally as $f(\omega)$. One can view ω as the result of some experiment and $f(\omega)$ as the result of a measurement made on the experiment. The experiment outcome ω is from an abstract space, e.g., real numbers, integers, ASCII characters, waveforms, sequences,

Chinese characters, etc. The resulting value of the measurement or random variable $f(\omega)$, however, must be "concrete" in the sense of being a real number, e.g., a meter reading. The randomness is really all in the original probability space and not in the random variable; that is, once the ω is selected in a "random" way, the output value or sample value of the random variable is determined.

EXAMPLES OF RANDOM VARIABLES

Before giving a formal definition of a random variable—which will require an additional technical assumption to be introduced shortly—we list several simple examples of random variables. In every case we are given a probability space $(\Omega, \mathbf{F}, P)$. However, for the moment we will concentrate on the sample space Ω and the random variable that is defined functionally on that space. Note that the function must be defined for *every* value in the sample space if it is to be a valid function. On the other hand, the function does not have to assume every possible value in its range.

As you will see, there is nothing magic about the names of the random variables. So far we have used the lower-case letter f. On occasion we will use other lower-case letters such as g and h. However, we will often use upper case letters far down in the alphabet, such as X, Y, Z, U, V, and W, as well.

[4.1]

Let $\Omega = \mathbf{R}$, the real line, and define the random variable $X: \Omega \to \Omega$ by $X(\omega) = \omega^2$ for all $\omega \in \Omega$. Thus the random variable is the square of the sample point. Note that since the square of a real number is always nonnegative, we could replace the range Ω by the range space $[0, \infty)$ and consider X as a mapping $X: \Omega \to [0, \infty)$. Other random variables mapping Ω into itself are $Y(\omega) = |\omega|$, $Z(\omega) = \sin(\omega)$, $U(\omega) = 3 \times \omega + 321.5$, and so on. We can also consider the identity mapping as a random variable; that is, we can define a random variable $W: \Omega \to \Omega$ by $W(\omega) = \omega$.

[4.2]

Let $\Omega = \mathbf{R}$ as in example [4.1] and define the random variable $f: \Omega \to \{-V, V\}$ by

$$f(r) = \begin{cases} +V & \text{if } r \geq 0 \\ -V & \text{if } r < 0 \,. \end{cases}$$

This example models what is called a *binary quantizer* of a real input. Observe that it maps a continuous space into a discrete space.

So far we have used ω exclusively to denote the argument of the random variable. However, we can obviously use any letter to denote the dummy variable or argument of the function, provided that we specify its domain; that is, we do not need to use ω all the time to specify elements of Ω: r, x, or any other dummy variable will do. We will, however, as a convention, always use *only lower-case letters* to denote dummy variables.

When referring to a function, we will use several methods of specification. Sometimes we will only give its name, say f; sometimes we will specify its domain and range, as in $f: \Omega \to A$; sometimes we will provide a specific dummy variable, as in $f(r)$; and sometimes we will provide the dummy variable and its domain, as in $f(r)$; $r \in \Omega$. Finally, functions can be shown with a place for the dummy variable marked by a period to avoid annointing any particular dummy variable as being somehow special, as in $f(.)$. These various notations are really just different means of denoting the same thing while emphasizing certain aspects of the functions. The only real danger of this notation is the same as that of calculus and trigonometry: If one encounters a function, say $\sin t$, does this mean the sine of a *particular t* (and hence a real number) or does it mean the entire waveform of $\sin t$ for *all t*? The distinction should be clear from the context, but the ambiguity can be removed, for example, by defining something like $\sin t_0$ to mean a particular value and $\{\sin t; t \in \mathbf{R}\}$ or $\sin(.)$ to mean the entire waveform.

[4.3]

Let U be as in example [4.1] and f as in [4.2]. Then the function $g: \Omega \to \Omega$ defined by $g(\omega) = f(U(\omega))$ is also a random variable. This relation is often abbreviated by dropping the explicit dependence on ω to write $g = f(U)$. More generally, any function of a function is another function, called a "composite" function. Thus a function of a random variable is another random variable. Similarly, one can consider a random variable formed by a complicated combination of other random variables—for example, $g(\omega) = \dfrac{1}{\omega}\sinh^{-1}[\pi \times e^{\cos(|\omega|^{3.4})}]$.

[4.4]

Let $\Omega = \mathbf{R}^k$, k-dimensional Euclidean space. Occasionally it is of interest to focus attention on the random variable which is defined as a particular coordinate of a vector $\omega = (x_0, x_1, \ldots, x_{k-1}) \in \mathbf{R}^k$. Toward this end we can define for each $i = 0, 1, \ldots, k-1$ a *sampling function* (or *coordinate function*) $\Pi_i: \mathbf{R}^k \to \mathbf{R}$ as the following random variable:

$$\Pi_i(\omega) = \Pi_i((x_0, \ldots, x_{k-1})) = x_i .$$

The sampling functions are also called "projections" of the higher dimensional space onto the lower. (This is the reason for the choice of Π (Greek P)—not to be confused with the product symbol $\prod$—to denote the functions.)

Similarly, we can define a sampling function for any product space, e.g., for sequence and waveform spaces.

[4.5]

Given a space A, an index set $\boldsymbol{I}$, and the product space $A^{\boldsymbol{I}}$, define as a random variable, for any fixed $t\in\boldsymbol{I}$, the sampling function $\Pi_t\colon A^{\boldsymbol{I}}\to A$ as follows: Since any $\omega\in A^{\boldsymbol{I}}$ is a vector or function of the form $\{x_s;\, s\in\boldsymbol{I}\}$, define for each t in $\boldsymbol{I}$ the mapping

$$\Pi_t(\omega) = \Pi_t(\{x_s;\, s\in\boldsymbol{I}\}) = x_t \;.$$

Thus, for example, if Ω is a one-sided binary sequence space $\prod_{i\in\mathbf{Z}_+}\{0,1\}_i = \{0,1\}^{\mathbf{Z}_+}$, and hence every point has the form $\omega = (x_0,x_1,\ldots)$, then $\Pi_3((0,1,1,0,0,0,1,0,1,\ldots)) = 0$. As another example, if for all t in the index set $\mathbf{R}_t$ is a replica of $\mathbf{R}$ and Ω is the space

$$\mathbf{R}^{\mathbf{R}} = \prod_{t\in\mathbf{R}}\mathbf{R}_t$$

of all real-valued waveforms $\{x(t);\, t\in(-\infty,\infty)\}$, then for $\omega = \{\sin t;\, t\in\mathbf{R}\}$, the sampling function at the particular time $t = 2\pi$ is

$$\Pi_{2\pi}(\{\sin t;\, t\in\mathbf{R}\}) = \sin 2\pi = 0 \;.$$

This is illustrated in Figure 4.1. Also shown are sampling functions for several other ω waveforms and sampling times.

[4.6]

Suppose that we have a one-sided binary sequence space $\{0,1\}^{\mathbf{Z}_+}$. For any $n\in\{1,2,\ldots\}$, define the random variable Y_n by $Y_n(\omega) = Y_n((x_0,x_1,x_2,\ldots)) = $ the index (time) of occurrence of the nth 1 in ω. For example, $Y_2((0,0,0,1,0,1,1,0,1,\ldots)) = 5$ because the second sample to be 1 is x_5.

[4.7]

Say we have a one-sided sequence space $\Omega = \prod_{i\in\mathbf{Z}_+}\mathbf{R}_i$, where $\mathbf{R}_i$ is a replica of the real line for each i in the index set. Since every ω in this space has the form $\{x_0,x_1,\ldots\} = \{x_i;\, i\in\mathbf{Z}_+\}$, we can define for each positive integer n the random variable, depending on n,

$$S_n(\omega) = S_n(\{x_i;\, i\in\mathbf{Z}_+\}) = n^{-1}\sum_{i=0}^{n-1} x_i \;,$$

the arithmetic average or "mean" of the first n coordinates of the infinite sequence.

For example, if $\omega = \{1,1,1,1,1,1,1,\ldots\}$, then $S_n = 1$. This average is also called a *Cèsaro mean* or *sample average* or *time average* since the index being summed over often corresponds to time; viz., we are adding the outputs at times 0 through $n-1$ in the preceding equation. Such arithmetic means will play a fundamental role in describing the long-term average behavior of random processes when we treat the law of large numbers and ergodic theorems in later chapters.

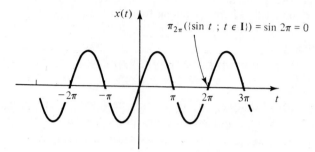

(a) $x(t) = \sin t$, $\mathbf{I} = (-\infty, \infty)$, $s = 2\pi$

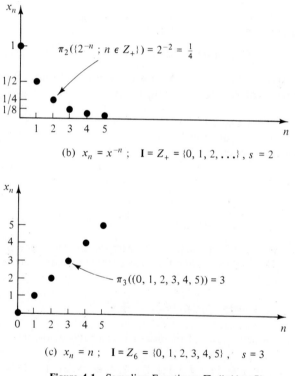

(b) $x_n = x^{-n}$; $\mathbf{I} = Z_+ = \{0, 1, 2, \ldots\}$, $s = 2$

(c) $x_n = n$; $\mathbf{I} = Z_6 = \{0, 1, 2, 3, 4, 5\}$, $s = 3$

Figure 4.1 Sampling Functions: $\Pi_s (\{x(t); t \in \mathbf{I}\})$.

We also write the arithmetic mean using coordinate functions as

$$S_n(\omega) = n^{-1} \sum_{i=0}^{n-1} \Pi_i(\omega), \tag{4.1a}$$

which we abbreviate to

$$S_n = n^{-1} \sum_{i=0}^{n-1} \Pi_i \qquad\qquad (4.1b)$$

by suppressing the dummy variable or argument ω. Equation (4.1b) is shorthand for (4.1a) and says the same thing: The arithmetic average of the first n terms of a sequence is the sum of the first n coordinates or samples of the sequence.

[4.8]

As a generalization of the sample average we can consider weighted averages of sequences. Such weighted averages occur in the convolutions of linear system theory. Let $\Omega = \prod_{i\in\mathbf{Z}}\mathbf{R}_i$, where the $\mathbf{R}_i$ are all copies of the real line. Suppose that we have a fixed sequence of real numbers $\{h_k;\ k = 0,1,2,...\}$ that we are going to use to form a weighted average of the coordinates of $\omega \in \Omega$. Since each ω in this space has the form $\omega = (...,x_{-1},x_0,x_1,...) = \{x_i;\ i\in\mathbf{Z}\}$, we can define for each integer n the random variable

$$Y_n(\omega) = \sum_{k=0}^{\infty} h_k x_{n-k} \ .$$

Thus the random variable Y_n is formed as a linear combination of the coordinates of the sequence constituting the point ω in the double-sided sequence space. This is a discrete time convolution of an input sequence with a linear weighting. In linear system theory the weighting is called a *pulse response* (or *Kronecker delta response* or *δ response*), and it is the discrete time equivalent of an impulse response. Note that we could also use the sampling function notation to write Y_n as a weighted sum of the sample random variables.

[4.9]

In a similar fashion, we can define complicated random variables on waveform spaces. For example, let $\Omega = \prod_{t\in\mathbf{R}}\mathbf{R}_t$, the space of all real-valued functions of time such as voltage-time waveforms. For each T, we can define the sample average function

$$Y_T(\omega) = Y_T(\{x(t);\ t\in\mathbf{R}\}) = T^{-1}\int_0^T x(t)dt \ ,$$

or given the impulse response $h(t)$ of a causal, linear time-invariant system, we could define the weighted average

$$W_T(\omega) = \int_0^{\infty} h(t)x(T-t)dt \ .$$

Are these also random variables? They are certainly functions defined on the underlying sample space, but as one might suspect, the sample space of all real-valued waveforms is quite large and contains some fairly weird waveforms. For example, the waveforms can be sufficiently pathological to preclude the existence of the integrals cited (see chapter 3 for a discussion of this point). These examples are sufficiently complicated to force us now to look a bit closer at a proper definition of a random variable and to develop a technical condition that constrains the generality of our definition but ensures that the definition will lead to a useful theory. It should be pointed out, however, that this difficulty is no accident and is not easily solved: Waveforms are truly more complicated than sequences because of the wider range of possible waveforms, and hence continuous time random processes are more difficult to deal with rigorously than are discrete time processes. One can write equations such as the integrals and then find that the integrals do not make sense, even in the general Lebesgue integral sense. Often fairly advanced mathematics are required to properly patch up the problems. Hence, in this book, for simplicity we usually concentrate on sequences (and hence the discrete time situation) rather than waveforms, and we gloss over the technical problems when we consider continuous time examples.

RANDOM VARIABLES

We now develop the promised technical condition for random variables and present a precise definition of a random variable. As you might guess, the technical condition for random variables is required because of certain subtle pathological problems that have to do with the ability to determine probabilities for the random variable. To arrive at the precise definition, we start with the informal definition of a random variable that we have already given and then show the inevitable difficulty that results without the technical condition. We have informally defined a random variable as being a function on a sample space. Suppose we have a probability space $(\Omega, \boldsymbol{F}, P)$. Let $f: \Omega \rightarrow \boldsymbol{R}$ be a function mapping the sample space into the real line so that f is a candidate for a random variable. Since the selection of the original sample point ω is random, that is, governed by a probability measure, so should be the output of our measurement or random variable $f(\omega)$. That is, we should be able to find the probability of an "output event" such as the event "the outcome of the random variable f was between a and b," that is, the event $F \subset \boldsymbol{R}$ given by $F = (a,b)$. Observe that there are two different kinds of events being considered here: (1) output events or members of the event space of the range or range space of the random variable, that is, events consisting of subsets of possible output values of the random variable; and (2) input events or Ω events, events in the original sample space of the original probability space. Can we find the probability of this output event? That is, can we make mathematical sense out of the quantity "the probability that f assumes a value in an event $F \subset \mathbf{R}$"? On reflection it seems clear that we can. The probability that f assumes a value in some set of values must be the probability of all values in the original sample space that result in a value of f in the given set. We will make this concept more precise shortly. To save writing we will abbreviate such English statements to the form $\Pr(f \in F)$, or $\Pr(F)$, that is, when the notation $\Pr(F)$ is encountered it should be interpreted as shorthand for the English statement for "the probability of an event F" or "the probability that the event F will occur" and not as a precise mathematical quantity.

Recall from chapter 3 that for a subset F of the real line $\mathbf{R}$ to be an event, it must be in a sigma field or event space of subsets of $\mathbf{R}$. Recall also that we adopted the Borel field $\boldsymbol{B}(\mathbf{R})$ as our basic event space for the real line. Hence it makes sense to require that our output event F be a Borel set.

Thus we can now state the question as follows: Given a probability space $(\Omega, \boldsymbol{F}, P)$ and a function $f: \Omega \to \mathbf{R}$, is there a reasonable and useful precise definition for the probability $\Pr(f \in F)$ for any $F \in \boldsymbol{B}(\mathbf{R})$, the Borel field or event space of the real line? Since the probability measure P sits on the original measurable space $(\Omega, \boldsymbol{F})$ and since f assumes a value in F if and only if $\omega \in \Omega$ is chosen so that $f(\omega) \in F$, the desired probability is obviously $\Pr(f \in F) = P(\{\omega: f(\omega) \in F\}) = P(f^{-1}(F))$. In other words, the probability that a random variable f takes on a value in a Borel set F is the probability (defined in the original probability space) of the set of all (original) sample points ω that yield a value $f(\omega) \in F$. This, in turn, is the probability of the inverse image of the Borel set F under the random variable f. This idea of computing the probability of an output event of a random variable using the original probability measure of the corresponding inverse image of the output event under the random variable is depicted in Figure 4.2.

This natural definition of the probability of an output event of a random variable indeed makes sense if and only if the probability $P(f^{-1}(F))$ makes sense, that is, if the subset $f^{-1}(F)$ of Ω corresponding to the output event F is itself an event, in this case an input event or member of the event space $\boldsymbol{F}$ of the original sample space. This, then, is the required technical condition: A function f mapping the sample space of a probability space $(\Omega, \boldsymbol{F}, P)$ into the real line $\mathbf{R}$ is a random variable if and only if the inverse images of all Borel sets in $\mathbf{R}$ are members of $\boldsymbol{F}$, that is, if all of the Ω sets corresponding to output events (members of $\boldsymbol{B}(\mathbf{R})$) are input events (members of $\boldsymbol{F}$). Unlike some of the other pathological conditions that we have met, it is easy to display some trivial examples where the technical condition is not met (as we will see in example [4.11]). We now formalize the definition:

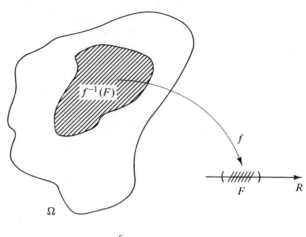

$$(\Omega, F, P) \quad \xrightarrow{\quad f \quad} \quad (R, B(R))$$

$$\Pr(f \in F) = P(\{\omega : f(\omega) \in F\})$$
$$= P(f^{-1}(F)) \qquad \text{if} \quad f^{-1}(F) \in F$$

Figure 4.2   Probability computation by inverse image technique.

Given a probability space $(\Omega, \boldsymbol{F}, P)$, a (real-valued) *random variable* is a function $f: \Omega \to \mathbf{R}$ with the property that if $F \in \boldsymbol{B}(\mathbf{R})$, then also $f^{-1}(F) = \{\omega: f(\omega) \in F\} \in \boldsymbol{F}$.

In some cases one may wish to consider a random variable with a more limited range space than the real line, e.g., when the random variable is binary. (Recall from chapter 2 that the range space of f is the image of Ω.) If so, $\mathbf{R}$ can be replaced in the definition by the appropriate subset, say $A \subset \mathbf{R}$. This is really just a question of semantics since the two definitions are equivalent. One or the other view may, however, be simpler to deal with for a particular problem.

A function meeting the condition in the definition we have given is said to be *measurable*. This is because such functions inherit a probability measure on their output events (specifically a probability measure in our context; in other contexts more general measures can be defined). Thus a random variable is simply a measurable function defined on a probability measure space.

One must know the event space being considered in order to determine whether or not a function is a random variable. While we will virtually always assume the usual event spaces (that is, the power set for discrete spaces, the Borel field for the real line or subsets of the real line, and the corresponding product event spaces for product sample spaces), it is useful to consider some other examples to help clarify the basic definition.

[4.10]

First consider $(\Omega, \boldsymbol{F}, P)$ where Ω is itself a discrete subset of the real line $\mathbf{R}$, e.g., $\{0,1\}$ or $\mathbf{Z}_+$. If, as usual, we take $\boldsymbol{F}$ to be the power set, then *any* function $f: \Omega \rightarrow \mathbf{R}$ is a random variable. This follows since the inverse image of any Borel set in $\mathbf{R}$ must be a subset of Ω and hence must be in the collection of all subsets of Ω.

Thus with the usual event space for a discrete sample space—the power set—*any* function defined on the probability space is a random variable. This is why all of the structure of event spaces and random variables is not seen in elementary texts that consider only discrete spaces: There is no need.

It should be noted that for any Ω, discrete or not, if $\boldsymbol{F}$ is the power set, then *all* functions defined on Ω are random variables. This fact is useful, however, only for discrete sample spaces since the power set is not a useful event space in the continuous case (since we cannot endow it with useful probability measures).

If, however, $\boldsymbol{F}$ is *not* the power set, some functions defined on Ω are *not* random variables, as the following simple example shows:

[4.11]

Let Ω be discrete, as in example [4.10], but let $\boldsymbol{F}$ be the trivial sigma field $\{\Omega, \varnothing\}$. On this space it is easy to construct functions that are not random variables (and hence are nonmeasurable functions). For example, let $\Omega = \{0,1\}$ and define $f(\omega) = \omega$, the identity function. Then $f^{-1}(\{0\}) = \{0\}$ is not in $\boldsymbol{F}$, and hence this simple function is not a random variable. In fact, it is obvious that any function that assigns different values to 0 and 1 is not a random variable. Note, however, that some functions are random variables.

The problem illustrated by this example is that the input event space is not big enough or "fine" enough to contain all input sets corresponding to output events. This apparently trivial example suggests an important technique for dealing with advanced random process theory, especially for continuous time random processes: If the event space is not large enough to include the inverse image of all Borel sets, then enlarge the event space to include all such events, viz., by using the power set as in example [4.10]. Alternatively, we might try to force $\boldsymbol{F}$ to contain *all* sets of the form $f^{-1}(F)$, $F \in \boldsymbol{B}(\mathbf{R})$; that is, make $\boldsymbol{F}$ the sigma field generated by such sets. Further treatment of this subject is beyond the scope of the book. However, it is worth remembering that if a sigma field is not big enough to make a function a random variable, it can often be enlarged to be big enough. This is not idle twiddling; such a procedure is required for important applications, e.g., to make integrals over time defined on a waveform space into random variables.

On a more hopeful tack, if the probability space $(\Omega,\boldsymbol{F},P)$ is chosen with $\Omega = \mathbf{R}$ and $\boldsymbol{F} = \boldsymbol{B}(\mathbf{R})$, then all functions f normally encountered in the real world are in fact random variables. For example, continuous functions, polynomials, step functions, trigonometric functions, limits of measurable functions, maxima and minima of measurable functions, and so on are random variables. It is, in fact, extremely difficult to construct functions on Borel spaces that are *not* random variables. The same statement holds for functions on sequence spaces. The difficulty is comparable to constructing a set on the real line that is not a Borel set and is beyond the scope of the book.

So far we have considered abstract philosophical aspects in the definition of random variables. We are now ready to develop the properties of the defined random variables.

Suppose we have a probability space $(\Omega,\boldsymbol{F},P)$ with a random variable, X, defined on the space. The random variable X takes values on its range space which is some subset A of $\mathbf{R}$ (possibly $A = \mathbf{R}$). The range space A of a random variable is often called the *alphabet* of the random variable. Because X is a random variable, we know that all subsets of Ω of the form $X^{-1}(F) = \{\omega: X(\omega)\epsilon F\}$, with $F\epsilon\boldsymbol{B}(A)$, must be members of $\boldsymbol{F}$ by definition. Thus the set function P_X defined by

$$P_X(F) = P(X^{-1}(F)) = P(\{\omega: X(\omega)\epsilon F\}); \quad F\epsilon\boldsymbol{B}(A) \tag{4.2}$$

is well defined and assigns probabilities to *output* events involving the random variable in terms of the original probability of *input* events in the original experiment. The three written forms in equation (4.2) are all read as $\Pr(X\epsilon F)$ or "the probability that the random variable X takes on a value in F." Furthermore, since inverse images preserve all set-theoretic operations (see exercise 2.12), P_X satisfies the axioms of probability as a probability measure on $(A,\boldsymbol{B}(A))$—it is nonnegative, $P_X(A) = 1$, and it is countably additive. Thus P_X is a probability measure on the measurable space $(A,\boldsymbol{B}(A))$. Therefore, given a probability space and a random variable X, we have constructed a new probability space $(A,\boldsymbol{B}(A),P_X)$ where the events describe outcomes of the random variable. The probability measure P_X is called the *distribution* of X (as opposed to a "cumulative distribution function," which will be introduced later).

If two random variables have the same distribution, then they are said to be *equivalent* since they have the same probabilistic description, whether or not they are defined on the same underlying space or have the same functional form (exercise 4.17).

DISTRIBUTIONS OF RANDOM VARIABLES

A substantial part of probability theory is devoted to determining the distributions of random variables. One begins with a probability space. A random variable is defined on that space. The distribution of the random variable is then derived, and this results in a new probability space. This topic is called variously "derived distributions" or "transformations of random variables" and is often developed in the literature as a sequence of apparently unrelated subjects. We shall emphasize that all such examples are just applications of the basic inverse image formula (4.2) and form a unified whole. In fact, this formula, with its vector analog, is one of the most important in probability theory. Its specialization to discrete spaces using sums and to continuous spaces using integrals will be seen and used often throughout this book.

It is useful to bear in mind both the mathematical and the intuitive concepts of a random variable when studying them. Mathematically, a random variable, say X, is a "nice" (= measurable) real-valued function defined on the sample space of a probability space $(\Omega, \boldsymbol{F}, P)$. Intuitively, a random variable is something that takes on values at random. The randomness is described by a distribution P_X, that is, by a probability measure on an event space of the real line. When doing computations involving random variables, it is usually simpler to concentrate on the probability space $(A, \boldsymbol{B}(A), P_X)$ instead of on the original underlying probability space $(\Omega, \boldsymbol{F}, P)$, where A is the range space of X. Many experiments can yield equivalent random variables, and the space $(A, \boldsymbol{B}(A), P_X)$ can be considered as a canonical description of the random variable that is often more useful for computation. Suppose, for example, that the original probability space consists of [0,1], its Borel sets, and a probability measure described by a uniform pdf. A random variable X is defined as 1 if $\omega < 1/2$ and 0 otherwise. The resulting range space, A, for the random variable is then a binary space described with a pmf placing probability $1/2$ on the outcomes 0 and 1, that is, the same space one would get flipping a single fair coin. If one is only interested in calculations involving the output values of X, then the second discrete probability space is easier and more direct to deal with.

The original space is important, however, for two reasons: First, all distribution properties of random variables are inherited from the original space. Therefore much of the theory of random variables is just the theory of probability spaces specialized to the case of real sample spaces. If we understand probability spaces in general, then we understand random variables in particular. Second, and more important, we will often have many interrelated random variables defined on the same probability space. Because of the interrelationships, we cannot consider the random variables independently with separate probability spaces and distributions. We must refer to the original space in order to study the dependencies among the various random variables.

Since a distribution is a special case of a probability measure, in many cases it can be induced or described by a probability function, i.e., a pmf or a pdf. If the range space of the random variable is discrete or, more generally, if there is a discrete subset of the range space A such that $P_X(A) = 1$, then there is a pmf, say p_X, corresponding to the distribution P_X. The two are related via the formulas

$$p_X(x) = P_X(\{x\}), \text{ all } x \in A \ , \tag{4.3}$$

where A is the range space or alphabet of the random variable, and

$$P_X(F) = \sum_{x \in F} p_X(x); \quad F \in \boldsymbol{B}(A) \ . \tag{4.4}$$

In (4.3) both quantities are read as $\Pr(X = x)$.

The pmf and the distribution imply each other from (4.3) and (4.4), and hence either formula specifies the random variable.

If the range space of the random variable is continuous and if a pdf f_X exists, then we can write the integral analog to (4.4):

$$P_X(F) = \int_F f_X(x)dx; \quad F \in \boldsymbol{B}(A) \ . \tag{4.5}$$

There is no direct analog of (4.3) since a pdf is not a probability. It is desirable, however, to have a pair of results like (4.3) and (4.4) that show how to go both ways, that is, to get the probability function from the distribution as well as vice versa. From considerations of elementary calculus it seems that we should somehow differentiate both sides of (4.5) to yield the pdf in terms of the distribution. This is not immediately possible, however, because F is a set and not a real variable. Instead, to find a pdf from a distribution, we use the intermediary of a cumulative distribution function or *cdf*. We pause to give the formal definition:

Given a random variable X with distribution P_X, the *cumulative distribution function* or *cdf* F_X is defined by

$$F_X(\alpha) = P_X((-\infty, \alpha]) = P_X(\{x : x \le \alpha\}); \quad \alpha \in \boldsymbol{R} \ .$$

The cdf is seen to represent the *cumulative* probability of all values of the random variable in the infinite interval from minus infinity up to and including the real number argument of the cdf. In English, $F_X(\alpha) = \Pr(X \le \alpha)$. If the random variable X is defined on the probability space $(\Omega, \boldsymbol{F}, P)$, then by definition

$$F_X(\alpha) = P(X^{-1}((-\infty, \alpha])) = P(\{\omega : X(\omega) \le \alpha\}) \ .$$

If a distribution possesses a pdf, then the cdf and pdf are related through the distribution and (4.5) by

$$F_X(\alpha) = P_X((-\infty,\alpha]) = \int_{-\infty}^{\alpha} f_X(x)dx; \ \alpha\epsilon\mathbf{R} \ . \tag{4.6}$$

The relationship between the pdf and cdf is illustrated in Figure 4.3. The motivation for the definition of the cdf in terms of our previous discussion is now obvious. Since integration and differentiation are mutually inverse operations, the pdf is determined from the cdf (and hence the distribution) by

$$f_X(\alpha) = \frac{dF_X(\alpha)}{d\alpha}; \ \ \alpha\epsilon\mathbf{R} \ , \tag{4.7}$$

where, as is customary, the right-hand side is shorthand for

$$\frac{dF_X(x)}{dx}\Big|_{x=\alpha} \ ,$$

the derivative evaluated at α. Alternatively, (4.7) also follows from the fundamental theorem of calculus and the observation that

$$P_X((a,b]) = \int_a^b f_X(x)dx = F_X(b)-F_X(a) \ . \tag{4.8}$$

Thus (4.6) and (4.7) together show how to find a pdf from a distribution and hence provide the continuous analog of (4.3). Equation (4.7) is useful, however, only if the derivative, and hence the pdf, exists. Observe that the cdf is always well defined (because the semi-infinite interval is a Borel set and therefore an event), regardless of whether or not the pdf exists in both the continuous and the discrete alphabet cases. For example, if X is a discrete alphabet random variable with alphabet $\mathbf{Z}$ and pmf p_X, then the cdf is

$$F_X(x) = \sum_{k=-\infty}^{x} p_X(k) \ , \tag{4.9}$$

the analogous sum to the integral of (4.6). Furthermore, for this example, the pmf can be determined from the cdf (as well as the distribution) as

$$p_X(x) = F_X(x)-F_X(x-1) \ , \tag{4.10}$$

a difference analogous to the derivative of (4.7).

It is desirable to use a single notation for the discrete and continuous cases whenever possible. This is accomplished for expressing the distribution in terms of the probability functions by using a Stieltjes integral, which is defined as follows:

$$P_X(F) = \int_F dF_X(x) = \begin{cases} \sum_{x\epsilon F} p_X(x) & \text{if } X \text{ is discrete} \\ \int_F f_X(x)dx & \text{if } X \text{ has a pdf} \ . \end{cases} \tag{4.11}$$

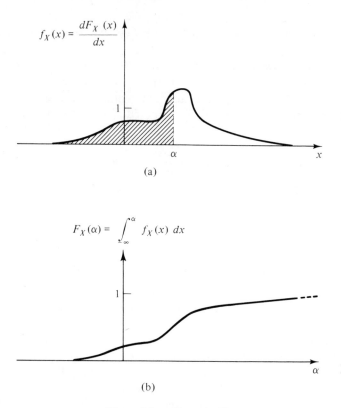

(a)

(b)

Figure 4.3   pdf's and cdf's.

Thus (4.11) is a combination of both (4.4) and (4.5).

More generally, we may have a random variable that has both discrete and continuous aspects and hence is not describable by either a pmf alone or a pdf alone. For example, we might have a probability space $(\mathbf{R}, \boldsymbol{B}(\mathbf{R}), P)$ where P is described by a Gaussian pdf $f(\omega)$; $\omega \in \mathbf{R}$. $\omega \in \mathbf{R}$ is input to a soft limiter with output $X(\omega)$—a device with the characteristic shown in Figure 4.4a. As long as $|\omega| \leq 1$, $X(\omega) = \omega$. But for values outside this range, the output is set equal to unity. Thus all of the probability density outside the limiting range "piles up" on the ends so that $\Pr(X(\omega) = 1) = \int_{\omega \geq 1} f(\omega) d\omega$ is not zero. This is illustrated in (b) and (c) of Figure 4.4 by showing the input pdf and the output probabilistic description as having two parts: a pdf for the continuous portion and a pmf for the discrete portion.

Random variables of this type can be described by a distribution that is the weighted sum of two other distributions—a discrete distribution and a continuous distribution. The weighted sum is an example of a mixture distribution, that is, a mixture of probability measures as in example

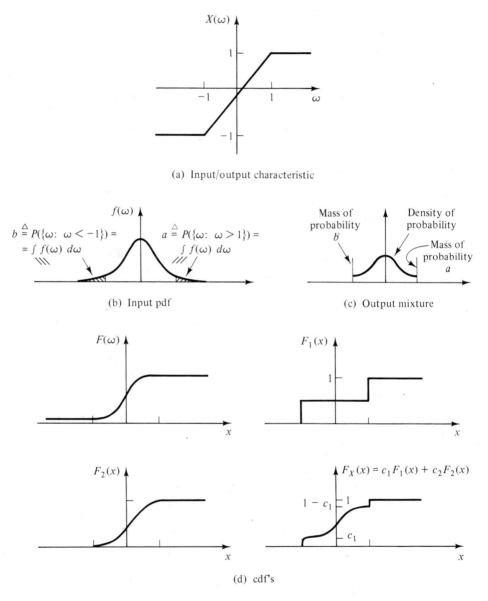

(a) Input/output characteristic

$b \overset{\triangle}{=} P(\{\omega: \ \omega < -1\}) =$
$= \int f(\omega) \ d\omega$

$a \overset{\triangle}{=} P(\{\omega: \ \omega > 1\}) =$
$\int f(\omega) \ d\omega$

(b) Input pdf

(c) Output mixture

(d) cdf's

Figure 4.4   Soft limiter.

[3.18].   Specifically, let P_1 be a discrete distribution with corresponding pmf p, and let P_2 be a continuous distribution described by a pdf f. For any positive weights c_1, c_2 with $c_1 + c_2 = 1$, the following mixture distribution P_X is defined:

$$P_X(F) = c_1 P_1(F) + c_2 P_2(F)$$

$$= c_1 \sum_{k \in F} p(k) + c_2 \int_F f(x)dx; \quad F \in \boldsymbol{B}(\boldsymbol{R}) . \tag{4.12}$$

For the example illustrated in Figure 4.4, the pmf places probability one half on ± 1. The pdf is Gaussian-shaped for magnitudes less than unity (i.e., it is a truncated Gaussian pdf normalized so that the pdf integrates to one over the range $(-1,1)$). The constant c_1 is the integral of the shaded area in Figure 4.4 and $c_2 = 1 - c_1$. Observe that the cdf for a random variable with a mixture distribution is

$$F_X(\alpha) = c_1 \sum_{k \leq \alpha} p(k) + c_2 \int_{-\infty}^{\alpha} f(x)dx$$

$$= c_1 F_1(\alpha) + c_2 F_2(\alpha) , \tag{4.13}$$

where F_1 and F_2 are the cdf's corresponding to P_1 and P_2, respectively. Figure 4.4c shows the cdf's for the discrete and continuous pieces and for the entire output for the example. The combined notation for discrete and continuous alphabets using the Stieltjes integral notation of (4.11) also can be used as follows: Given a random variable with a mixture distribution of the form (4.12), then

$$P_X(F) = \int_F dF_X(x); \quad F \in \boldsymbol{B}(\boldsymbol{R}) , \tag{4.14a}$$

where

$$\int_F dF_X(x) = c_1 \sum_{x \in F} p(x) + c_2 \int_F f(x)dx . \tag{4.14b}$$

Observe that (4.14) includes (4.11) as a special case where either c_1 or c_2 is 0. Equation (4.14) provides a general means for finding the distribution of a random variable X given its cdf, provided the distribution has the form of (4.12) or, equivalently, that the cdf has the form of (4.13).

All random variables can be described by a cdf. But, more subtly, do all random variables have a cdf of the form (4.13)? The answer is almost yes. Certainly all of the random variables encountered in this course and in engineering practice have this form. It can be shown, however, that the most general cdf has the form of a mixture of three cdf's: a continuous piece induced by a pdf, a discrete piece induced by a pmf, and a third pathological piece. The third piece is an odd beast wherein the cdf is something called a singular function—the cdf is continuous (it has no jumps as in the discrete case), and the cdf is differentiable almost everywhere (here "almost everywhere" means that the cdf is differentiable at all points except some set F for which $\int_{x \in F} dx = 0$), but this derivative is 0 almost everywhere and hence it cannot be integrated to find a probability! Thus for this third piece, one cannot use pmf's or pdf's to compute probabilities. The construction of such a cdf is beyond the scope of this text, but we can point out for the curious that the typical example involves placing probability measures on what is called a Cantor set. At any rate, as such examples never arise in practice, we shall ignore them and henceforth consider only random variables for which (4.14) holds.

While the general mixture distribution random variable has both discrete and continuous pieces, for pedagogical purposes it is usually simplest to treat the two pieces separately—i.e., to consider random variables that have either a pdf or a pmf. Hence we will rarely consider mixture distribution random variables and will almost always focus on those that are described either by a pmf or by a pdf and not both.

To summarize our discussion, we will define a random variable to be a *discrete*, *continuous*, or *mixture* random variable depending on whether it is described probabilistically by a pmf, pdf, or mixture as in (4.14) with $c_1, c_2 > 0$.

We note in passing that some texts endeavor to use a uniform approach to mixture distributions by permitting pdf's to possess Dirac delta or impulse functions. The purpose of this approach is to permit the use of the continuous ideas in discrete cases. As an example of this procedure, refer again to the example of Figure 4.4. Differentiate the cdf. Note that a legitimate pdf results (without the need for a pmf) if a delta function is allowed at the two discontinuities of the cdf. As a general practice we prefer the Stieltjes notation, however, because of the added notational clumsiness resulting from using pdf's to handle inherently discrete problems. For example, compare the notation for the geometric pmf with the corresponding pdf that is written using Dirac delta functions.

We next consider a few simple examples. Other examples are given in the exercises.

EXAMPLES OF DERIVED DISTRIBUTIONS

[4.12]

Let $(\Omega, \boldsymbol{F}, P)$ be a discrete probability space with Ω a discrete subset of the real line and $\boldsymbol{F}$ the power set. Let p be the pmf corresponding to P, that is,

$$p(\omega) = P(\{\omega\}), \text{ all } \omega \epsilon \Omega .$$

(*Note*: There is a very subtle possibility for confusion here. $p(\omega)$ could be considered to be a random variable because it satisfies the definition for a random variable. We do not use it in this sense, however; we use it as a pmf for evaluating probabilities in the context given. In addition, no confusion should result because we never use lower-case letters for random variables.) Let X be a random variable defined on this space. Since the domain of X is discrete, its range space, A, is also discrete (refer to the definition of a function to understand this point). Thus the probability measure P_X must also correspond to a pmf, say p_X;

that is, (4.3) and (4.4) must hold. Thus we can derive either the distribution P_X or the simpler pmf p_X in order to complete a probabilistic description of X. Using (4.2) yields

$$p_X(x) = P_X(\{x\}) = P(X^{-1}(\{x\})) = \sum_{\omega:\ X(\omega)\ =\ x} p(\omega) \ . \tag{4.15}$$

Equation (4.15) provides a formula for computing the pmf and hence the distribution of any random variable defined on a discrete probability space. As a specific example, consider a discrete probability space $(\Omega, \boldsymbol{F}, P)$ with $\Omega = \mathbf{Z}_+$, $\boldsymbol{F}$ the power set of Ω, and P the probability measure induced by the geometric pmf. Define a random variable Y on this space by

$$Y(\omega) = \begin{cases} 1 & \text{if } \omega \text{ even} \\ 0 & \text{if } \omega \text{ odd} \end{cases}$$

where we consider 0 (which has probability zero under the geometric pmf) to be even. Thus we have a random variable $Y\colon \mathbf{Z}_+ \to \{0,1\}$. Using the formula (4.15) for the pmf for $Y(\omega) = 1$ results in

$$p_Y(1) = \sum_{\omega:\ \omega \text{ even}} p(\omega) = \sum_{k\ =\ 2,4,\dots} (1-p)^{k-1} p$$

$$= \frac{p}{(1-p)} \sum_{k\ =\ 1}^{\infty} ((1-p)^2)^k = p(1-p) \sum_{k\ =\ 0}^{\infty} ((1-p)^2)^k$$

$$= p\frac{(1-p)}{1-(1-p)^2} = \frac{1-p}{2-p} \ ,$$

where we have used the standard geometric series summation formula. We can calculate the remaining point in the pmf from the axioms of probability: $p_Y(0) = 1 - p_Y(1)$. Thus we have found a nonobvious derived distribution by computing a pmf via (4.15), a special case of (4.2). Of course, given the pmf, we could now calculate the distribution from (4.4) for all four sets in the power set of $\{0,1\}$.

[4.13]

Say we have a probability space $(\mathbf{R}, \boldsymbol{B}(\mathbf{R}), P)$ where P is described by a pdf g; that is, g is a nonnegative function of the real line with total integral 1 and

$$P(F) = \int_{r\epsilon F} g(r) dr; \qquad F \epsilon \boldsymbol{B}(\mathbf{R}) \ .$$

Suppose that we have a random variable $X\colon \mathbf{R} \to \mathbf{R}$. We can use (4.2) and (4.5) to write a general formula for the distribution of X:

$$P_X(F) = P(X^{-1}(F)) = \int_{r:\ X(r)\epsilon F} g(r) dr \ .$$

Ideally, however, we would like to have a simpler description of X. In particular, if X is a "reasonable function" it should have either a discrete range space (e.g., a quantizer) or a continuous range space (or possibly both, as in the general mixture case). If the range space is discrete, then X can be described by a pmf, and the preceding formula (with the requisite change of dummy variable) becomes

$$p_X(x) = \int_{r:\, X(r)\, =\, x} g(r)dr \ .$$

If, however, the range space is continuous, then there should exist a pdf for X, say f_X, such that (4.5) holds. How do we find this pdf? As previously discussed, to find a pdf from a distribution, we first find the cdf F_X. Then we differentiate the cdf with respect to its argument to obtain the pdf. As a nontrivial example, suppose that we have a probability space $(\mathbf{R}, \mathbf{\mathit{B}}(\mathbf{R}), P)$ with P the probability measure induced by the Gaussian pdf. Define a random variable $W: \mathbf{R} \rightarrow \mathbf{R}$ by $W(r) = r^2;\ r \in \mathbf{R}$. Following the described procedure, we first attempt to find the cdf F_W for W:

$$F_W(w) = \Pr(W \le w) = P(\{\omega:\ W(\omega) = \omega^2 \le w\})$$

$$= P([-w^{1/2}, w^{1/2}]);\ \text{if } w \ge 0 \ .$$

The cdf is clearly 0 if $w < 0$. Since P is described by a Gaussian pdf, say g (the specific Gaussian form is not yet important), then

$$F_W(w) = \int_{-w^{1/2}}^{w^{1/2}} g(r)dr \ .$$

If one should now try to plug in the specific form for the Gaussian density, one would quickly discover that no closed form solution exists. Happily, however, the integral does not have to be evaluated explicitly—we need only its derivative. Therefore we can use the following handy formula from elementary calculus for differentiating the integral:

$$\frac{d}{dw} \int_{a(w)}^{b(w)} g(r)dr = g(b(w))\frac{db(w)}{dw} - g(a(w))\frac{da(w)}{dw} \ . \qquad (4.16)$$

Application of the formula yields

$$f_W(w) = g(w^{1/2})(\frac{w^{-1/2}}{2}) - g(-w^{1/2})(\frac{-w^{-1/2}}{2}) \ .$$

The final answer is found by plugging in the Gaussian form of g. For simplicity we do this only for the special case where $m = 0$. Then g is symmetric; that is, $g(w) = g(-w)$ so that

$$f_W(w) = w^{-1/2}g(w^{1/2});\ w \in [0, \infty) \ ,$$

and finally

$$f_W(w) = \frac{w^{-1/2}}{\sqrt{2\pi\sigma^2}} e^{-\frac{w}{2\sigma^2}} \; ; \; w \in [0, \infty) \; .$$

(This pdf happens to be called a chi-squared pdf with one degree of freedom.) Observe that the functional form of the pdf is valid only for the given domain. By implication the pdf is zero outside the given domain—in this example, negative values of W cannot occur. One should always specify the domain of the dummy variable of a pdf; otherwise the description is incomplete.

It can be seen that although the details may vary from application to application, all derived distribution problems are solved by the general formula (4.2). In some cases the solution will result in a pmf; in others the solution will result in a pdf. To review the general philosophy, one uses the inverse image formula to compute the probability of an output event. This is accomplished by finding the probability with respect to the original probability measure of all input events that result in the given output event. In the discrete case one concentrates on output events of the form $X = x$ and thereby finds a pmf. In the continuous case, one concentrates on output events of the form $X \leq x$ and thereby finds a cdf. The pdf is then found by differentiating. If a random variable has a pmf or a pdf with a specific name, then the name is often applied also to the random variable; e.g., a continuous random variable with a Gaussian pdf is called a Gaussian random variable.

[4.14]

As a final example, suppose that we are given a probability space $(\Omega, \boldsymbol{B}(\Omega), P)$ with $\Omega \subset \boldsymbol{R}$. Define the identity mapping $X : \Omega \to \Omega$ by $X(\omega) = \omega$. The identity mapping on the real line with the Borel field is always a random variable because the measurability requirement is automatically satisfied. Obviously the distribution P_X is identical to the original probability measure P. Thus all probability spaces with real sample spaces provide examples of random variables through the identity mapping. A random variable described in this form instead of as a general function (not the identity mapping) on an underlying probability space is called a "directly given" random variable.

RANDOM VECTORS

In this chapter we consider random variables, random vectors, and random processes. Thus far we have considered only random variables, scalar functions on a sample space that assume real values. In some cases we may wish to model processes or measurements with complex values. Complex outputs can be considered as two-dimensional real vectors with the components being the real and imaginary parts or, equivalently, the magnitude and phase. More generally, we may have k-dimensional real vector outputs. Given that a random variable is a real-valued function of a sample space (with a technical condition), that is, a function mapping a sample space into the real line **R**, the obvious random vector definition is a vector-valued function definition. Under this definition, a random vector is a vector of random variables, a function mapping the sample space into $\mathbf{R}^k$ instead of **R**. The following is a more precise definition of the idea.

Given a probability space $(\Omega, \boldsymbol{F}, P)$, a finite collection of random variables $\{X_i; i = 0, 1, ..., k-1\}$ is called a *random vector*. We will often denote a random vector in boldface as **X**. Thus a random vector is a vector-valued function $\mathbf{X}: \Omega \to \mathbf{R}^k$ defined by $\mathbf{X} = (X_0, X_1, ..., X_{k-1})$ with each of the components being a random variable.

Since a random vector takes values in a space $\mathbf{R}^k$, analogous to random variables one might expect that the events in this space, that is, the members of the event space $\boldsymbol{B}(\mathbf{R})^k$, should inherit a probability measure from the original probability space. This is in fact true. Also analogous to the case of a random variable, the probability measure is called a *distribution* and is defined as

$$P_{\mathbf{X}}(F) = P(\mathbf{X}^{-1}(F)) = P(\{\omega : \mathbf{X}(\omega) \epsilon F\})$$

$$= P(\{\omega : (X_0(\omega), X_1(\omega), ..., X_{k-1}(\omega)) \epsilon F\}) , \qquad (4.17)$$

$$F \epsilon \boldsymbol{B}(\mathbf{R})^k ,$$

where the various forms are equivalent and all stand for $\mathrm{Pr}(\mathbf{X} \epsilon F)$. Equation (4.17) is the vector generalization of the equation (4.2) for random variables. Hence (4.17) is the fundamental formula for deriving vector distributions, that is, probability distributions describing random vector events.

By definition the distribution given by (4.2) is valid because the individual components of the vector are random variables. This does not immediately imply that the distribution given by (4.17) for events on all components together is valid. As in the case of a random variable, the distribution will be valid if the output events $F \epsilon \boldsymbol{B}(\mathbf{R})^k$ have inverse images under **X** that are input events, that is, if $\mathbf{X}^{-1}(F) \epsilon \boldsymbol{F}$ for every $F \epsilon \boldsymbol{B}(\mathbf{R})^k$. From the discussion following example [3.11] we can at least resolve the issue for certain types of output events, viz., events that are rectangles. Rectangles are special events in that the values assumed by any component in the event are not constrained by any of the other components (compare a two-dimensional rectangle with a circle, as in exercise 3.22). Specifically, $F \epsilon \boldsymbol{B}(\mathbf{R})^k$ is a rectangle if it has the form

$$F = \{\mathbf{x} : x_i \epsilon F_i; i = 0, 1, ..., k-1\} = \bigcap_{i=0}^{k-1} \{\mathbf{x} : x_i \epsilon F_i\} = \prod_{i=0}^{k-1} F_i ,$$

where all $F_i \epsilon \boldsymbol{B}(\mathbf{R})$; $i = 0, 1, ..., k-1$ (refer to Figure 3.3 for a two-dimensional illustration of such a rectangle). Because inverse images preserve set operations (exercise 2.12), the inverse image of F can be specified as the intersection of the inverse images of the individual events:

$$\mathbf{X}^{-1}(F) = \{\omega: X_i(\omega)\epsilon F_i; \; i = 0,1,\ldots,k-1\} = \bigcap_{i=0}^{k-1} X_i^{-1}(F_i)$$

Since the X_i are each random variables, the inverse images of the individual events $X_i^{-1}(F_i)$ must all be in $\boldsymbol{F}$. Since $\boldsymbol{F}$ is an event space, the intersection of events must also be an event, and hence $\mathbf{X}^{-1}(F)$ is indeed an event.

Thus we conclude that the distribution is well defined for rectangles. As to more general output events, we simply observe that a result from measure theory ensures that if (1) inverse images of rectangles are events and (2) rectangles are used to generate the output event space, then the inverse images of all output events are events. These two conditions are satisfied by our definition. Thus the distribution of the random vector $\mathbf{X}$ is well defined. Although a detailed proof of the measure theoretical result will not be given, the essential concept can be given: Any event in $\boldsymbol{F}$ can be approximated arbitrarily closely by finite unions of rectangles (e.g., a circle can be approximated by lots of very small squares). The union of the rectangles is an event. Finally, the limit of the events as the approximation gets better must also be an event.

Given a probability space $(\Omega,\boldsymbol{F},P)$ and a random vector $\mathbf{X}: \Omega \rightarrow \mathbf{R}^k$, we have seen that there is a probability measure $P_{\mathbf{X}}$ that the random vector inherits from the original space. With the new probability measure we define a new probability space $(\mathbf{R}^k,\boldsymbol{B}(\mathbf{R})^k,P_{\mathbf{X}})$. As in the scalar case, the distribution can be described by probability functions, that is, cdf's and either pmf's or pdf's (or both). If the random vector has a discrete range space, then the distribution can be described by a multidimensional pmf $p_{\mathbf{X}}(\mathbf{x}) = P_{\mathbf{X}}(\{\mathbf{x}\}) = \Pr(\mathbf{X} = \mathbf{x})$ as

$$P_{\mathbf{X}}(F) = \sum_{\mathbf{x}\epsilon F} p_{\mathbf{X}}(\mathbf{x})$$

$$= \sum_{(x_0,x_1,\ldots,x_{k-1})\epsilon F} p_{X_0,X_1,\ldots,X_{k-1}}(x_0,x_1,\ldots,x_{k-1}) \; ,$$

where the last form points out the economy of the vector notation of the previous line. If the random vector $\mathbf{X}$ has a continuous range space, then in a similar fashion its distribution can be described by a multidimensional pdf $f_{\mathbf{X}}$ with

$$P_{\mathbf{X}}(F) = \int_F f_{\mathbf{X}}(\mathbf{x})dx \; .$$

In order to derive the pdf from the distribution, as in the scalar case, we introduce the notion of a cdf:

Given a k-dimensional random vector $\mathbf{X}$, define its *cumulative distribution function* $F_{\mathbf{X}}$ by

$$F_{\mathbf{X}}(\alpha) = F_{X_0, X_1, \ldots, X_{k-1}}(\alpha_0, \alpha_1, \ldots, \alpha_{k-1})$$

$$= P_{\mathbf{X}}(\{\mathbf{x}: x_i \le \alpha_i; \ i = 0, 1, \ldots, k-1\}) .$$

In English, $F_{\mathbf{X}}(\mathbf{x}) = \Pr(X_i \le x_i; \ i = 0, 1, \ldots, k-1)$. Note that the cdf for any value of its argument is the probability of a special kind of rectangle. For example, if we have a two-dimensional random vector (X, Y), then the cdf $F_{X,Y}(\alpha, \beta) = \Pr(X \le \alpha, Y \le \beta)$ is the probability of the semi-infinite rectangle depicted in Figure 4.5.

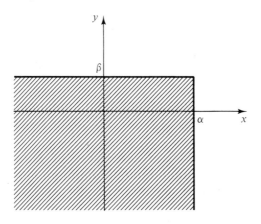

$F_{X,Y}(\alpha_1 \beta) = P_{X,Y}((-\infty, \alpha] \times (-\infty, \beta])$
$= P_{X,Y}(\text{shaded region}) = \Pr(X \le \alpha \text{ and } Y \le \beta)$

Figure 4.5   Two-dimensional cdf.

Observe that we can also write this probability in several other ways, e.g.,

$$F_{\mathbf{X}}(\mathbf{x}) = P_{\mathbf{X}}\left(\prod_{i=0}^{k-1}(-\infty, x_i]\right) = P(\{\omega: X_i(\omega) \le x_i; \ i = 0, 1, \ldots, k-1\})$$

$$= P\left(\bigcap_{i=0}^{k-1} X_i^{-1}((-\infty, x_i])\right) .$$

Since integration and differentiation are inverses of each other, it follows that

$$f_{X_0, X_1, \ldots, X_{k-1}}(x_0, x_1, \ldots, x_{k-1}) =$$

$$\frac{\partial^k}{\partial x_0 \, \partial x_1 \, ... \partial x_{k-1}} F_{X_0,X_1,...,X_{k-1}}(x_0,x_1,...,x_{k-1}) \; .$$

As with random variables, random vectors can, in general, have discrete and continuous parts with a corresponding mixture distribution. We will concentrate on random vectors that are described completely by either pmf's or pdf's. Also as with random variables, we can always unify notation using a multidimensional Stieltjes integral to write

$$P_{\mathbf{X}}(F) = \int_F dF_{\mathbf{X}}(\mathbf{x}); \; F \epsilon \boldsymbol{B}(\mathbf{R})^k \; ,$$

where the integral is defined as the usual integral if $\mathbf{X}$ is described by a pdf, as a sum if $\mathbf{X}$ is described by a pmf, and by a weighted average if $\mathbf{X}$ has both a discrete and a continuous part. Random vectors are said to be continuous, discrete, or mixture random vectors in accordance with the above analogy to random variables.

MARGINAL AND JOINT DISTRIBUTIONS

By definition a random vector $\mathbf{X}$ is a collection of random variables $(X_0,X_1,...,X_{k-1})$ in vector form that takes on values randomly as described by a probability distribution $P_{\mathbf{X}}$. $P_{\mathbf{X}}$ in turn may be induced by a pmf $p_{\mathbf{X}}$ or a pdf $f_{\mathbf{X}}$. From any of these probabilistic descriptions we can find a probabilistic description for any of the component random variables. For example, given a value of i in $\{0,1,...,k\text{-}1\}$, the distribution of the random variable X_i is found by evaluating the distribution $P_{\mathbf{X}}$ for the random vector on one-dimensional rectangles where only the component X_i is constrained to lie in some set—the rest of the components can take on any value. That is, $P_{\mathbf{X}}$ is evaluated on rectangles of the form $\{\mathbf{x} = (x_0,...,x_{k-1}): x_i \epsilon G\}$ for each $G \epsilon \boldsymbol{B}(\mathbf{R})$ as

$$P_{X_i}(G) = P_{\mathbf{X}}(\{\mathbf{x}: x_i \epsilon G\}), \; G \epsilon \boldsymbol{B}(\mathbf{R}) \; . \tag{4.18}$$

Alternatively, we can consider this a derived distribution problem on the vector probability space $(\mathbf{R}^k, \boldsymbol{B}(\mathbf{R})^k, P_{\mathbf{X}})$ using a sampling function $\Pi_i : \mathbf{R}^k \to \mathbf{R}$ as in example [4.4]. Specifically, let $\Pi_i(\mathbf{X}) = X_i$. Using (4.2) we write

$$P_{\Pi_i}(G) = P_{\mathbf{X}}(\Pi_i^{-1}(G)) = P_{\mathbf{X}}(\{\mathbf{x}: x_i \epsilon G\}) \; .$$

These two formulas demonstrate that Π_i and X_i are equivalent random variables, and indeed they correspond to the same physical events—the outputs of the ith coordinate of the random vector $\mathbf{X}$. For any value of ω they provide equal outputs, but they are not identical functions: They are defined on different spaces. X_i is defined on the original sample space Ω, while Π_i is defined on the vector space $\mathbf{R}^k$. They are related through the relation $\Pi_i(\mathbf{X}(\omega)) = X_i(\omega)$. Intuitively, the two random variables provide different models of the same thing. As usual, which is "better" depends on which is the simpler model to handle for a given problem.

Instead of finding the distribution for each component from the distribution of the random vector $\mathbf{X}$ we could also find the distribution directly by using the underlying probability measure P on the original probability space $(\Omega, \mathbf{F}, P)$. Through the usual derived distribution formula of (4.2):

$$P_{X_i}(G) = P(X_i^{-1}(G)), \quad G \in \mathbf{B}(\mathbf{R}) .$$

Even though this may seem more direct, it is often more convenient to find the distributions for the individual random variables—which are called *marginal distributions*—from the distribution for the random vector—which is called a *joint distribution*. For the cases where the distributions are induced by pmf's (marginal pmf's and joint pmf's) or pdf's (marginal pdf's or joint pdf's), the relation becomes, respectively,

$$p_{X_i}(\alpha) =$$

$$\sum_{x_0,x_1,\ldots,x_{i-1},x_{i+1},\ldots,x_{k-1}} p_{X_0,X_1,\ldots,X_{k-1}}(x_0,x_1,\ldots,x_{i-1},\alpha,x_{i+1},\ldots,x_{k-1})$$

or

$$f_{X_i}(\alpha) =$$

$$\int_{x_0,\ldots,x_{i-1},x_{i+1},\ldots,x_{k-1}} f_{X_0,\ldots,X_{k-1}}(x_0,\ldots,x_{i-1},\alpha,x_{i+1},\ldots,x_{k-1})dx_0\ldots dx_{i-1}dx_{i+1}\ldots dx_{k-1} .$$

That is, one sums or integrates over all of the dummy variables corresponding to the unwanted random variables in the vector to obtain the pmf or pdf for the random variable X_i. The two formulas look identical except that one sums for discrete random variables and the other integrates for continuous ones. For emphasis, we repeat the fact that both formulas are simple consequences of the fact that for any $G \in \mathbf{B}(\mathbf{R})$,

$$P_{X_i}(\{x: x \in G\}) = P_{\mathbf{X}}(\{x: x_i \in G, x_j \in \mathbf{R}; \ j \neq i\}) = P(\{\omega: X_i(\omega) \in G\}) .$$

One can also use (4.18) to derive the cdf of X_i by setting $G = (-\infty, \alpha]$. The cdf is

$$F_{X_i}(\alpha) = F_{\mathbf{X}}(\infty, \infty, \ldots, \infty, \alpha, \infty, \ldots, \infty) ,$$

where the α appears in the ith position. In English, this equation states that $\Pr(X_i \leq \alpha) = \Pr(X_i \leq \alpha$ and $X_j \leq \infty$, all $j \neq i)$. The expressions for pmf's and pdf's also can be derived from the expression for cdf's.

The sheer details of notation with k random variables can cloud the meaning of the relations we are discussing. Therefore we rewrite them for the special case of $k = 2$ to emphasize the essential form. Say we have a random vector (X,Y). Then the marginal distribution of X is obtained from the joint distribution of X and Y by leaving Y unconstrained, i.e., as in equation (4.18):

$$P_X(F) = P_{X,Y}(\{(x,y): x \in F\}); \quad F \in \boldsymbol{B}(\mathbf{R}) .$$

Furthermore, the marginal cdf of X is

$$F_X(\alpha) = F_{X,Y}(\alpha, \infty) .$$

If the range space of the vector (X,Y) is discrete, the marginal pmf of X is

$$p_X(x) = \sum_y p_{X,Y}(x,y) .$$

If the range space of the vector (X,Y) is continuous and the cdf is differentiable, the marginal pdf of X is

$$f_X(x) = \int_{-\infty}^{\infty} f_{X,Y}(x,y)dy ,$$

with similar expressions for the distribution and probability functions for the random variable Y.

In summary, given a probabilistic description of a random vector, we can always determine a probabilistic description for any of the component random variables of the random vector. It is important to note that the opposite statement is not true. Given all the marginal distributions of the component random variables, we cannot find the joint distribution of the random vector formed from the components. This is true because the marginal distributions provide none of the information about the interrelationships of the components that is contained in the joint distribution. As an example, consider a random vector (X,Y) with range space $\{0,1\}^2$, the space of all binary pairs, and assume that $P_{X,Y}$ is induced by the pmf

$$p_{X,Y}(x,y) = \begin{cases} 1/2 & \text{if } (x,y) = (1,1) \\ 1/2 & \text{if } (x,y) = (0,0) \\ 0 & \text{otherwise} . \end{cases}$$

that is, $X = Y$ with probability one. Intuitively, one can consider the two coins to be soldered together in this example so that they always show the same face. One finds quickly that $p_X(x) = p_Y(x) = 1/2$ for $x = 0$ and $x = 1$. But these marginals are the same as those that result if $P_{X,Y}$ is described by a product distribution, that is, if $p_{X,Y}(x,y) = p_X(x)p_Y(y) = 1/4$ for all (x,y). Thus the same set of marginals can result from two quite different joint distributions!

In a similar manner we can deduce the distributions or probability functions of "subvectors" of a random vector, that is, if we have the distribution for $\mathbf{X} = (X_0, X_1, ..., X_{k-1})$ and if k is big enough, we can find the distribution for the random vector (X_1, X_2) or the random vector (X_5, X_{10}, X_{15}), and so on. Writing the general formulas is, however, tedious and adds little insight. One important fact should be noted. One always starts with a probability space $(\Omega, \mathbf{F}, P)$. From this point one can proceed in several directions. But no matter how one proceeds, the same answer must result for the same quantity. For example, after finding the distribution of a random vector $\mathbf{X}$, the marginal distribution for the specific component X_i can be found from the joint distribution. This marginal distribution must agree with the marginal distribution obtained for X_i directly from the probability space. As another possibility, one might first find a distribution for a subvector containing X_i, say the vector $\mathbf{Y} = (X_{i-1}, X_i, X_{i+1})$. This distribution can be used to find the marginal distribution for X_i. All answers must be the same since all can be expressed in the form $P(X^{-1}(F))$ using the original probability measure. The point is that all of the distributions derived on a common probability space must be *consistent* in the sense that they agree with one another on events.

We now give examples of the computation of marginal probability functions from joint probability functions.

[4.15]

Say that we are given a pair of random variables X and Y such that the random vector (X, Y) has a pmf of the form

$$p_{X,Y}(x,y) = r(x)q(y) ,$$

where r and q are both valid pmf's. In other words, $p_{X,Y}$ is a product pmf. Then it is easily seen that

$$p_X(x) = \sum_y p_{X,Y}(x,y) = \sum_y r(x)q(y)$$
$$= r(x)\sum_y q(y) = r(x) .$$

Thus in the special case of a product distribution, knowing the marginal pmf's is enough to know the joint distribution. A similar result and conclusion hold for more general product distributions. For example, if both r and q are pmf's assigning probability 1/2 to the points 0 and 1 and hence provide a model of a flip of a single fair coin, then the product distribution assigns probability 1/4 to each possible pair of outcomes $(0,0),(0,1),(1,0)$, and $(1,1)$ and provides a model for a single flip of two fair coins (or two flips in a row of one fair coin).

[4.16]

Consider flipping two fair coins, but now we connect them by a piece of rubber that is fairly flexible. Unlike the example where the coins were soldered together, it is not certain that they will show the same face; it is, however, more probable. To quantify the pmf, say that the probability of the pair (0,0) is .4, the probability of the pair (1,1) is .4, and the probabilities of the pairs (0,1) and (1,0) are each .1. As with the soldered-coins case, this is clearly not a product distribution, but a simple computation shows that as in example [4.15], p_X and p_Y both place probability 1/2 on 0 and 1. Thus this distribution, the soldered-coins distribution, and the product distribution of example [4.15] all yield the same marginal pmf's. The point again is that the marginal probability functions are not enough to describe a vector experiment, we need the joint probability function to describe the interrelations or dependencies among the random variables.

[4.17]

A gambler has a pair of very special dice: the sum of the two dice comes up as seven on every roll. All combinations have equal probability; e.g., the probability of a one and a six has the same probability as a three and a four. Although the two dice are identical, we will distinguish between them by number for the purposes of assigning two random variables. The outcome of the roll of the first die is denoted X and the outcome of the roll of the second die is called Y so that (X,Y) is a random vector taking values in $\mathbf{I}^2 = \{(x,y):x,y = 1,2,3,4,5,6\}$. The joint pmf of X and Y is

$$p_{X,Y}(x,y) = C, \ x+y = 7, \ (x,y)\epsilon\mathbf{I}^2 \ ,$$

where C is a constant to be determined. The pmf of X is determined by summing the pmf with respect to y. However, for any given $X\epsilon\mathbf{I}$, the value of Y is determined; viz., $Y = 7-X$. Therefore the pmf of X is

$$p_X(x) = C, \ x\epsilon\mathbf{I} \ .$$

The value of C is determined by summing over $x\epsilon\mathbf{I}$. Performing this summation results in $C = 1/6$, or

$$p_X(x) = 1/6, \ x\epsilon\mathbf{I} \ .$$

Note that this pmf is the same as one would derive for the roll of a single unbiased die! Note also that the pmf for Y is identical with that for X. Obviously, then, it is impossible to tell that the gambler is using unfair dice *as a pair* from looking at outcomes of the rolls of each die alone. Again note that the joint pmf cannot be deduced from the marginal pmf's alone.

[4.18]

Let (X,Y) be a random vector with a pdf that is constant on the unit disk in the XY plane; i.e.,

$$f_{X,Y}(x,y) = C, \ x^2+y^2 \leq 1 \ .$$

This pdf is illustrated in Figure 4.6. The constant C is determined by the requirement that the pdf integrate to 1; i.e.,

$$\int_{x^2+y^2\leq1} C\,dxdy = 1 \,.$$

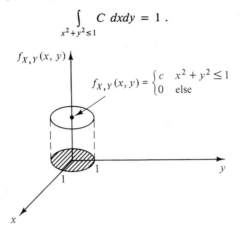

$$f_{X,Y}(x, y) = \begin{cases} c & x^2 + y^2 \leq 1 \\ 0 & \text{else} \end{cases}$$

Figure 4.6 Example 4.18.

Since this integral is just the area of a circle times C, we have immediately that $C = 1/\pi$. For the moment, however, we leave the joint pdf in terms of C and determine the pdf of X in terms of C by integrating with respect to y:

$$f_X(x) = \int_{-(1-x^2)^{1/2}}^{+(1-x^2)^{1/2}} Cdy = 2C\,(1-x^2)^{1/2},\ x^2\leq1 \,.$$

Observe that we could now also find C by a second integration:

$$\int_{-1}^{+1} 2C\,(1-x^2)^{1/2}dx = \pi C = 1 \,,$$

or $C = \pi^{-1}$. Thus the pdf of X is

$$f_X(x) = 2\pi^{-1}(1-x^2)^{1/2}\,,\ x^2 \leq 1 \,.$$

By symmetry Y has the same pdf. Note that the marginal pdf is *not constant*, as the joint pdf is. Furthermore, it is obvious that it would be impossible to determine the joint density from the marginal pdf's alone.

[4.19]

Consider the two-dimensional Gaussian pdf of example [3.17] with $k = 2$, $\mathbf{m} = (0,0)$, and $\Lambda = \{\lambda(i,j):\lambda(1,1) = \lambda(2,2) = 1, \lambda(1,2) = \lambda(2,1) = \rho\}$. Since the inverse of the matrix

$$\begin{bmatrix} 1 & \rho \\ \rho & 1 \end{bmatrix}$$

is

$$\frac{1}{1-\rho^2}\begin{bmatrix} 1 & -\rho \\ -\rho & 1 \end{bmatrix},$$

the joint pdf for the random vector (X,Y) is

$$f_{X,Y}(x,y) = ((2\pi)^2(1-\rho^2))^{-1/2}e^{-\frac{1}{2(1-\rho^2)}(x^2+y^2-2\rho xy)}, \quad (x,y)\in\mathbf{R}^2.$$

ρ is called the "correlation coefficient" between X and Y and must satisfy $\rho^2<1$ for Λ to be positive definite. To find the pdf of X we complete the square in the exponent so that

$$f_{X,Y}(x,y) = ((2\pi)^2(1-\rho^2))^{-1/2}e^{-\frac{(y-\rho x)^2}{2(1-\rho^2)}-\frac{x^2}{2}}$$

$$= (2\pi(1-\rho^2))^{-1/2}e^{-\frac{(y-\rho x)^2}{2(1-\rho^2)}}(2\pi)^{-1/2}e^{-1/2x^2}.$$

The pdf of X is determined by integrating with respect to y on $(-\infty,\infty)$. To perform this integration we refer to the form of the one-dimensional Gaussian pdf with $m = \rho x$ (note that x is fixed while we integrate with respect to y) and $\sigma^2 = 1-\rho^2$. The first factor in the preceding equation has this form. Because the one-dimensional pdf must integrate to one, the pdf of X that results from integrating y out from the two-dimensional pdf is also a one-dimensional Gaussian pdf; i.e.,

$$f_X(x) = (2\pi)^{-1/2}e^{-1/2x^2}.$$

As in examples [4.16], [4.17], and [4.18], Y has the same pdf as X. Note that by varying ρ there is a whole family of joint Gaussian pdf's with the same marginal Gaussian pdf's.

EXAMPLES OF RANDOM VECTORS

We will now describe the two most important types of random vectors of arbitrary length.

[4.20] *Gaussian random vectors*

A random vector is said to be *Gaussian* if its density is Gaussian, that is, if its distribution is described by the multidimensional pdf explained in chapter 3. This is an obvious extension of example [4.19]. Note that the symmetric matrix Λ of the k-dimensional vector pdf has $k(k+1)/2$ parameters and that the vector m has k parameters. On the other hand, the k marginal pdf's together have only $2k$ parameters. Again we note the impossibility of constructing joint pdf's without more specification than the marginal pdf's alone.

[4.21] *i.i.d. random vectors*

A random vector is said to have independent and identically distributed components or, simply, to be *i.i.d.*, if its distribution has the form

$$P_{X_0,\ldots,X_{k-1}}\left(\prod_{i=0}^{k-1} F_i\right) = \prod_{i=0}^{k-1} P_{X_i}(F_i)$$

for all choices of $F_i \in \boldsymbol{B}(\mathbf{R})$, $i = 0,1,\ldots,k-1$, and if all the marginal distributions are the same, e.g., if there is a distribution P_X such that $P_{X_i}(F) = P_X(F)$; all $F \in \boldsymbol{B}(\mathbf{R})$ for all i. More generally, any distribution of this form (whether or not the marginal distributions are identical) is called a *product distribution*.

For example, a random vector will have a product distribution if it has a joint pdf or pmf that is a product pdf or pmf as described in example [3.16]. The random vector will be i.i.d. if it has a joint pdf with the form

$$f_{\mathbf{X}}(\mathbf{x}) = \prod_i f(x_i)$$

for some pdf f defined on $\mathbf{R}$ or if it has a joint pmf with the form

$$p_{\mathbf{X}}(\mathbf{x}) = \prod_i p(x_i)$$

for some pmf p defined on some discrete subset of the real line. Both of these cases are included in the following statement: A random vector will be i.i.d. if and only if its cdf has the form

$$F_{\mathbf{X}}(\mathbf{x}) = \prod_i F(x_i)$$

for some cdf F.

Note that, in contrast with the preceding examples, the specification "product distribution," along with the marginal pdf's or pmf's or cdf's, is sufficient to specify the joint distribution.

We will now elaborate on the meaning of the term *i.i.d.* The fact that the distributions of all of the component random variables are the same explains the "identically distributed" portion of the name. To explain the rest of the name, we turn to the concept of independence.

INDEPENDENCE

Given a probability space $(\Omega, \boldsymbol{F}, P)$, two events F and G are defined to be *independent* if $P(F \cap G) = P(F)P(G)$. A collection of events $\{F_i; i = 0,1,\ldots,k-1\}$ is said to be *independent* or *mutually independent* if

$$P(\bigcap_{i=0}^{k-1} F_i) = \prod_{i=0}^{k-1} P(F_i) .$$

The concept of independence in the probabilistic sense we have defined relates easily to the intuitive idea of independence of physical events. For example, if a fair die is rolled twice, one would expect the second roll to be unrelated to the first roll because there is no physical connection between the individual outcomes. Independence in the probabilistic sense is reflected in this experiment. The probability of any given outcome for either of the individual rolls is 1/6. The probability of any given pair of outcomes is $(1/6)^2 = 1/36$—the addition of a second outcome diminishes the overall probability by exactly the probability of the individual event, viz., 1/6. Note that the probabilities are not *added*—the probability of two successive outcomes cannot reasonably be greater than the probability of either of the outcomes alone. Do *not*, however, confuse the concept of independence with the concept of disjoint or mutually exclusive events. If you roll the die once, the event {the roll is a one} is not independent of the event {the roll is a six}. Given one event, the other cannot happen—they are neither physically nor probabilistically independent. These are mutually exclusive events.

The notion of independent events is in turn used to define independent random variables.

Two random variables X and Y defined on a probability space are *independent* if the events $X^{-1}(F)$ and $Y^{-1}(G)$ are independent for all F and G in $\mathbf{B}(\mathbf{R})$. A collection of random variables $\{X_i; i = 0,1,...,k-1\}$ is said to be *independent* or *mutually independent* if all collections of events of the form $\{X_i^{-1}(F_i); i = 0,1,...,k-1\}$ are independent for any $F_i \in \mathbf{B}(\mathbf{R})$; $i = 0,1,...,k-1$.

Thus two random variables are independent if and only if their output events correspond to independent input events. Translating this statement into distributions yields the following:

Random variables X and Y are independent if and only if

$$P_{X,Y}(F_1 \times F_2) = P_X(F_1)P_Y(F_2), \text{ all } F_1,F_2 \in \mathbf{B}(\mathbf{R}) .$$

(Recall that $F_1 \times F_2$ is an alternate notation for $\prod_{i=1}^{2} F_i$—we will frequently use the alternate notation when the number of product events is small.) Note that a *product* and not an *intersection* is used here. The reader should be certain that this is understood.

Random variables $X_0,...,X_{k-1}$ are mutually independent if and only if

$$P_{X_0,...,X_{k-1}}(\prod_{i=0}^{k-1} F_i) = \prod_{i=0}^{k-1} P_{X_i}(F_i) ;$$

for all $F_i \in \mathbf{B}(\mathbf{R})$; $i = 0,1,...,k-1$.

The general form for distributions can be specialized to pmf's, pdf's, and cdf's as follows:

Two discrete random variables X and Y are independent if and only if the joint pmf factors as

$$p_{X,Y}(x,y) = p_X(x)p_Y(y); \text{ all } x,y .$$

A collection of discrete random variables X_i; $i = 0,1,...,k-1$ is mutually independent if and only if the joint pmf factors as

$$p_{X_0,...,X_{k-1}}(x_0,...,x_{k-1}) = \prod_{i=0}^{k-1} p_{X_i}(x_i); \text{ all } x_i .$$

Similarly, if the random variables are continuous and described by pdf's, then two random variables are independent if and only if the joint pdf factors as

$$f_{X,Y}(x,y) = f_X(x)f_Y(y); \text{ all } x,y \in \mathbf{R} .$$

A collection of continuous random variables is independent if and only if the joint pdf factors as

$$f_{X_0,...,X_{k-1}}(x_0,...,x_{k-1}) = \prod_{i=0}^{k-1} f_{X_i}(x_i) .$$

Two general random variables (discrete, continuous, or mixture) are independent if and only if the joint cdf factors as

$$F_{X,Y}(x,y) = F_X(x)F_Y(y); \text{ all } x,y \in \mathbf{R} .$$

A collection of general random variables is independent if and only if the joint cdf factors as

$$F_{X_0,...,X_{k-1}}(x_0,...,x_{k-1}) = \prod_{i=0}^{k-1} F_{X_i}(x_i) ; \text{ all } (x_0,x_1,...,x_{k-1}) \in \mathbf{R}^k .$$

We have separately stated the two-dimensional case because of its simplicity and common occurrence. The student should be able to prove the equivalence of the general distribution form and the pmf form. If one does not consider technical problems regarding the interchange of limits of integration, then the equivalence of the general form and the pdf form can also be proved.

Given the definition of independence, it follows that an i.i.d. vector has mutually independent component random variables, completing the explanation of the name *i.i.d.* To understand the notion better, however, more motivation for the basic definition of independence is required. The notion of independence is further developed in terms of conditional probabilities in the next section. Since all of the definitions of independence derive from the definition of independence of events and because the generalization from two to many events is obvious, we will concentrate on two independent events.

ELEMENTARY CONDITIONAL PROBABILITY

Intuitively, independence of two events means that the occurrence of one event should not affect the occurrence of the other. For example, the knowledge of the outcome of the first roll of a die should not change the probabilities for the outcome of a second roll of the die. To be more precise, the notion of elementary conditional probability is required. Consider the following motivation. Say that an observer is told that an event G has occurred and is then asked to calculate the probability of another event F given this information. We will denote this probability of F given G by $P(F|G)$. Since the observer has been told that G has occurred and hence that $\omega \epsilon G$, clearly no sample point in F not also in G can have occurred. Thus a conditional probability measure given G occurred should put zero probability on the total of all points outside of G. At first glance this fact seems to contradict the statement that the first event should not affect the second if the events are to be independent. However, the statement is not really contradicted. The *probability* of the second event is at issue, not the microscopic details of the actual elementary sample points included in the probability calculation. Observe that there is no reason to change the *relative* probabilities within G. For example, in the discrete case, the relative probabilities of the points within G should remain the same; that is, if ω_1 was twice as probable as ω_2 before being told that G occurred, then the same should still be true after being told that G occurred, provided that both points are in G. Thus $P(F|G)$ must be proportional to $P(F \cap G)$, the probability of points common to both F and G. In fact, we must have $P(F|G) \geq P(F \cap G)$ because the occurrence of *both* events can have a probability that is no larger than the probability of a single event no matter how much or how little information is given about the single event. To determine the constant of proportionality, we tentatively set $P(F|G) = cP(F \cap G)$, where c is some constant (dependent on G but not F). A conditional probability for a fixed given event G is also a probability and hence must satisfy the axioms of probability. Thus for the particular event Ω, we must have

$$1 = P(\Omega|G) = cP(\Omega \cap G) = cP(G) \, ,$$

which implies the following definition:

Given a probability space $(\Omega, \boldsymbol{F}, P)$ and an event G for which $P(G) > 0$, the conditional probability of any event F given the event G is defined by $P(F|G) = P(F \cap G)/P(G)$.

The conditional probability can be interpreted as "cutting down" the original probability space to a probability space with the smaller sample space G and with probabilities equal to the renormalized probabilities of the intersection of events with the given event G on the original space.

With this notion of conditional probability we can observe that two events F and G, with $P(G) > 0$, are independent if $P(F) = P(F|G)$; that is, the probability of F given G is identical to the probability of F given no knowledge of G. Thus, as we mentioned before, independence implies the intuitive notion that the occurrence of one event does not influence the probability of another. Note, however, that this would not make as useful a definition of independence as the product definition since it is less general: it requires that one of the events have a nonzero probability—an obviously unnecessary qualification.

RANDOM PROCESSES

We have seen that a random vector is just a finite bunch of random variables all defined on the same probability space. We shall next see that a random process is also a bunch of random variables, just an infinite bunch with a slightly different notation! To make this idea more precise, note that we can use a different random vector notation without changing the essential definition of a random vector. A random vector $\mathbf{X} = (X_0, X_1, \ldots, X_{k-1})$ can be defined as an *indexed* family of random variables $\{X_i; \, i \in \mathbf{I}\}$ where $\mathbf{I}$ is the index set $\mathbf{Z}_k = \{0, 1, \ldots, k-1\}$. The index set in some examples will correspond to time; e.g., X_i is a measurement on an experiment at time i for k different times. We get a random process by using the same definition with an infinite index set, which almost always corresponds to time. This we now do.

A *random process* or *stochastic process* is an indexed family of random variables $\{X_t; \, t \in \mathbf{I}\}$ or, equivalently, $\{X(t); \, t \in \mathbf{I}\}$ defined on a common probability space $(\Omega, \mathbf{F}, P)$. The process is said to be *discrete time* if $\mathbf{I}$ is discrete, e.g., $\mathbf{Z}_+$ or $\mathbf{Z}$, and *continuous time* if the index set $\mathbf{I}$ is continuous, e.g., $\mathbf{R}$ or $[0, \infty)$. A discrete time random process is often called a *time series*. It is said to be *discrete alphabet* or *discrete amplitude* if all finite-length random vectors of random variables drawn from the random process are discrete random vectors. The process is said to be *continuous alphabet* or *continuous amplitude* if all finite-length random vectors of random variables drawn from the random process are continuous random vectors. The process is said to have a *mixed alphabet* if all finite-length random vectors of random variables drawn from the random process are mixture random vectors.

Thus a random process is a collection of random variables indexed by time into the indefinite future and sometimes into the infinite past as well. For each value of time t, X_t or $X(t)$ is a random variable. Both notations are used, but X_t or X_n is more common for discrete time processes whereas $X(t)$ is more common for continuous time processes. It is useful to recall that random variables are functions on an underlying sample space Ω and hence implicitly depend on $\omega \in \Omega$. Thus a random process (and a random vector) is actually a function of two arguments, written explicitly as $X(t,\omega)$; $t \in \boldsymbol{I}$, $\omega \in \Omega$ (or $X_t(\omega)$—we will use the first notation for the moment). Observe that for a fixed value of time, $X(t,\omega)$ is a random variable whose value depends probabilistically on ω. On the other hand, if we fix ω and allow t to vary deterministically, we have either a sequence ($\boldsymbol{I}$ discrete) or a waveform ($\boldsymbol{I}$ continuous). If we fix both t and ω, we have a number. Overall we can consider a random process as a two-space mapping $X: \Omega \times \boldsymbol{I} \to \mathbf{R}$ or as a one-space mapping $X: \Omega \to \mathbf{R}^{\boldsymbol{I}}$. Figure 4.7 is an illustration to aid in the visualization of this concept. Several arbitrarily shaped functions show the joint dependence on ω and t. The reader should also refer to Figure 4.1 and the sampling function and note the close connection to random processes.

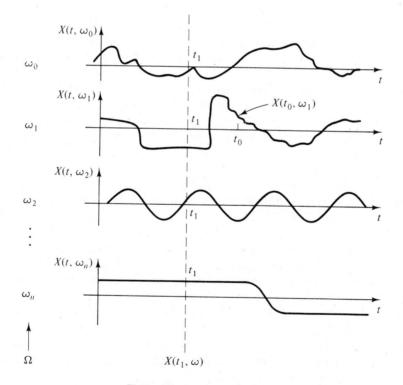

Figure 4.7 Sample functions.

There is a common bit of notational ambiguity and hence confusion when dealing with random processes. It is the same problem we encountered with functions in the context of random variables at the beginning of the chapter. The notation $X(t)$ or X_t usually means a sample of the random process at a specified time t, i.e., a random variable, just as $\sin t$ means the sine of a specified value t. Often in the literature, however, the notation is used as an abbreviation for $\{X(t); t \in \boldsymbol{I}\}$ or $\{X_t; t \in \boldsymbol{I}\}$, that is, for the entire random process or family of random variables. The abbreviation is the same as the common use of $\sin t$ to mean $\{\sin t; t \in (-\infty, \infty)\}$, that is, the entire waveform and not just a single value. Thus, in summary, the common (and sometimes unfortunate) ambiguity is in whether or not the dummy variable t means a *specific* value or is implicitly allowed to vary over its entire domain. Of course, as noted at the beginning of the chapter, the problem could be avoided by reserving a different notation to specify a fixed time value, say t_0, but this is usually not done to avoid a proliferation of notation. In this book we will attempt to avoid the potential confusion by using the abbreviations $\{X(t)\}$ and $\{X_t\}$ for the random processes when the index set is clear from context and reserving the notation $X(t)$ and X_t to mean the tth random variable of the process, that is, the sample of the random process at time t. The reader should beware in reading other sources, however, because this sloppiness will undoubtedly be encountered at some point in the literature; when this happens one can only hope that the context will make the meaning clear.

There is also an ambiguity regarding the alphabet of a random process. If $X(t)$ takes values in A_t, then strictly speaking the alphabet of the random process is $\prod_{t \in \boldsymbol{I}} A_t$, the space of all possible waveforms or sequences with coordinate t taking values in A_t. If all of the A_t are the same, say $A_t = A$, this process alphabet is $A^{\boldsymbol{I}}$. In this case, however, the alphabet of the process is commonly said to be simply A, the set of values from which all of the coordinate random variables are drawn. We will frequently use this convention.

SIMPLE EXAMPLES OF RANDOM PROCESSES

The following simple and somewhat artificial examples of random processes are chosen to illustrate the principles we have described. The reader should be very careful to understand the concepts involved, especially the relationship between the two spaces on which the random process is defined.

[4.22]

Consider the binary probability space $(\Omega, \boldsymbol{F}, P)$ with $\Omega = \{0,1\}$, $\boldsymbol{F}$ the usual event space, and P induced by the pmf $p(0) = \alpha$ and $p(1) = 1 - \alpha$, where α is some constant, $0 \le \alpha \le 1$. Define a random process on this space as follows:

$$X(t,\omega) = \cos(\omega t) = \begin{cases} \cos(t), \ t\in\mathbf{R} \text{ if } \omega = 1 \\ 1, \ t\in\mathbf{R} \text{ if } \omega = 0 \ . \end{cases}$$

Thus if a 1 occurs a cosine is sent forever, and if a 0 occurs a constant 1 is sent forever.

This process clearly has continuous time and at first glance it might appear to also have continuous amplitude, but only two waveforms are possible, as illustrated in Figure 4.8. Thus the alphabet at each time contains at most two values and these possible values change with time. Thus this process is in fact a discrete amplitude process and random vectors drawn from this source are described by pmf's. We can consider the alphabet of the process to be either $\mathbf{R}^I$ or $[-1,1]^I$, among other possibilities. Let's fix time at $t = \pi/2$. Then $X(\pi/2)$ is a random variable with pmf

$$p_{X(\pi/2)}(x) = \begin{cases} \alpha, \text{ if } x = 1 \\ 1-\alpha, \text{ if } x = 0 \ . \end{cases}$$

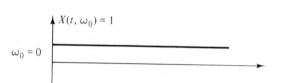

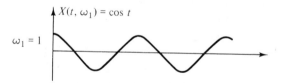

Figure 4.8   Example 4.22.

The reader should try other instances of time. What happens at $t = 0, 2\pi, 4\pi, ...$?

[4.23]

Consider a probability space $(\Omega, \mathbf{F}, P)$ with $\Omega = \mathbf{R}$, $\mathbf{F} = \mathbf{B}(\mathbf{R})$, the Borel field, and probability measure P induced by the pdf

$$f(r) = \begin{cases} 1 \text{ if } r\in[0,1] \\ 0 \text{ otherwise} \ . \end{cases}$$

Again define the random process $\{X(t)\}$ by $X(t,\omega) = \cos(\omega t)$; $t\in\mathbf{R}$.

Again the process is continuous time but now it has mixed alphabet because an uncountable infinity of waveforms is possible corresponding to all angular frequencies between 0 and 1 so that $X(t,\omega)$ is a continuous random variable except at $t = 0$. $X(0,\omega) = 1$ is a discrete random variable. If you calculate the pdf of the random variable $X(t)$ you see that it varies as a function of time (exercise 4.21).

[4.24]

Consider the probability space of example [4.23], but cut it down to the unit interval; that is, consider the probability space $([0,1],\boldsymbol{B}([0,1]),P)$ where P is the probability measure induced by the pdf $f(r) = 1$; $r\in[0,1]$. (So far this is just another model for the same thing.) Define for $n = 1,2,...$ $X_n(\omega) = b_n(\omega) = $ the nth digit in the binary expansion of ω, that is,

$$\omega = \sum_{n=1}^{\infty} b_n 2^{-n}$$

or, equivalently, $\omega = .b_1b_2b_3...$ in binary.

Then $\{X_n; n = 1,2,...\}$ is a one-sided discrete time discrete alphabet random process with alphabet $\{0,1\}^I$ or, simply, $\{0,1\}$. It is important to understand that nature has selected ω at the beginning of time, but the observer has no way of determining ω completely without waiting until the end of time. Nature only reveals one bit of ω per unit time, so the observer can only get an improved *estimate* of ω as time goes on. This is an excellent example of how a random process can be modeled by nature selecting only a single elementary outcome, yet the observer sees a process that evolves forever.

In this example our change in the sample space to [0, 1] from **R** was done for convenience. By restricting the sample space we did not have to define the random variable outside of the unit interval (as we would have had to do to provide a complete description).

At times it is necessary to extend the definition of a random process to include vector-valued functions of time so that the random process is a function of three arguments instead of two. The most important extension is to complex-valued random processes, i.e., vectors of length 2. We will not make such extensions frequently, but we will include an example at this time.

[4.25]

Given the same probability space as in example [4.24], define a complex-valued random process $\{\mathbf{X}_n\}$ as follows: Let α be a fixed real parameter and define

$$\mathbf{X}_n(\omega) = e^{jn\alpha}e^{j2\pi\omega} = e^{j(n\alpha+2\pi\omega)}; \quad n = 1,2,3,....$$

This process, called the random rotations process, is a discrete time continuous (complex) alphabet one-sided random process. Note that an alternative description of the same process would be to define Ω as the unit circle in the complex plane together with its Borel field and to define a process $Y_n(\omega) = c^n\omega$ for some fixed $c \in \Omega$; this representation points out that successive values of Y_n are obtained by rotating the previous value through an angle determined by c.

Note that the joint pdf of the complex components of X_n varies with time, n, as the pdf in example [4.23] does (exercise 4.22).

[4.26]

Again consider the probability space of example [4.24]. We define a random process recursively on this space as follows: Define $X_0 = \omega$ and

$$X_n(\omega) = 2X_{n-1}(\omega) \bmod 1 ,$$

where $r \bmod 1$ is the fractional portion of r. In other words, if $X_{n-1}(\omega) = x$ is in $[0,1/2)$, then $X_n(\omega) = 2x$. If $x \in [1/2,1)$, then $X_n(\omega) = 2x - 1$.

[4.27]

Given the same probability space as in the example [4.26], define $X(t,\omega) = \cos(t + 2\pi\omega)$, $t \in \mathbf{R}$. The resulting random process $\{X(t)\}$ is continuous time and continuous amplitude and is called a random phase process since all of the possible waveforms are shifts of one another. Note that the pdf of $X(t,\omega)$ does *not* depend on time (exercise 4.23).

[4.28]

Take any one of the foregoing (real) processes and quantize or clip it; that is, define a binary quantizer q by

$$q(r) = \begin{cases} a & \text{if } r \geq 0 \\ b & \text{if } r < 0 \end{cases}$$

and define the process $Y(t,\omega) = q(X(t,\omega))$, all t. (Typically $b = -a$ or 0.)

This process is discrete alphabet and is either continuous or discrete time, depending on the original X process. In any case, $Y(t)$ has a binary pmf that, in general, varies with time.

[4.29]

Say we have two random variables U and V defined on a common probability space $(\Omega, \mathbf{F}, P)$. Then

$$X(t) = U\cos(2\pi f_0 t + V)$$

defines a random process on the same probability space for any fixed parameter f_0. (Why?)

All the foregoing random processes are well defined. The values of the processes inherit probabilistic descriptions from the underlying probability space. The techniques of derived distributions can be used to compute probabilities involving the outputs. Several examples are explored in the exercises. We close this chapter with a more abstract example.

[4.30]
Suppose that we are given a probability space $(\Omega, \boldsymbol{F}, P)$ where $\Omega = A^I$ is a sequence or waveform space, e.g., the space of all real-valued sequences, the space of all binary sequences, the space of all real-valued waveforms. Let Π_t denote the sampling functions of example [4.5]; that is, $\Pi_t(\omega)$ samples the sequence or waveform ω at time t. Then $\{\Pi_t, t \in I\}$ is a random process. A random process given in this form, that is, as a sampling function on a probability space whose points are sequences or waveforms, is said to be *directly given*.

EXERCISES

4.1   Given the probability space $(\boldsymbol{R}, \boldsymbol{B}(\boldsymbol{R}), m)$, where m is the probability measure induced by the uniform pdf f on $[0,1]$ (that is, $f(r) = 1$ for $r \in [0,1]$ and is 0 otherwise), find the pdf's for the following random variables defined on this space:

   (a) $X(r) = r^2$,
   (b) $Y(r) = r^{1/2}$,
   (c) $Z(r) = \ln|r|$,
   (d) $V(r) = ar + b$, where a and b are fixed constants.

4.2   Do exercise 4.1 for an exponential pdf on the original sample space.

4.3   Do exercise 4.1a−d for a Gaussian pdf on the original sample space.

4.4   Use the properties of probability measures to prove the following facts about cdf's: If F is the cdf of a random variable, then

   (a) $F(-\infty) = 0$ and $F(\infty) = 1$.
   (b) $F(r)$ is a monotonically nondecreasing function, that is, if $x \geq y$, then $F(x) \geq F(y)$.
   (c) F is continuous from the right, that is, if ϵ_n, $n = 1, 2, \ldots$ is a sequence of positive numbers decreasing to zero, then

$$\lim_{n \to \infty} F(r + \epsilon_n) = F(r) .$$

   Note that continuity from the right is a result of the fact that we defined a cdf as the probability of an event of the form $(-\infty, r]$. If instead we had defined it as the probability of an event of the form $(-\infty, r)$ (as is often done in Eastern Europe), then cdf's would be continuous from the left instead of from the right. When is a cdf continuous from the left? When is it discontinuous?

4.5 Say we are given an arbitrary cdf F for a random variable and we would like to simulate an experiment by generating one of these random variables as input to the experiment. As is typical of computer simulations, all we have available is a uniformly distributed random variable U; that is, U has the pdf of exercise 4.1. This exercise explores a means of generating the desired random variable from U (this method is occasionally used in computer simulations). Given the cdf F, define the *inverse* cdf F^{-1} as follows: For $r \in [0,1]$ (the range space of a cdf), define $F^{-1}(r)$ as the smallest value of $x \in \mathbf{R}$ for which $F(x) = r$. We specify "smallest" to ensure a unique definition since F may have the same value for an interval of x. Find the cdf of the random variable Y defined by $Y = F^{-1}(U)$.

4.6 A probability space $(\Omega, \mathbf{F}, P)$ models the outcome of rolling two fair four-sided dice on a glass table and reading their down faces. Hence we can take $\Omega = \{1,2,3,4\}^2$, the usual event space (the power set or, equivalently, the Borel field), and a pmf placing equal probability on all 16 points in the space. On this space we define the following random variables: $W(\omega) = $ the down face on die #1; that is, if $\omega = (\omega_1, \omega_2)$, where ω_i denotes the down face on die #i, then $W(\omega) = \omega_1$. (We could use the sampling function notation here: $W = \Pi_1$.) Similarly, define $V(\omega) = \omega_2$, the down face on the second die. Define also $X(\omega) = \omega_1 + \omega_2$, the sum of the down faces, and $Y(\omega) = \omega_1 \omega_2$, the product of the down faces. Find the pmf and cdf for the random variables X, Y, W, and V. Find the pmf's for the random vectors (X,Y) and (W,V). Write a formula for the distribution of the random vector (W,V) in terms of its pmf.

Suppose that a greedy scientist has rigged the dice using magnets to ensure that the two dice always yield the same value; that is, we now have a new pmf on Ω that assigns equal values to all points where the faces are the same and zero to the remaining points. Repeat the calculations for this case.

4.7 Consider the two-dimensional probability space $(\mathbf{R}^2, \mathbf{B}(\mathbf{R})^2, P)$, where P is the probability measure induced by the pdf g, which is equal to a constant c in the square $\{(x,y): x \in [-1/2,1/2], y \in [-1/2,1/2]\}$ and zero elsewhere.

(a) Find the constant c.
(b) Find $P(\{x,y: x < y\})$.
(c) Define the random variable $U: \mathbf{R}^2 \to \mathbf{R}$ by $U(x,y) = x+y$. Find an expression for the cdf $F_U(u) = \Pr(U \le u)$.
(d) Define the random variable $V: \mathbf{R}^2 \to \mathbf{R}$ by $V(x,y) = xy$. Find the cdf $F_V(v)$.

(e)   Define the random variable $W: \mathbf{R}^2 \to \mathbf{R}$ by $W(x,y) = \max(x,y)$, that is, the larger of the two coordinate values. Thus $\max(x,y) = x$ if $x \geq y$. Find the cdf $F_W(w)$.

4.8   Let (X,Y) be a random vector with distribution $P_{X,Y}$ induced by the pdf $f_{X,Y}(x,y) = f_X(x)f_Y(y)$, where

$$f_X(x) = f_Y(x) = \lambda e^{-\lambda x}; \quad x \geq 0 ,$$

that is, (X,Y) is described by a product pdf with exponential components.

(a)   Find the pdf for the random variable $U = X + Y$.

(b)   Let the "max" function be defined as in exercise 4.7 and define the "min" function as the smaller of two values; that is, $\min(x,y) = x$ if $x \leq y$. Define the random vector (W,V) by $W = \min(X,Y)$ and $V = \max(X,Y)$. Find the pdf for the random vector (W,V).

4.9   Let (X,Y) be a random vector with distribution $P_{X,Y}$ induced by a product pdf $f_{X,Y}(x,y) = f_X(x)f_Y(y)$ with $f_X(x) = f_Y(y)$ equal to the Gaussian pdf with $m = 0$. Consider the random vector as representing the real and imaginary parts of a complex-valued measurement. It is often useful to consider instead a magnitude-phase representation vector (R,Θ), where R is the magnitude $(X^2 + Y^2)^{1/2}$ and $\Theta = \tan^{-1}(Y/X)$. Find the joint pdf of the random vector (R,Θ). Find the marginal pdf's of the random variables R and Θ. The pdf of R is called the Rayleigh pdf. Are R and Θ independent?

4.10   Let $(\Omega,\mathbf{F},P)$ be a probability space and consider events F, G, and H for which $P(F) > P(G) > P(H) > 0$. Events F and G form a partition of Ω, and events F and H are independent. Can events G and H be disjoint?

4.11   Given a probability space $(\Omega,\mathbf{F},P)$, let F, G, and H be events such that $P(F \cap G | H) = 1$. Which of the following statements are true? Why or why not?

(a) $P(F \cap G) = 1$

(b) $P(F \cap G \cap H) = P(H)$

(c) $P(F^c | H) = 0$

(d) $H = \Omega$

4.12   Let f be the uniform pdf f on $[0, 1]$, as in exercise 4.1. Let (X,Y) be a random vector described by a joint pdf

$$f_{X,Y}(x,y) = f(y)f(x - y), \text{ all } x,y .$$

(a)   Find the marginal densities f_X and f_Y. Are X and Y independent?

(b)   Find $P(X \geq 1/2 | Y \leq 1/2)$.

4.13 In example [4.24] of the binary random process formed by taking the binary expansion of a uniformly distributed number on [0,1], find the pmf for the random variable X_n for a fixed n. Find the pmf for the random vector (X_n, X_k) for fixed n and k. Consider both the cases where $n = k$ and where $n \neq k$. Find the probability $\Pr(X_5 = X_{12})$.

4.14 In example [4.27] of the random phase process, find $\Pr(X(t) \geq 1/2)$.

4.15 Evaluate the pmf $p_{Y(t)}(y)$ for the quantized process of example [4.28] for each possible case. (Choose $b = 0$ if the process is nonnegative and $b = -a$ otherwise.)

4.16 Let $([0,1], B([0,1]), P)$ be a probability space with pdf $f(\omega) = 1$; $\omega \in [0,1]$. Find a random vector $\{X_t; \epsilon \in \{1,2,...,n\}\}$ such that $\Pr(X_t = 1) = \Pr(X_t = 0) = 1/2$ and $\Pr(X_t = 1$ and $X_{t-1} = 1) = 1/8$, for all relevant t.

4.17 Give an example of two equivalent random variables (that is, two random variables having the same distribution) that

(a)   are defined on the same space but are not equal for any $\omega \in \Omega$,

(b)   are defined on different spaces and have different functional forms.

4.18 Let $(R, B(R), m)$ be the probability space of exercise 4.1.

(a)   Define the random process $\{X(t); t \in [0, \infty)\}$ by

$$X(t,\omega) = \begin{cases} 1 \text{ if } 0 < t \leq \omega \\ 0 \text{ otherwise .} \end{cases}$$

Find $\Pr(X(t) = 1)$ as a function of t.

(b)   Define the random process $\{X(t); t \in [0, \infty)\}$ by

$$X(t,\omega) = \begin{cases} t/\omega \text{ if } 0 < t \leq \omega \\ 0 \text{ otherwise .} \end{cases}$$

Find $\Pr(X(t) > x)$ as a function of t for $x \in (0,1)$.

4.19 Two continuous random variables X and Y are described by the pdf

$$f_{X,Y}(x,y) = \begin{cases} c \text{ if } |x| + |y| \leq r \\ 0 \text{ otherwise ,} \end{cases}$$

where r is a fixed real constant and c is a constant. In other words, the pdf is uniform on a square whose side has length $\sqrt{2}\, r$.

(a)   Evaluate c in terms of r.

(b)   Find $f_X(x)$.

(c)   Are X and Y independent random variables? (Prove your answer.)

(d)   Define the random variable $Z = (|X| + |Y|)$. Find the pdf $f_Z(z)$.

4.20 You are given a random variable U described by a pdf that is 1 on $[0, 1]$. Describe and make a labeled sketch of a function g such that the random variable $Y = g(U)$ has a pdf $\lambda e^{-\lambda x}$; $x \geq 0$.

4.21 Find the pdf of $X(t)$ in example [4.23] as a function of time. Find the joint cdf of the vector $(X(1), X(2))$.

4.22 Find the joint cdf of the complex components of $X_n(\omega)$ in example [4.25] as a function of time.

4.23 Find the pdf of $X(t)$ in example [4.27].

4.24 A certain communication system outputs a discrete time series $\{X_n\}$ where X_n has pmf $p_X(1) = p_X(-1) = 1/2$. Transmission noise in the form of a random process $\{Y_n\}$ is added to X_n to form a random process $\{Z_n = X_n + Y_n\}$. Y_n has a Gaussian distribution with $m = 0$ and $\sigma = 1$.

  (a) Find the pdf of Z_n.

  (b) A receiver forms a random process $\{R_n = \text{sgn}(Z_n)\}$ where sgn is the sign function $\text{sgn}(x) = 1$, if $x \geq 0$, $\text{sgn}(x) = -1$, if $x < 0$. R_n is output from the receiver as the receiver's estimate of what was transmitted. Find the pmf of R_n and the probability of detection (i.e., $Pr(R_n = X_n)$).

4.25 If X is a Gaussian random variable, find the marginal pdf $f_{Y(t)}$ for the random process $Y(t)$ defined by

$$Y(t) = X\cos(2\pi f_0 t); \quad t \in \mathbf{R} \ ,$$

where f_0 is a known constant frequency.

4.26 Let X and Z be the random variables of exercises 4.1 through 4.3. For each assumption on the original density find the cdf for the random vector (X, Z), $F_{X,Z}(x,z)$. Try to find the corresponding pdf $f_{X,Z}(x,z)$. Does the appropriate derivative exist? Is it a valid pdf?

4.27 Let N be a random variable giving the number of molecules molecules of hydrogen in a spherical region of radius r and volume $V = 4\pi r^3/3$. Assume that N is described by a Poisson pmf

$$p_N(n) = \frac{e^{-\rho V}(\rho V)^n}{n!} \ , \quad n = 0,1,2,\dots.$$

where ρ can be viewed as a limiting density of molecules in space. Say we choose an arbitrary point in deep space as the center of our coordinate system. Define a random variable X as the distance from the origin of our coordinate center to the nearest molecule. Find the pdf of the random variable X, $f_X(x)$.
Hint: Find $Pr(X > x)$ in terms of p_N.

4.28 Let V be a random variable with a uniform pdf on $[0,a]$. Let W be a random variable, independent of V, with an exponential pdf with parameter λ, that is,

$$f_W(w) = \lambda e^{-\lambda w} \; ; \; w \in [0, \infty) \, .$$

Let $p(t)$ be the pulse with value 1 when $0 \le t \le 1$ and 0 otherwise. Define the random process $\{X(t); \, t \in [0, \infty)\}$ by

$$X(t) = V p(t - W) \, .$$

(This is a model of a square pulse that occurs randomly in time with a random amplitude.) Find for a fixed time $t > 1$ the cdf $F_{X(t)}(\alpha) = \Pr(X(t) \le \alpha)$. You must specify the values of the cdf for all possible real values α. Show that there exists a pmf p with a corresponding cdf F_1, a pdf f with a corresponding cdf F_2, and a number $\beta_t \in (0,1)$ such that

$$F_{X(t)}(\alpha) = \beta_t F_1(\alpha) + (1 - \beta_t) F_2(\alpha).$$

Give expressions for p, f, and β_t.

4.29 Suppose that (X, Y) is a two-dimensional random vector with joint pdf

$$f_{X,Y}(x,y) = \begin{cases} c(x^2 + y^2) ; & x^2 + y^2 \le 1, x \ge 0, y \ge 0 \\ 0 ; & \text{otherwise.} \end{cases}$$

(a) Find the constant c.

(b) Find the marginal densities of X and Y. Are X and Y independent?

(c) Define the random variables $R = \sqrt{X^2 + Y^2}$ and $\Theta = \tan^{-1}(Y/X)$. Find the joint pdf $f_{R,\Theta}(r,\theta)$. Are R and Θ independent?

Suppose that the random vector (R, Θ) is quantized in two dimensions using the polar quantizer $q(r,\theta)$ given below to produce a new random variable $Q = q(R, \Theta)$.

$$q(r,\theta) = \begin{cases} i \text{ when } 0 \le r \le (1/2)^{1/4} \, , \; (i-1)\pi/8 \le \theta \le i\pi/8 \; ; \; 1 \le i \le 4 \\ i+4 \text{ when } (1/2)^{1/4} \le r < 1 \, , \; (i-1)\pi/8 \le \theta \le i\pi/8 \; ; \; 1 \le i \le 4 \end{cases}$$

(d) Find the pmf for the random variable Q.

5

SPECIFICATION
OF RANDOM PROCESSES

In chapter 4 we studied the definitions and probability distributions of random variables, vectors, and processes. The probability distributions for random variables and vectors are relatively easy to specify because they are defined on $\mathbf{R}^k$ for finite k. For random processes, however, the specification of a distribution for the whole process at once, i.e., the distribution of an infinite-dimensional vector, is generally not possible or useful except for a few simple examples such as those at the end of chapter 4. Instead, we are generally interested in the distribution of finite-dimensional subsets of the random process, e.g., the distribution of random vectors from the discrete time random process formed by sampling on finite discrete time intervals. In this chapter the various distributions induced by a random process on such finite-dimensional collections of sample times are studied. The principal result of this chapter is the Kolmogorov extension theorem, one of the results that form the foundation for the theory of random processes. Roughly speaking, this theorem states that a family of finite-dimensional cdf's, pmf's, or pdf's defined on all possible finite-dimensional time-sampled subsets of a random process suffices to completely describe or specify a random process provided that the family is consistent—consistent in the sense that all probabilities of finite-dimensional events on the random process can be calculated and are unique. Alternatively, if we can describe in a consistent fashion the probabilistic behavior of the outputs of a random process for finite collections of sample times, then we have uniquely characterized the random process. This provides a form of canonical model for random processes, as will be seen. A proof of the Kolmogorov extension theorem is beyond the scope of this book. However, the theorem is quite believable and is presented with a discussion of its meaning and implications.

PROCESS DISTRIBUTIONS

A random process was defined in chapter 4 as an infinite family of random variables $\{X_t; t \in \mathbf{I}\}$, where $\mathbf{I}$ is an index set corresponding to the possible values of time. The family is defined as a set of functions on a common probability space $(\Omega, \mathbf{F}, P)$. In principle, given this underlying probability space and the family of functions on the space, we should be able to compute the probability of any output event of the random process. In other words, the underlying probability space, together with the family of random variables constituting the random process, induces a probability measure on events in the sequence space or waveform space of possible outputs of the random process. This follows from the appropriate generalization of the basic derived distribution formula for random variables and random vectors. To be more precise, recall that we can form an event space for the space of all possible output sequences or waveforms of the random process by using the product event space. In particular, the random process outputs are members of the product space $\mathbf{R}^{\mathbf{I}}$ = {all real-valued $\{x_t; t \in \mathbf{I}\}\}$. The corresponding event space is $\mathbf{B}(\mathbf{R})^{\mathbf{I}}$, the event space generated by all one-dimensional events of the form $\{\{x_t; t \in \mathbf{I}\}: x_s \in F\}$ for $s \in \mathbf{I}$ and $F \in \mathbf{B}(\mathbf{R})$. The derived distribution for the random process is the probability measure, m, on the measurable space $(\mathbf{R}^{\mathbf{I}}, \mathbf{B}(\mathbf{R})^{\mathbf{I}})$ defined by

$$m(F) = P(\omega: \{X_t(\omega); t \in \mathbf{I}\} \in F); \quad F \in \mathbf{B}(\mathbf{R})^{\mathbf{I}}, \qquad (5.1)$$

that is, as usual, the probability of a sequence or waveform event F is the probability under the original probability space of all ω yielding the output event. (m could also be written using an inverse image notation. We do not do this, however, because it is a bit clumsy to denote the inverse image of an entire sequence or waveform.) The space of possible sequences or waveforms produced by the process is sometimes called the *ensemble*.

As with random variables and random vectors, formula (5.1) makes sense only if the given ω set is indeed an event in the original event space $\mathbf{F}$. As with random vectors, we simply state without proof that this is indeed the case. Proofs can be constructed that rely on the same principles for random processes as for random vectors. The reasoning starts with the observation that, by definition, the ω sets corresponding to one-dimensional output events are indeed input events (since such sets are inverse images of random variables that are defined with this property).

This observation, together with the establishment that one-dimensional output events generate the output event space, forms the proof.

Thus the probability measure m of (5.1) is well defined. It is called the *distribution* of the random process or, more specifically, the *process distribution*.

We can now form a probability space for any random process that deals directly with the space of possible output sequences or waveforms of the random process in place of the less direct underlying probability space. That is, the original probability space $(\Omega, \boldsymbol{F}, P)$, together with the random process $\{X_t; t \in \boldsymbol{I}\}$, induces the new probability space $(\mathbf{R}^{\boldsymbol{I}}, \boldsymbol{B}(\mathbf{R})^{\boldsymbol{I}}, m)$. The triplet forming this new probability space includes the probability measure of (5.1) defined on the event space whose elements are collections of sample sequences or waveforms of the random process. As with random variables or vectors, we can compute probabilities of output events using either the distribution m or the original probability measure P, as convenient. In practice, the derived distribution is almost always used for such computations.

We shall consider two random processes to be the *same* or to be *equivalent* if they have the same process distributions. They are said to be equivalent because all possible output events will have the same probability for either random process even though their underlying probability spaces or their functional dependence on their defining spaces are not the same. (This is the same principle used with equivalent random variables in chapter 4; see exercise 4.17.) It is possible that two quite different models will yield equivalent processes. For example, define two random processes on the space of infinite coin flips. One random process is just the sampling function on the sequences. The second shifts the sequence exactly one flip to the right in time and then samples; i.e., one random process is the time-delayed version of the other. Obviously, these are equivalent random processes. More to the point, by construction, the random process $\{X_t; t \in \boldsymbol{I}\}$ defined on the probability space $(\Omega, \boldsymbol{F}, P)$ has exactly the same process distribution as the process $\{\Pi_t; t \in \boldsymbol{I}\}$ of sampling random variables Π_t defined on the sequence or waveform probability space $(\mathbf{R}^{\boldsymbol{I}}, \boldsymbol{B}(\mathbf{R})^{\boldsymbol{I}}, m)$. The two models can be viewed as different models for the same process since both yield the same process distributions. The first model may be more basic, perhaps showing the physical origins of the process as measurements on some underlying experiment. The second model, however, is more direct in the sense that it shows directly the measure on possible sequences or waveforms produced by the process. For this reason, the sequence space or waveform model of a process is often referred to as a *directly given* model, which can be considered a canonical model for a random process.

Quite different physical, as well as mathematical, models can result in identical directly given models. As a simple example that we will later consider more carefully, we observe that choosing a number in $[0,1)$ according to a uniform pdf and then putting out the binary expansion of that number forever yields a binary process with the same process distribution as that obtained by flipping a fair coin forever.

While m describes the probabilities of events involving the entire family of random variables making up the process—that is, events described by what the process does over all time—it is usually of more interest to concentrate on events that involve only a finite number of samples of the random process. In addition, in practice it is frequently not easy to derive the "whole" m, but only that part of m which deals with certain subsets of the event space, viz., finite numbers of samples. As we will see subsequently, the knowledge of probabilities on all combinations of finite numbers of samples is enough to build up to, or approximate, m. To move in that direction, we consider probabilities for one sample. Given a specific sample time $t_0 \in I$ and a one-dimensional event of the form $X_{t_0} \in F$ for some $F \in \boldsymbol{B}(\mathbf{R})$, we can evaluate the probability $\Pr(X_{t_0} \in F)$ using the basic derived distribution formula as

$$P_{X_{t_0}}(F) = P(X_{t_0}^{-1}(F)); \; F \in \boldsymbol{B}(\mathbf{R})$$

since for a fixed time t_0, X_{t_0} is just a random variable defined on the underlying probability space. Alternatively, we could find the same probability using the process distribution as

$$P_{X_{t_0}}(F) = m(\{x_t; \; t \in \boldsymbol{I}\}: x_{t_0} \in F) = m(\Pi_{t_0}^{-1}(F)) \;,$$

where Π_{t_0} is the sampling function at time t_0. In other words, $P_{X_{t_0}}(F)$ is the probability of all sequences having the t_0^{th} coordinate in F. Note that the distribution $P_{X_{t_0}}(F)$ for the random variable X_{t_0} defined on the probability space $(\Omega, \boldsymbol{F}, P)$ is the same as the distribution $P_{\Pi_{t_0}}(F)$ of the random variable Π_{t_0} defined on the probability space $(\mathbf{R}^I, \boldsymbol{B}(\mathbf{R})^I, m)$. This again points out that we can compute probabilities using either the original underlying probability space or the directly given probability space induced by the original space and a random process.

More generally, we can also find the probability of any output event involving a finite number of sample values of the output using the same principle, since such a finite collection constitutes a random vector defined on the space. Say we are given a finite collection of sample times $t_0, t_1, ..., t_{k-1}$, all in $\boldsymbol{I}$. For example, if $\boldsymbol{I} = \mathbf{Z}$ or $\boldsymbol{I} = \mathbf{Z}_+$, an interesting collection of sample times would be the first k possible times starting at time zero, that is, $t_0 = 0$, $t_1 = 1,...,$ and $t_{k-1} = k-1$. If $\boldsymbol{I} = \mathbf{R}$, we might be interested in a sequence of consecutive samples taken every second or every minute or every hour. Given these fixed times and an arbitrary event in $\boldsymbol{B}(\mathbf{R})^k$ we can find the probability that the collection of samples at the given times lies in the event by using the vector derived distribution formula: For any $F \in \boldsymbol{B}(\mathbf{R})^k$

$$\Pr((X_{t_0}, X_{t_1}, \ldots, X_{t_{k-1}}) \in F) = P_{X_{t_0}, X_{t_1}, \ldots, X_{t_{k-1}}}(F)$$

$$= m(\{x_t; \ t \in \boldsymbol{I}\}: (x_{t_0}, \ldots, x_{t_{k-1}}) \in F)$$

$$= P(\omega: (X_{t_0}(\omega), \ldots, X_{t_{k-1}}(\omega)) \in F) . \qquad (5.2)$$

Of particular importance is the specialization to k-dimensional *rectangles*—the events, introduced in chapter 3, in which the separate samples are confined to take on values in separate one-dimensional events. For rectangles, the general event $F \in \boldsymbol{B}(\mathbf{R})^k$ becomes an event of the form

$$\mathop{\times}_{i=0}^{k-1} F_i = \{\text{all } (r_0, r_1, \ldots, r_{k-1}): r_i \in F_i; \ i = 0, 1, \ldots, k-1\}$$

for some collection $F_i; \ i = 0, 1, \ldots, k-1$ of events in $\boldsymbol{B}(\mathbf{R})$. In this case formula (5.2) specializes to

$$\Pr(X_{t_i} \in F_i; \ i = 0, 1, \ldots, k-1) = P_{X_{t_0}, X_{t_1}, \ldots, X_{t_{k-1}}}(\mathop{\times}_{i=0}^{k-1} F_i)$$

$$= m(\{x_t; \ t \in \boldsymbol{I}\}: x_{t_i} \in F_i; \ i = 0, 1, \ldots, k-1)$$

$$= P(\{\omega: (X_{t_0}(\omega), \ldots, X_{t_{k-1}}(\omega)) \in \mathop{\times}_{i=0}^{k-1} F_i\})$$

$$= P(\mathop{\cap}_{i=0}^{k-1} X_{t_i}^{-1}(F_i)); \qquad (5.3)$$

$$F_i \in \boldsymbol{B}(\mathbf{R}); \ i = 0, 1, \ldots, k-1 .$$

Thus, given a random process, we can write down formulas for the distributions at any given finite collection of sample times. If the process is discrete, then these distributions will in turn be induced by pmf's. If the process is continuous and the appropriate pdf's exist, then these distributions are completely characterized by the corresponding pdf's. In other words, *given a random process, we have a family of finite-dimensional distributions: one for each choice of dimension k and each set of sample times $t_0, \ldots, t_{k-1}$.*

Obviously, these distributions all agree with one another; that is, they are *consistent* in the sense that no matter which distribution in the family we use to compute a given event, we get the same answer. This is the same idea as consistency of distributions of random vectors: Higher-dimensional distributions must agree with lower-dimensional distributions. For example, we could use any of the distributions P_{X_0, X_1, X_2} or P_{X_0, X_1} or P_{X_1, X_2} to compute $P_{X_1}(F)$ since, for example,

$$P_{X_1}(F) = \Pr(X_1 \in F)$$

$$= \Pr(X_1 \in F \text{ and } X_2 \in \mathbf{R}) = P_{X_1,X_2}(F \times \mathbf{R})$$

$$= \Pr(X_0 \in \mathbf{R} \text{ and } X_1 \in F \text{ and } X_2 \in \mathbf{R})$$

$$= P_{X_0,X_1,X_2}(\mathbf{R} \times F \times \mathbf{R}) .$$

In English, all the above formulas translate as $\Pr(X_1 \in F$ and $X_i \in \mathbf{R}$ for other indices i). Since $X_i \in \mathbf{R}$ is true for all ω, this is just $\Pr(X_1 \in F)$. To be more precise, all the versions are computed using the derived distribution formula to yield $P(X_1^{-1}(F))$ or, equivalently, $m(\{x_t; \ t \in \mathbf{I}\}: x_1 \in F)$, and hence they all produce the same number.

Thus, given a random process, the family of finite-dimensional distributions that yields the probability of finite-dimensional events is automatically consistent. On the other hand, not all families of distributions are consistent. To emphasize the requirement of consistent distributions, we point out two simple examples of inconsistent distributions. First, suppose we are told that an engineer has modeled a binary {0,1} random process and found, among other things, that the distribution for sample times 0 and 1 is described by a pmf $p_{X_0,X_1}(x,y)$ that is 1/4 for all four possible values of (x,y). The engineer also tells us that the marginal distribution for the sample at time 0 is described by a pmf $p_{X_0}(x)$ that is 5/6 for $x = 0$ and 1/6 for $x = 1$. While both the joint and the marginal pmf's are valid pmf's, they are inconsistent: If we compute the probability $\Pr(X_0 = 1)$ we get 1/2 using the joint pmf and 1/6 using the marginal. In other words, something is wrong with the engineer's model. There cannot be a random process with the distributions stated because they give inconsistent results: The same physical event has different probabilities, depending on which distribution is used for computation.

As a second and even more flagrant example of inconsistency, suppose that the engineer told us that the distribution for the pair of samples X_{t_0}, X_{t_1} placed probability 1 on the event $\{(1,1)\}$ and that the marginal distribution for X_{t_0} placed probability 1 on the event $\{0\}$. Again, these distributions do not agree with each other. No process that would yield these distributions could exist.

For reference and precision we will next give a formal definition of *consistency*. The reader should keep in mind the informal definition in order to see the basic idea through the messy notation: A family of finite-dimensional distributions is *consistent* if the distributions agree with each other, that is, if higher dimensional distributions agree with lower dimensional distributions on events where they both are defined. In other words, if we compute the probability of a given event using any of the distributions which can be used, then we must get the same answer. Furthermore, if we are *given* a random process, then the family of finite-dimensional distributions *must* be consistent since the probability of *any* finite-dimensional event can be *uniquely* determined using the original probability measure of the underlying probability space. The formal definition of consistency follows.

Let I be infinite index set and $\{P_{X_{t_0},X_{t_1},...,X_{t_{k-1}}}$; all $k < \infty$; all $K = \{t_0,t_1,...,t_{k-1}\} \subset I \}$ be a family of finite-dimensional distributions. The family of finite-dimensional distributions is *consistent* on I if for every dimension $k < \infty$, for *every* choice of sample times $K = \{t_0,...,t_{k-1}\}$, and for *every* $J \subset K$,

$$P_{X_{s_0},...,X_{s_{j-1}}}(\{(x_{s_0},...,x_{s_{j-1}}): x_{s_i} \in F_{s_i}; \, i = 0,1,...,j-1\}) = \qquad (5.4a)$$

$$P_{X_{t_0},...,X_{t_{k-1}}}(\{(x_{t_0},...,x_{t_{k-1}}): x_{t_i} \in F_{t_i} \text{ if } t_i \in J; \, x_{t_i} \in \mathbf{R} \text{ otherwise}; \, i = 0,1,...,k-1\}) \, ,$$

for *all* choices of $F_{s_i} \in \mathbf{B}(\mathbf{R})$.

Equivalently, the family of distributions is consistent if the corresponding probability functions satisfy similar relations: If the alphabet of the process is $A^I \subset \mathbf{R}^I$, then for every dimension $k < \infty$, for *every* choice of sample times $K = \{t_0,...,t_{k-1}\}$, and for *every* $J \subset K$, the cdf's satisfy

$$F_{X_{s_0},...,X_{s_{j-1}}}(x_{s_0},...,x_{s_{j-1}}) =$$

$$F_{X_{t_0},X_{t_1},...,X_{t_{k-1}}}(x_{t_0},...,x_{t_{k-1}}); \, x_{t_i} = \infty \text{ if } t_i \notin J \, , \qquad (5.4b)$$

for *all* choices of $x_{s_i} \in A$. Similar descriptions can be given in terms of pdf's and pmf's. We now turn, however, to specific examples of consistency relations.

The general idea of consistency is likely best demonstrated by giving important examples of consistency relations for cdf's, pmf's, and pdf's. These relations are more important for applications than the general abstract definition, but they are all simple consequences of the general definition. Say we have a family of consistent finite-dimensional distributions. If the corresponding family of cdf's is $\{F_{X_{t_0},...,X_{t_{k-1}}}\}$, then (5.4) implies

$$F_{X_{t_0},...,X_{t_{k-2}},X_{t_{k-1}}}(x_0,x_1,...,x_{k-2},\infty) =$$

$$F_{X_{t_0},...,X_{t_{k-2}}}(x_0,x_1,...,x_{k-2}); \text{ all } x_i \in \mathbf{R} \, , \qquad (5.5a)$$

and

$$F_{X_{t_0},X_{t_1},...,X_{t_{k-1}}}(\infty,x_1,...,x_{k-1}) =$$

$$F_{X_{t_1},...,X_{t_{k-1}}}(x_1,...,x_{k-1}); \text{ all } x_i \epsilon \mathbf{R} . \tag{5.5b}$$

Observe that the first statement translates as $\Pr(X_i \leq x_i , i=0,1,...,k-2,$ and $X_{k-1} \leq \infty) = \Pr(X_i \leq x_i, i=0,1,...,k-2)$, hardly a profound observation!

These two relations are not the only consistency relations for cdf's, but they are the most important since they show how to find probabilistic descriptions of short vectors from long vectors which contain them. Similarly, we can apply either the basic consistency of the (5.4) definition or the cdf relationships of (5.5) to get similar relations for pmf's and pdf's:

$$\sum_{x_{k-1} \epsilon A} p_{X_{t_0},...,X_{t_{k-2}},X_{t_{k-1}}}(x_0,...,x_{k-2},x_{k-1}) =$$

$$p_{X_{t_0},...,X_{t_{k-2}}}(x_0,...,x_{k-2}); \text{ all } x_i \epsilon A , \tag{5.6a}$$

and

$$\sum_{x_0 \epsilon A} p_{X_{t_0},X_{t_1},...,X_{t_{k-1}}}(x_0,x_1,...,x_{k-1}) =$$

$$p_{X_{t_1},...,X_{t_{k-1}}}(x_1,...,x_{k-1}); \text{ all } x_i \epsilon A , \tag{5.6b}$$

where $A^{\boldsymbol{I}}$ is the alphabet of the process. The analogous pdf result is

$$\int_A f_{X_{t_0},...,X_{t_{k-2}},X_{t_{k-1}}}(x_0,...,x_{k-2},x_{k-1})dx_{k-1} =$$

$$f_{X_{t_0},...,X_{t_{k-2}}}(x_0,...,x_{k-2}); \text{ all } x_i \epsilon A , \tag{5.7a}$$

and

$$\int_A f_{X_{t_0},X_{t_1},...,X_{t_{k-1}}}(x_0,x_1,...,x_{k-1})dx_0 =$$

$$f_{X_{t_1},...,X_{t_{k-1}}}(x_1,...,x_{k-1}) , \text{ all } x_i \epsilon A . \tag{5.7b}$$

It should be kept in mind that these relations hold for all $k=0,1,2,...$ and all choices of sample times $t_i \epsilon \boldsymbol{I}; i=0,1,...,k-1$. The consistency relations of (5.5) through (5.7) for probability functions are special cases of the general relation in that we did not specifically consider all subvectors of each vector—we only considered what happens at the ends of vectors.

The foregoing discussion does not contain anything surprising or really new. Effectively, we have only made a definition and collected together a lot of bookkeeping information that states what is fairly obvious: If we have a random process, that is, a family of random variables defined on a common probability space, then we can compute a family of finite-dimensional distributions or, equivalently, a family of joint probability functions, and these distributions or probability functions are consistent in the sense we have defined. A perhaps surprising and definitely useful fact is that we can go in the opposite direction: *If we have a family of consistent finite-dimensional distributions, then there exists a random process described by this family. In addition, all random processes described by these distributions or probability functions are equivalent.*

This result is of extreme practical importance, since it implies that one has a complete model of a random process if one can describe its family of consistent *finite* dimensional cdf's, pmf's, or pdf's. Thus one can completely characterize or *specify* a random process by its finite sample behavior. This is important because in most applications one does not have an *a priori* model of a random process in terms of an underlying probability space and a family of functions on the space as was the case in all of the examples of chapter 4. Instead one has finite sample probability functions with which to describe directly the probabilistic behavior of the output sequences or waveforms of the random process. The result—called the Kolmogorov extension theorem, to be stated concisely shortly—implies that one need only provide a consistent description of the finite sample behavior of the process, that is, a rule for constructing the distribution for all possible finite-dimensional random vectors produced by the process. Providing such a rule is referred to as *specifying* the process, and a description of the finite-dimensional distributions via a formula for the finite-dimensional pmf's or pdf's is called a *specification* of the process. From this Kolmogorov specification one could, in concept, work up a (nonunique) underlying probability space and family of functions to describe the random process. This would almost never be done, as we are normally interested in the behavior of the sample functions of the random process themselves and not the underlying structure. Furthermore, in many instances, the complete model is too global: One really does not know a complete specification; one only knows enough about the finite sample distributions to solve the problem at hand.

Next we formally state the Kolmogorov extension theorem for reference and discuss its more familiar forms in terms of cdf's, pmf's, and pdf's. We then present two common and important examples of direct random process specification through the description of their finite-dimensional distributions. These random processes are *not* described as functions on an underlying probability space as in chapter 4.

The Kolmogorov Extension Theorem.   Given a consistent family of finite-dimensional distributions

$$\{P_{X_{t_0},X_{t_1},...,X_{t_{k-1}}}; \text{ all } k=0,1,2,...; \text{ all } t_i \in \boldsymbol{I}, \ i=0,1,...,k-1\} \ ,$$

there exists a random process $\{X_t; \ t \in \boldsymbol{I}\}$ described by these distributions, that is, having a process distribution m that induces the given finite-dimensional distributions. Furthermore, the process distribution m is unique. Thus if two random processes have the same finite-dimensional distributions, then they have the same process distribution and hence are equivalent random processes.

We now have two means of constructing a mathematical model for a random process: The first is to define explicitly a probability space and a family of measurable functions on it. The second is to ignore the underlying probability space and to provide a formula for a family of consistent finite-dimensional joint probability functions. The first technique is the most important from a conceptual, global point of view and is frequently used in the literature in theorem statements and proofs of general properties of random processes. The second technique is by far the more important for engineering applications. We gave some examples of the first technique in chapter 4. In this chapter we shall give two examples of the second or specification technique.

EXAMPLES OF SPECIFICATION

[5.1] i.i.d. random processes

A discrete time random process $\{X_n; \ n \in \boldsymbol{I}\}$ (where usually $\boldsymbol{I} = \mathbf{Z}$ or $\mathbf{Z}_+$) is said to be *independent identically distributed* or *i.i.d.* if all vectors formed by a finite number of samples of the process are i.i.d. random vectors (as defined in example [4.21]). Equivalently, the process is i.i.d. if there is a marginal cdf $F(r)$; $r \in \mathbf{R}$ such that the joint cdf's are given by

$$F_{X_{t_0},X_{t_1},...,X_{t_{k-1}}}(x_0,x_1,...,x_{k-1}) = \prod_{i=0}^{k-1} F(x_i); \text{ all } x_i \in \mathbf{R} \ , \qquad (5.8a)$$

for all $k=1,2,...$ and all choices of distinct $t_i \in \boldsymbol{I}$; $i=0,1,...,k-1$.

It is instructive to specialize the definition to the pmf and pdf cases. A discrete time discrete alphabet random process is i.i.d. if there is a marginal pmf $p(x)$; $x \in A$ such that the joint pmf's satisfy

$$p_{X_{t_0},X_{t_1},...,X_{t_{k-1}}}(x_0,x_1,...,x_{k-1}) = \prod_{i=0}^{k-1} p(x_i); \text{ all } \mathbf{x} \in A^k \ . \qquad (5.8b)$$

A discrete time continuous alphabet random process is i.i.d. if there is a marginal pdf $f(x)$; $x \in \mathbf{R}$ such that the joint pdf's satisfy

$$f_{X_{t_0},X_{t_1},\ldots,X_{t_{k-1}}}(x_0,x_1,\ldots,x_{k-1}) = \prod_{i=0}^{k-1} f(x_i); \text{ all } \mathbf{x} \in \mathbf{R}^k . \tag{5.8c}$$

Thus, given any marginal cdf, pmf, or pdf, we can specify an i.i.d. process by (5.8). It is trivial to verify that the rules of (5.8) indeed satisfy the consistency condition (5.5) to (5.7). The more general consistency relations follow similarly. Hence the Kolmogorov extension theorem guarantees the existence of the process. We observe in passing that i.i.d. processes are also referred to as *discrete time memoryless processes*. Note that i.i.d. processes are defined only for discrete time. While there is a continuous time analog (called continuous time white noise), it is far more difficult to define. We will consider white noise in a subsequent chapter.

We have compiled a fairly large number of marginal cdf's, pmf's, or pdf's in our earlier material. Therefore, the i.i.d. example immediately provides a large number of examples of random processes. The name *i.i.d.* is used for the same reasons as in the i.i.d. vector case: If a process is i.i.d., then its output samples are mutually independent and identically distributed.

I.i.d. processes play a fundamental role in the theory and applications of discrete time random processes. They are, in a sense, both the simplest nontrivial processes and the most "random," since the past outputs provide no information about the future. We shall see, however, that i.i.d. processes can serve as building blocks for more complicated processes with memory. This is accomplished by filtering and performing other operations on the i.i.d. processes.

One example of an i.i.d. process that is sufficiently common and important enough to merit its own name is the Bernoulli random process: A *Bernoulli random process* is simply a binary alphabet i.i.d. random process. The Bernoulli random process is a model for such physical phenomena as binary data communications and flipping a (possibly) biased coin. The complete random process is defined by the marginal pmf and hence, since the process is binary, by the single number $p_{X_i}(1) = p_X(1)$, where, as is common, we have made the time independence notationally explicit. The number, often denoted by p, is called the *parameter* of the Bernoulli process and corresponds to the probability of a 1 occurring in a binary data stream or a *head* occurring in a single coin flip.

By extending exercise 4.13, it follows that a one-sided Bernoulli process with parameter 1/2 is an equivalent model for the random process formed as in example [4.24] from the binary expansion of a number in [0, 1) selected at random according to a uniform pdf (see exercise 5.3). This is an excellent example of how the same process can be obtained in very different manners. In one construction the random process is a sequence of random variables formed as a function on a sample space (the terms in the binary expansion on the unit interval). In the other construction the random process is formed by specifying finite-dimensional distributions. Since both constructions yield the same finite-dimensional distributions, the Kolmogorov extension theorem implies that the two random processes must be the same. It can also be shown that example [4.26] also yields an equivalent model to the one-sided Bernoulli process with parameter 1/2.

The next example provides a large and useful class of both discrete and continuous time random processes having continuous alphabets. Without doubt, it is the single most important class of random processes, providing an accurate model for many physically occurring signal, noise, and other processes.

[5.2] *Gaussian random processes*

A random process $\{X_t;\ t\in\mathbf{I}\}$ is said to be a *Gaussian random process* if all finite collections of samples of the process $(X_{t_0}, X_{t_1},\ldots, X_{t_{k-1}})$ are Gaussian random vectors.

Given a family of Gaussian pdf's, it is easy to *verify* consistency, as we will demonstrate shortly. However, unlike the i.i.d. example, it is not trivial to *prove* the consistency relations we will pose, because of the complicated form of the joint pdf's. Clearly the $\mathbf{m}$ vectors and Λ matrices required by the joint pdf's must be related if the family of pdf's is to be consistent. We summarize what this relation must be, but we state without proof that the resulting pdf's are indeed consistent. The persistent student interested in a proof can accomplish it with an understanding of changing variables in multidimensional integrals (using Jacobians), a good book of integral tables, and a great deal of stamina. Alternatively, the proof can be accomplished somewhat more easily using the transform methods that we will discuss in chapter 9.

A Gaussian random process $\{X_t;\ t\in\mathbf{I}\}$ is completely described by a real-valued function $m(t); t\in\mathbf{I}$ (called the *mean function* for reasons to be seen later) and a symmetric positive definite function $\Lambda(t,s); t,s\in\mathbf{I}$. That is, by symmetry we mean $\Lambda(t,s) = \Lambda(s,t)$, all $t,s\in\mathbf{I}$. By positive definite we mean that for any dimension k, any collection of sample times $t_0,t_1,\ldots,t_{k-1}$, and any nonzero real vector $(r_0,r_1,\ldots,r_{k-1})$ we have

$$\sum_{i=0}^{k-1}\sum_{j=0}^{k-1} r_i\Lambda(t_i,t_j)r_j > 0.$$

($\Lambda(t,s)$ is called the *covariance function* for reasons to be seen later.) These functions are used to define the joint pdf's as follows: Given any dimension k and collection of sample times $t_0,t_1,\ldots,t_{k-1}$, define

$$f_{X_{t_0},X_{t_1},\ldots,X_{t_{k-1}}}(\mathbf{x}) = \frac{e^{-(\mathbf{x}-\mathbf{m})^t\Lambda^{-1}(\mathbf{x}-\mathbf{m})/2}}{(2\pi)^{k/2}(\det\Lambda)^{1/2}};\quad \mathbf{x}\in\mathbf{R}^k ,$$

where

$$\mathbf{m} = (m(t_0), m(t_1), \ldots, m(t_{k-1}))^t$$

and

$$\Lambda = \begin{bmatrix} \Lambda(t_0, t_0) & \Lambda(t_0, t_1) & \cdots & \Lambda(t_0, t_{k-1}) \\ \Lambda(t_1, t_0) & \Lambda(t_1, t_1) & \cdots & \Lambda(t_1, t_{k-1}) \\ \cdot & \cdot & \cdot & \\ \cdot & \cdot & \cdot & \\ \cdot & \cdot & \cdot & \\ \Lambda(t_{k-1}, t_0) & \Lambda(t_{k-1}, t_1) & \cdots & \Lambda(t_{k-1}, t_{k-1}) \end{bmatrix}.$$

This family of probability density functions can be shown to be consistent. That is, any mean function and any symmetric positive definite function yields a consistent family of Gaussian pdf's. The most complicated aspect of Gaussian random processes is their definition or specification: The formulas just given are a mess! The basic idea is that we have an m function and a Λ function, and for any collection of sample times, the random vector has a Gaussian density with $\mathbf{m}$ vector and Λ matrix formed by sampling these functions. The form of the joint pdf is, however, admittedly complicated.

As a simple example of a Gaussian process, consider a discrete time Gaussian process with a mean function $m(t) = m$ for all t, where m is a fixed constant, and a covariance function $\Lambda(t,s) = \sigma^2$ for $t = s$ and $\Lambda(t,s) = 0$ for $t \neq s$. Then the matrix Λ is simply σ^2 times the identity matrix. Therefore $\det \Lambda = \sigma^{2k}$ and Λ^{-1} is σ^{-2} times the identity matrix, and hence the joint pdf is

$$f_{X_{t_0}, X_{t_1}, \ldots, X_{t_{k-1}}}(x_0, x_1, \ldots, x_{k-1}) = \frac{e^{-\frac{1}{2\sigma^2}\sum_{i=0}^{k-1}x_i^2}}{(2\pi\sigma^2)^{k/2}} = \prod_{i=0}^{k-1} \frac{e^{-\frac{1}{2\sigma^2}x_i^2}}{(2\pi\sigma^2)^{1/2}}.$$

This is exactly an i.i.d. random process with Gaussian marginals! This process is called the i.i.d. Gaussian process or the memoryless Gaussian process. More general Gaussian processes are formed by using more complicated Λ and m functions.

EXERCISES

5.1   Describe the joint distribution of the random vector $(X(t), X(s))$ for random process examples [4.24] and [4.26] through [4.28].

5.2   Suppose that a discrete time discrete alphabet random process $\{X_n; n \in \mathbf{Z}+\}$ has a finite alphabet A with a letters and a pmf such that for all $k=1,2,...$ and all $\mathbf{x} = (x_0,x_1,...,x_{k-1})$ with $x_i \in A$ all i,

$$p_{X_0,X_1,...,X_{k-1}}(x_0,x_1,...,x_{k-1}) = c^k$$

for some constant c. Is the process necessarily i.i.d.? What is c?

5.3 Show that the binary example [4.24] of chapter 4 is a Bernoulli process with parameter 1/2.

5.4 Suppose we have a function $q: \mathbf{R} \to \mathbf{R}$. Show that if a random process $\{X_n; n \in \mathbf{I}\}$ is i.i.d., then so is the random process $\{q(X_n); n \in \mathbf{I}\}$. Thus, for example, a quantized i.i.d. process is also i.i.d.

5.5 Say $\{X_n; n \in \mathbf{Z}\}$ is a Bernoulli process (with alphabet $\{0,1\}$). For fixed n, define the random variable

$$Y_n = \sum_{i=0}^{n-1} X_i .$$

Find the pmf $p_{Y_n}(y)$. Is the random process $\{Y_n; n=1,2,...\}$ i.i.d.?

5.6 Let $\{X_n; n=0,1,2,...\}$ be a Bernoulli process, as in exercise 5.5. Define the random variable Z by $Z=j$ if j is the first index greater than zero for which $X_n=1$. In other words, we can define Z on the same probability space as the Bernoulli process by $Z(\omega)=j$ if and only if $X_n(\omega)=0$ for $n=1,2,...,j-1$ and $X_j(\omega)=1$. Find the pmf of Z.

5.7 Let $\{X_n; n \in \mathbf{Z}_+\}$ be an i.i.d. random process with marginal pmf $p_X(k)=c2^{-k}$, $k=0,1,...$. Evaluate the constant c. Given a fixed integer m, find the probabilities $\Pr(X_n > m)$ and $\Pr(X_n=k|X_n>m)$ for all k.

5.8 Let $\{X_n; n=0,1,2,...\}$ be an i.i.d. process with alphabet $\{1,2,3,4,5,6\}$ and marginal pmf $p_X(k)=1/6$; $k \in \{1,2,3,4,5,6\}$. Find the pmf's for the following random variables:

(a) $Y=X_1+X_2$.

(b) W defined as 1 if X_0 is even and 0 otherwise.

(c) $V = \min(X_{10},X_{30})$.

(d) Find an expression for the cdf of the random variable $U = \max(X_0,X_1,...,X_{k-1})$.

5.9 Let $\{X_n\}$ be a Bernoulli process. Find the pmf's for the random vector (X_n+X_m,X_nX_m) and for the random variables (X_n+X_m) and X_nX_m. Are the latter two random variables independent?

5.10 Consider the following random process $\{Y_n\}$: At the beginning of time, Mother Nature selects a random bias p at random according to a uniform pdf on (0,1) and then sends a Bernoulli process with parameter p forever. In other words, nature chooses a coin at random with the bias equally likely to be any number in (0,1) and then flips that coin forever. Is the resulting process i.i.d.? *Hint:* Consider a two-dimensional joint pmf and the corresponding one-dimensional pmf's. (This process distribution of the Y process is a continuous mixture analogous to the countable or finite mixture of example [3.18].)

5.11 Suppose that $\{X_n;\ n\in\mathbf{Z}\}$ is an i.i.d. process with marginal distribution described by a doubly exponential pdf. Let $q:\mathbf{R}\to\{-a,+a\}$ denote a binary quantizer defined by $q(r)=+a$ if $r\geq0$ and $q(r)=-a$ otherwise. Find the pmf for the random vector $(q(X_0),q(X_1),...,q(X_{k-1}))$. Is the process $\{q(X_n);\ n\in\mathbf{Z}_+\}$ i.i.d.?

5.12 Let $\{X_n;\ n\in\mathbf{Z}\}$ be an i.i.d. process with marginal distribution described by a Poisson pmf. Define two random processes $\{U_n\}$ and $\{W_n\}$ by $U_n=\min(X_n,X_{n-1})$ and $W_n=\max(X_n,X_{n-1})$. Find the marginal cdf's for these two processes. Is either of them i.i.d.?

5.13 Let $\{X_n\}$, $\{Y_n\}$, and $\{V_n\}$ be three mutually independent Bernoulli processes with parameters p_1, p_2, and α, respectively; e.g., $p_{X_n}(1)=p_1$. Define a new process W_n

by
$$W_n = \begin{cases} X_n \text{ if } V_n=1 \\ Y_n \text{ if } V_n=0 . \end{cases}$$

The process $\{W_n\}$ is called a *switched* process or *composite* process since it can be thought of as the output of a commutator or switch that randomly connects to one of the two processes $\{X_n\}$ or $\{Y_n\}$.

(a) Find $p_{W_n}(w)$.

(b) Find $p_{W_1,W_2,...,W_n}(w_1,...,w_n)$. Is $\{W_n\}$ a Bernoulli process?

5.14 Let $\{X_n\}$ and $\{Y_n\}$ be as in exercise 5.13, and let λ be a fixed number in $(0,1)$. Let m_1 and m_2 denote the process distributions of the X and Y processes, respectively. Suppose that $\{V_n\}$ is a directly given random process with process distribution given by the finite mixture

$$\bar{m} = \lambda m_1+(1-\lambda)m_2 .$$

Find the marginal pmf p_{V_1} and the joint pmf p_{V_1,V_2}. Is the V process i.i.d.? This is a model for a sequence of coin flips in which the coin itself is selected at random from one of two choices at the beginning of time.

6

EXPECTATION AND
LAWS OF LARGE NUMBERS

In engineering practice we are often interested in the average behavior of measurements on random processes. The goal of this chapter is to link the two distinct types of averages that are used—long-term time averages taken by calculations on an actual physical realization of a random process and averages calculated theoretically by probabilistic averages at some given instant of time, averages that are sometimes called *expectations*. As we shall see, both computations often (but by no means always) give the same answer. Such results are called *laws of large numbers* or *ergodic theorems*.

At first glance from a conceptual point of view, it seems unlikely that long-term time averages and instantaneous probabilistic averages would be the same. Figure 6.1 shows the situation graphically. You see several possible realizations of the random process $\{X(t,\omega)\}$ for different values of ω. If we take a long-term time average of a particular realization of the random process, we are averaging for a *particular* ω—an ω which we cannot know or choose; we do not use probability in any way and we are ignoring what happens with other values of ω. If, on the other hand, we take an instantaneous probabilistic average, say at the time t_0 shown in Figure 6.1, we are ignoring what happens with time and taking a probabilistically weighted average across all values of ω. Thus we have two averages, one along the time axis with ω fixed, the other along the ω axis with time fixed. It seems that there should be no reason for the answers to agree. Taking a more practical point of view, however, it seems that the time and probabilistic averages must be the same in many situations. For example, suppose that you measure the percentage of time that a particular noise voltage exceeds 10 volts. If you make the measurement over a sufficiently long period of time, the result should be a reasonably good estimate of the probability that the noise voltage exceeds 10 volts at any given instant of time—a probabilistic average value.

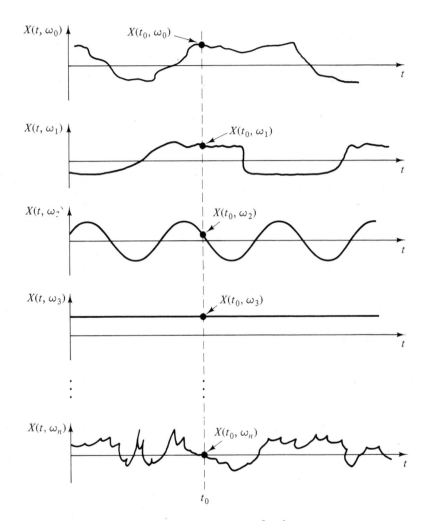

Figure 6.1   Fixed ω vs. fixed t.

To proceed further, for simplicity we concentrate on a discrete alphabet discrete time random process. Other cases are considered by converting appropriate sums into integrals. Let $\{X_n\}$ be an arbitrary discrete alphabet discrete time process. Since the process is random, we cannot predict accurately its instantaneous or short-term behavior—we can only make probabilistic statements. Based on experience with coins, dice, and roulette wheels, however, one expects that the long-term average behavior can be characterized with more accuracy. For example, if one flips a fair coin, short sequences of flips are unpredictable. However, if one flips long enough, one would expect to have an average of about 50% of the flips result in *heads*. This is a time average of an instantaneous function of a random process—a type of counting function that we will consider extensively. It is obvious that there are many functions that we can average, i.e., the average value, the average power, etc. We will proceed by defining one particular average, the sample average value of the random process, which is formulated as

$$S_n = n^{-1} \sum_{i=0}^{n-1} X_i; \; n = 1,2,3,\ldots$$

We will investigate the behavior of S_n for large n, i.e., for a long-term time average. Thus, for example, if the random process $\{X_n\}$ is the coin-flipping model, the binary process with alphabet $\{0,1\}$, then S_n is the number of 1's divided by the total number of flips—the fraction of flips that produced a 1. As noted before, S_n should be close to 50% for large n if the coin is fair.

Note that, as in example [4.7], for each n, S_n is a random variable that is defined on the same probability space as the random process $\{X_n\}$. This is made explicit by writing the ω dependence:

$$S_n(\omega) = n^{-1} \sum_{i=0}^{n-1} X_i(\omega) \; .$$

In more direct analogy to example [4.7], we can consider the $\{X_n\}$ as coordinate functions on a sequence space, say $(\mathbf{R}^\mathbf{Z}, \boldsymbol{B}(\mathbf{R})^\mathbf{Z}, m)$, where m is the distribution of the process, in which case S_n is defined directly on the sequence space. The form of definition is simply a matter of semantics or convenience. Observe, however, that in any case $\{S_n; \; n = 1,2,\ldots\}$ is itself a random process since it is an indexed family of random variables defined on a probability space.

For the discrete alphabet random process that we are considering, we can rewrite the sum in another form by grouping together all equal terms:

$$S_n(\omega) = \sum_{a \in A} a r_a^{(n)}(\omega) \tag{6.1}$$

where A is the range space of the discrete alphabet random variable X_n and $r_a^{(n)}(\omega) = n^{-1}$[number of occurrences of the letter a in $\{X_i(\omega), \; i = 0,1,2,\ldots,n-1\}$]. The random variable $r_a^{(n)}$ is called the nth-order *relative frequency* of the symbol a. Note that for the binary coin-flipping example we have considered, $A = \{0,1\}$, and $S_n(\omega) = r_1^{(n)}(\omega)$, the average number of heads in the first n flips. In other words, for the binary coin-flipping example, the sample average and the relative frequency of heads are the same quantity. More generally, the reader should note that $r_a^{(n)}$ can always be written as the sample average of the indicator function for a, $1_a(x)$:

$$r_a^{(n)} = n^{-1} \sum_{i=0}^{n-1} 1_a(X_i) \ ,$$

where

$$1_a(x) = \begin{cases} 1 & \text{if } x=a \\ 0 & \text{otherwise .} \end{cases}$$

Note that $1_{\{a\}}$ is a more precise, but more clumsy, notation for the indicator function of the singleton set $\{a\}$. We shall use the shorter form.

Let us now assume that all of the marginal pmf's of the given process are the same, say $p_X(x)$, $x \in A$. Based on intuition and gambling experience, one might suspect that as n goes to infinity, the relative frequency of a symbol a should go to its probability of occurrence, $p_X(a)$. To continue the example of binary coin flipping, the relative frequency of *heads* in n tosses of a fair coin should tend to $1/2$ as $n \to \infty$. If these statements are true, that is, if in some sense,

$$r_a^{(n)} \underset{n \to \infty}{\to} p_X(a) \ , \tag{6.2}$$

then it follows that in a similar sense

$$S_n \underset{n \to \infty}{\to} \sum_{a \in A} a p_X(a) \ , \tag{6.3}$$

the same expression as (6.1) with the relative frequency replaced by the pmf. Limiting statements of the form of (6.2) and (6.3) are called *laws of large numbers* or *ergodic theorems*. They relate long-run sample averages or time average behaviors to probabilistic calculations made at any given instant of time. It is obvious that such laws or theorems do not always hold. If the coin we are flipping wears in a known fashion with time so that the probability of a head changes, then one could hardly expect that the relative frequency of heads would equal the probability of heads at time zero.

In order to make precise statements and to develop conditions under which the laws or theorems do hold, we first need to develop the properties of the quantity on the right-hand side of (6.3)—called an *expectation*—and to make precise what the limits in (6.2) and (6.3) mean.

In particular, we cannot at this point make any sense out of a statement like " $\lim_{n \to \infty} S_n = \sum_{a \in A} a p_X(a)$," since we have no definition for such a limit of random variables or functions of random variables. It is obvious, however, that the usual definition of a limit used in calculus will not do, because S_n is a random variable, albeit a random variable whose "randomness" decreases in some sense with increasing n. Thus the limit must be defined in some fashion that involves probability.

EXPECTATION

Given a discrete alphabet random variable X specified by a pmf p_X, define the *expectation* (or *expected value, probabilistic average,* or *mean*) of X by

$$EX = \sum_{a \in A} a p_X(a) .$$

The expectation is also denoted by $E(X)$ or by an overbar, as $\bar{X}$. The expectation is also sometimes called an *ensemble average* to denote averaging across the ensemble of sequences that is generated for different values of ω at a given instant of time, as illustrated in Figure 6.1. To summarize the discussion of the preceding section, we are interested in conditions under which the time average of one realization of a random process converges to its ensemble average.

The mean of a random variable is simply a weighted average of the possible values of the random variable with the pmf used as a weighting. Before continuing, observe that we can define an analogous quantity for a continuous random variable possessing a pdf: If the random variable X is described by a pdf f_X, then we define the *expectation* of X by

$$EX = \int_{\mathbf{R}} x f_X(x) dx ,$$

where we have replaced the sum by an integral. While the integral does not have the intuitive motivation involving a relative frequency converging to a pmf that the earlier sum did, we shall see that it plays the analogous role in the laws of large numbers. Roughly speaking, this is because continuous random variables can be approximated by discrete random variables arbitrarily closely by very fine quantization. Through this procedure, the integrals with pdfs are approximated by sums with pmf's. Hence the discrete alphabet results imply the continuous alphabet results by taking appropriate limits. Because of the direct analogy, we shall develop the properties of expectations for continuous random variables along with those for discrete alphabet random variables. We note in passing that, analogous to using the Stieltjes integral as a unified notation for sums and integrals when computing probabilities, we can do the same thing for expectations. Hence, if F_X is the cdf of a random variable X, we make the definition

$$EX = \int x \, dF_X(x) = \begin{cases} \sum x p_X(x) \text{ if } X \text{ is discrete} \\ \int x f_X(x) dx \text{ if } X \text{ has a pdf.} \end{cases}$$

In a similar manner, we can define the expectation of a mixture random variable having both continuous and discrete parts in a manner analogous to (4.14).

Examples

The following examples provide some typical expectation computations.

[6.1]

As a slight generalization of the fair coin flip, consider the more general binary pmf with parameter p; that is, $p_X(1) = p$ and $p_X(0) = 1 - p$. In this case

$$EX = \sum_{i=0}^{1} x p_X(x) = 0(1-p) + 1p = p .$$

It is interesting to note that in this example, as is generally true, EX is not in the alphabet of the random variable, i.e., $EX \neq 0$ or 1 unless $p = 0$ or 1.

[6.2]

A more complicated discrete example is a geometric random variable. In this case

$$EX = \sum_{k=1}^{\infty} k p_X(k) = \sum_{k=1}^{\infty} kp(1-p)^{k-1} .$$

One may have access to a book of tables including this sum, but a useful trick can be used to evaluate the sum from the well-known result for summing a geometric series. Set $q = 1 - p$ and observe that we can replace the lower index of the sum by 0 without changing its value. Hence we wish to evaluate the sum

$$\sum_{k=0}^{\infty} k q^{k-1} .$$

Since $kq^{k-1} = \dfrac{d}{dq} q^k$ and since we can interchange differentiation and summation,

$$\sum_{k=0}^{\infty} k q^{k-1} = \frac{d}{dq} \sum_{k=0}^{\infty} q^k = \frac{d}{dq}(1-q)^{-1} ,$$

where we have used the standard geometric series sum formula. Thus we have that

$$\sum_{k=0}^{\infty} kq^{k-1} = \frac{1}{(1-q)^2} .$$

Applying this formula to the problem at hand, where $q = 1 - p$, we have that

$$EX = \frac{1}{p} .$$

Similar tricks often work for integrals, e.g., for the exponential density. Integrals, however, are more likely to be found easily in books of tables than are sums.

[6.3]

As an example of a continuous random variable, assume that X is a uniform random variable on [0,1], that is, that its density is one on [0,1]. Here

$$EX = \int_0^1 x f_X(x) dx = \int_0^1 x dx = 1/2 .$$

In some cases expectations can be found virtually by inspection. For example, if X has an even pdf f_X—that is, if $f_X(-x) = f_X(x)$ for all $x \in \mathbf{R}$—then if the integral exists, $EX = 0$, since $x f_X(x)$ is an odd function and hence has a zero integral. The assumption that the integral exists is necessary because not all even functions are integrable. For example, suppose that we have a pdf $f_X(x) = c/x^2$ for all $|x| \geq 1$, where c is a normalization constant. Then it is not true that EX is zero, even though the pdf is even, because the Riemann integral

$$\int_{x:\, |x| \geq 1} \frac{x}{x^2} dx$$

does not exist. (The puzzled reader should review the definition of indefinite Riemann integrals. Their existence requires that the limit

$$\lim_{T \to \infty} \lim_{S \to \infty} \int_{-T}^{S} x f_X(x) dx$$

exists regardless of how T and S tend to infinity; in particular, the existence for the limit with the constraint $T = S$ is not sufficient for the existence of the integral. These limits do not exist for the given example because $1/x$ is not integrable on $[1, \infty)$.) Nonetheless, it is conventional to set EX to 0 in this example because of the obvious intuitive interpretation.

Sometimes the pdf is an even function about some nonzero value, that is, $f_X(-(x-m)) = f_X(x-m)$, where m is some constant. When this happens, if the expectation exists, then $EX = m$, as the reader can quickly verify by a change of variable in the integral defining the expectation. The most important example of this is the Gaussian pdf, which is even about the constant m.

The same conclusions also obviously hold for an even pmf.

Expectations of Functions of Random Variables

In addition to the expectation of a given random variable, we will often be interested in the expectations of other random variables formed as functions of the given one. In the beginning of the chapter we introduced the relative frequency function, $r_a^{(n)}$, which counts the relative number of occurrences of the value a in a sequence of n terms. We are interested in its expected value and in the expected value of the indicator function that appears in the expression for $r_a^{(n)}$. More generally, given a random variable X and a function $g: \mathbf{R} \rightarrow \mathbf{R}$, we might wish to find the expectation of the random variable $Y = g(X)$. If X corresponds to a voltage measurement and g is a simple squaring operation, $g(X) = X^2$, then $g(X)$ provides the instantaneous energy across a unit resistor. Its expected value, then, represents the probabilistic average energy. We consider two ways of finding the expectation EY. First, we can find the expectation of Y by using derived distribution techniques to find the probability function for Y and then use the definition of expectation to evaluate EY. Specifically, if X is discrete, the pmf for Y is found as before as

$$p_Y(y) = \sum_{x:\, g(x)=y} p_X(x) \, .$$

EY is then found as

$$EY = \sum_y y p_Y(y) \, .$$

Although it is straightforward to find the probability function for Y, it can be a nuisance if it is being found only as a step in the evaluation of the expectation $EY = Eg(X)$. A second and easier method of finding EY is normally used. Looking at the formula for EX, it seems intuitively obvious that $E(g(X))$ should result if x is replaced by $g(x)$. This can be proved by the following simple procedure. Starting with the pmf for Y, then substituting for its expression in terms of the pmf of X and reordering the summation, the expectation of Y is found directly from the pmf for X as claimed:

$$EY = \sum_y y p_Y(y) = \sum_y y \{ \sum_{x:\, g(x)=y} p_X(x) \} = \sum_x p_X(x) g(x) \, .$$

This little bit of manipulation is given the fancy name of the *fundamental theorem of expectation*. It is a very useful formula in that it allows the computation of expectations of functions of random variables without the necessity of performing the (usually more difficult) derived distribution operations.

A similar change of variables argument with integrals in place of sums yields the analogous pdf result for continuous random variables. As is customary, however, we have only provided the proof for the simple discrete case. For the details of the continuous case, we refer the reader to books on integration or analysis. The reader should be aware that such integral results will have additional technical assumptions (almost always satisfied) required to guarantee the existence of the various integrals. We summarize the results below.

The Fundamental Theorem of Expectation.   Let a random variable X be described by a cdf F_X, which is in turn described by either a pmf p_X or a pdf f_X. Given any measurable function $g: \mathbf{R} \to \mathbf{R}$, the resulting random variable $Y = g(X)$ has expectation

$$EY = E(g(X)) = \int_y y \, dF_{g(X)}(y)$$

$$= \int_x g(x) \, dF_X(x) = \begin{cases} \sum_x g(x) p_X(x) \\ \text{or} \\ \int_x g(x) f_X(x) \, dx \ . \end{cases}$$

The qualification "measurable" is needed in the theorem to guarantee the existence of the expectation. Measurability is satisfied by almost any function that you can think of and, for all practical purposes, can be neglected.

Examples

[6.4]

As a simple example of the use of this formula, consider a random variable X with a uniform pdf on $[-1/2, 1/2]$. Define the random variable $Y = X^2$, that is, $g(r) = r^2$. We can use the derived distribution formula of chapter 4 to write

$$f_Y(y) = y^{-1/2} f_X(y^{1/2}); \quad y \geq 0 \ ,$$

and hence

$$f_Y(y) = y^{-1/2}; \quad y \in [0, 1/4] \ ,$$

where we have used the fact that $f_X(y^{1/2})$ is 1 only if the nonnegative argument is less than 1/2 or $y \leq 1/4$. We can then find EY as

$$EY = \int y f_Y(y) \, dy = \int_0^{1/4} y^{1/2} \, dy = \frac{(1/4)^{3/2}}{3/2} = \frac{1}{12} \ .$$

Alternatively, we can use the theorem to write

$$EY = E(X^2) = \int_{-1/2}^{1/2} x^2 dx = 2\frac{(1/2)^3}{3} = \frac{1}{12} \,.$$

Note that the result is the same for each method. However, the second calculation is much simpler, especially if one considers the work which has already been done in chapter 4 in deriving the density formula for the square of a random variable.

[6.5]

A second example shows that expectations can be used to express probabilities. Recall that the indicator function of an event F is defined by

$$1_F(x) = \begin{cases} 1 \text{ if } x \in F \\ 0 \text{ otherwise} . \end{cases}$$

The probability of the event F can be written in the following form which is convenient in certain computations:

$$E1_F(X) = \int 1_F(x)dF_X(x) = \int_F dF_X(x) = P_X(F) , \qquad (6.4)$$

where we have used the universal Stieltjes integral representation of (4.11) to save writing out both sums of pmf's and integrals of pdf's (the reader who is unconvinced by (6.4) should write out the specific pmf and pdf forms). As a graphic aid to understanding, Figure 6.2 shows the indicator function and an arbitrary pdf for the particular $F=(1,2)$. You can see that the expectation of the indicator function is just the integral of the product of the indicator function and the pdf, which in turn equals the integral of the pdf between 1 and 2, which equals $P_X((1,2))$.

It is obvious from the fundamental theorem of expectation that the expected value of any function of a random value can be calculated from its probability distribution. The preceding example demonstrates that the converse is also true: The probability distribution can be calculated from a knowledge of the expectation of a large enough set of functions of the random variable. The example provides the result for the set of all indicator functions. The choice is not unique, as shown by the following example:

[6.6]

Let $g(x)$ be the complex function e^{jux} where u is an arbitrary constant. For a cdf F_X, define

$$E(g(X)) = E(e^{juX}) = \int e^{jux}dF_X(x).$$

This expectation is a function of u, and it is called the *characteristic function* or *transform* of the cdf. We denote it by $M_X(ju)$. The reader can verify that for a continuous random variable, the characteristic function of the cdf specializes to the Fourier transform of the pdf. For a discrete random variable, the characteristic

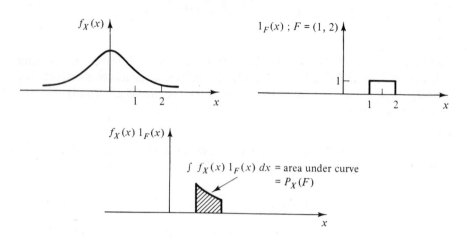

Figure 6.2 Expectation of indicator functions.

function is the discrete Fourier transform of the pmf. In any case, as we know from linear system theory, if the transform is known for all values of u, the transform can be inverted to yield F_X. The distribution P_X can, as usual, then be found from the cdf.

If ju is replaced by u in example [6.6], the expectation is called the *moment-generating function* of X because it can be used to find the *moments of X*, $\{E(X^n); n = 1,2,...\}$ (exercise 6.22).

Example [6.6] illustrates an obvious extension of the fundamental theorem of expectation. In [6.6] the complex function is actually a vector function of length 2. Thus it is seen that the theorem is valid for vector functions, $\mathbf{g}(x)$, as well as for scalar functions, $g(x)$. The expectation of a vector is simply the vector of expected values of the components.

FUNCTIONS OF SEVERAL RANDOM VARIABLES

As an analog to the expectation of a function of a single random variable, we will often wish to find the expectation of a function of a random vector, that is, a function of a bunch of random variables such as $Y = g(U,V)$, where (U,V) is a random vector. The fundamental theorem of expectation has a natural extension (which is proved in the same way). It is stated as follows: Given random variables $X_0, X_1, ..., X_{k-1}$ described by a cdf $F_{X_0, X_1, ..., X_{k-1}}$ and given a measurable function $g: \mathbf{R}^k \to \mathbf{R}$,

$$Eg(X_0, ..., X_{k-1}) = \int g(x_0, ..., x_{k-1}) dF_{X_0, ..., X_{k-1}}(x_0, ..., x_{k-1})$$

$$= \begin{cases} \displaystyle\sum_{x_0,\dots,x_{k-1}} g(x_0,\dots,x_{k-1})p_{X_0,\dots,X_{k-1}}(x_0,\dots,x_{k-1}) \\ \text{or} \\ \displaystyle\int g(x_0,\dots,x_{k-1})f_{X_0,\dots,X_{k-1}}(x_0,\dots,x_{k-1})dx_0\dots dx_{k-1} \ . \end{cases}$$

The theorem can also be extended to vector functions **g** of random vectors as well.

PROPERTIES OF EXPECTATION

Expectation possesses several basic properties that will prove useful. We now present these properties and prove them for the discrete case. The continuous results follow by using integrals in place of sums.

Property 1.   If X is a random variable such that $\Pr(X \geq 0) = 1$, then $EX \geq 0$.

Proof. $\Pr(X \geq 0) = 1$ implies that the pmf $p_X(x)=0$ for $x<0$. If $p_X(x)$ is nonzero only for nonnegative x, then the sum defining the expectation contains only terms $xp_X(x) \geq 0$, and hence the sum and EX are nonnegative. Note that property 1 parallels axiom 1 of probability. That is, the nonnegativity of probability measures implies property 1.

Property 2.   If X is a random variable such that for some fixed number r, $\Pr(X = r) = 1$, then $EX = r$. Thus the expectation of a constant equals the constant.

Proof. $\Pr(X = r) = 1$ implies that $p_X(r)=1$. Thus the result follows from the definition of expectation. Observe that property 2 parallels axiom 2 of probability. That is, the normalization of the total probability to 1 leaves the constant unscaled in the result. If total probability were different from 1, the expectation of a constant as defined would be a different, scaled value of the constant.

Property 3.   Expectation is linear; that is, given two random variables X and Y and two real constants a and b,

$$E(aX + bY) = aEX + bEY \ .$$

Proof

$$E(aX + bY) = \sum_{x,y}(ax + by)p_{X,Y}(x,y)$$

$$= a\sum_{x,y}xp_{X,Y}(x,y) + b\sum_{x,y}yp_{X,Y}(x,y)$$

$$= a\sum_{x}x\sum_{y}p_{X,Y}(x,y) + b\sum_{y}y\sum_{x}p_{X,Y}(x,y)$$

$$= a\sum_{x}xp_{X}(x) + b\sum_{y}yp_{Y}(y)$$

$$= aEX + bEY$$

Observe that the linearity of expectation follows from the additivity of probability. That is, from axiom 4 of probability, the addition of joint probabilities yields marginal probability functions.

The alert reader will likely have noticed the method behind the presentation of the properties of expectation: Each follows directly from the corresponding axiom of probability. Furthermore, using (6.4), the converse is true: That is, instead of starting with the axioms of probability, suppose we start by using the properties of expectation as the axioms of expectation. Then the axioms of probability become the derived properties of probability. Thus the first three axioms of probability and the first three properties of expectation are dual; one can start with either and get the other. One might suspect that to get a useful theory based on expectation, one would require a property analogous to axiom 4 of probability, that is, a limiting form of expectation property 3. This is, in fact, the case, and the fourth basic property of expectation is the countably infinite version of property 3. When dealing with expectations, however, the fourth property is more often stated as a continuity property, that is, in a form analogous to axiom 4 of probability given in equation (3.8b). For reference we state the property below without proof:

Property 4. Given an increasing sequence of nonnegative random variables $X_n; n = 0,1,2,\ldots,$ that is, $X_n \geq X_{n-1}$ for all n (i.e., $X_n(\omega) \geq X_{n-1}(\omega)$ for all $\omega \in \Omega$), then

$$E(\lim_{n \to \infty} X_n) = \lim_{n \to \infty} EX_n .$$

Thus as with probabilities, one can in certain cases exchange the order of limits and expectation. The cases include but are not limited to those of property 4. Property 4 is called the *monotone convergence theorem* and is one of the basic properties of integration as well as expectation. In fact, the four properties of expectation can be taken as a definition of an integral (viz., the Stieltjes integral) and used to develop the general Lebesgue theory of integration. That is, the theory of expectation is really just a specialization of the theory of integration. The duality between probability and expectation is just a special case of the duality between measure theory and the theory of integration.

CORRELATION

We next develop a property of expectation that is special to independent random variables. A weak form of this property provides a weak form of independence that will later be useful in characterizing certain random processes. Suppose we have two independent random variables X and Y and we have two functions or measurements on these random variables, say $g(X)$ and $h(Y)$, where $g: \mathbf{R} \rightarrow \mathbf{R}$ and $h: \mathbf{R} \rightarrow \mathbf{R}$ and $E[g(X)]$ and $E[h(Y)]$ exist and are finite. Consider the expected value of the product of these two functions. Applying the two-dimensional vector case of the fundamental theorem of expectation to discrete random variables results in

$$
\begin{aligned}
E(g(X)h(Y)) &= \sum_{x,y} g(x)h(y)p_{X,Y}(x,y) \\
&= \sum_{x} \sum_{y} g(x)h(y)p_X(x)p_Y(y) \\
&= \left[\sum_{x} g(x)p_X(x) \right] \left[\sum_{y} h(y)p_Y(y) \right] \\
&= (E(g(X)))(E(h(Y))) \,.
\end{aligned}
$$

A similar manipulation with integrals shows the same to be true for random variables possessing pdf's. Thus we have proved the following result, which we state formally as a lemma.

Lemma 6.1.   For any two independent random variables X and Y,

$$
E(g(X)h(Y)) = (Eg(X))(Eh(Y)) \tag{6.5}
$$

for all measurable functions g and h with finite expectations.

The usual remarks regarding the measurability condition apply. Without measurability, the expectations are not defined. Measurability is satisfied by almost all functions so that the qualification can be ignored for all practical purposes.

To cite the most important example, if g and h are identity functions $(h(r)=g(r)=r)$, then we have

$$
E(XY) = (EX)(EY) \,. \tag{6.6}
$$

Equation (6.6) is implied by independence and is certainly what one would anticipate from two independent random variables: The expectation of the product of two random variables behaves as if the two were independent in any sense of the word; that is, the expectation of the product is the product of the expectations. In general, the expected value of the product of two random variables X and Y, $E(XY)$, is called the *correlation* of the two random variables, and two random variables satisfying (6.6) are said to be *uncorrelated*. In terms of this definition, we have shown that if two discrete random variables are independent, then they are also uncorrelated. Note that independence implies not only that two random variables are uncorrelated but also that all *functions* of the random variables are uncorrelated—a much stronger property. In particular, two uncorrelated random variables need not be independent. For example, consider two random variables X and Y with the joint pmf

$$p_{X,Y}(x,y) = \begin{cases} 1/4 \text{ if } (x,y)=(1,1) \text{ or } (-1,1) \\ 1/2 \text{ if } (x,y)=(0,0) . \end{cases}$$

A simple calculation shows that

$$E(XY) = 1/4(1-1)+1/2(0) = 0$$

and

$$(EX)(EY) = (0)(1/2) = 0 ,$$

and hence the random variables are uncorrelated. They are not, however, independent. For example, $\Pr(X=0 \mid Y=0) = 1$ while $\Pr(X=0) = 1/2$. As another example, consider the case where $p_X(x) = 1/3$ for $x = -1,0,1$ and $Y=X^2$. X and Y are uncorrelated but not independent.

Thus uncorrelation does not imply independence. If, however, all possible functions of the two random variables are uncorrelated—that is, if (6.5) holds—then they must be independent. To see this in the discrete case, just consider all possible functions of the form $1_a(x)$, that is, indicator functions of all of the points. ($1_a(x)$ is 1 if $x=a$ and zero otherwise.) Let $g = 1_a$ and $h = 1_b$ for a in the range space of X and b in the range space of Y. It follows from (6.5) and (6.4) that

$$p_{X,Y}(a,b) = p_X(a)p_Y(b) .$$

Obviously the result holds for all a and b. Thus the two random variables are independent. It can now be seen that (6.5) provides a necessary and sufficient condition for independence, a fact we formally state as a theorem.

Theorem 6.1.   Two random variables X and Y are independent if and only if $g(X)$ and $h(Y)$ are uncorrelated for all measurable functions g and h with finite expectations, that is, if (6.5) holds.

This theorem is useful as a means of showing that two random variables are not independent: If we can find *any* functions g and h such that

$E(g(X)h(Y)) \neq (Eg(X))(Eh(Y))$, then the random variables are not independent.

Uncorrelation provides a weak notion of independence. In fact, in some texts uncorrelated random variables are referred to as weakly independent or second-order independent (meaning that $E(X^n Y^m) = E(X^n)E(Y^m)$ for all nonnegative integers m, n such that $m+n \leq 2$). We shall not, however, use this terminology and shall stick to *uncorrelated* as the name for the property.

The idea of uncorrelation can be stated conveniently in terms of another quantity, which we now define. Given two random variables X and Y, define their *covariance*, $COV(X,Y)$ by

$$COV(X,Y) = E[(X-EX)(Y-EY)].$$

As you can see, the *covariance* of two random variables equals the *correlation* of the two "centralized" random variables, $X-EX$ and $Y-EY$, that are formed by subtracting the means from the respective random variables. Keeping in mind that EX and EY are *constants*, it is seen that centralized random variables are zero-mean random variables; i.e., $E(X-EX)=E(X)-E(EX)=EX-EX=0$. Expanding the product in the definition, the covariance can be written in terms of the correlation and means of the random variables. Again remembering that EX and EY are constants, we get

$$COV(X,Y) = E[XY - YEX - XEY + (EX)(EY)]$$

$$= E(XY) - (EY)(EX) - (EX)(EY) + (EX)(EY)$$

$$= E(XY) - (EX)(EY) . \tag{6.7}$$

Thus the covariance is the correlation minus the product of the means. Using this fact and the definition of uncorrelated, we have the following statement:

Corollary 6.1.   Two random variables X and Y are uncorrelated if and only if their covariance is zero; that is, if $COV(X,Y) = 0$.

If we set $Y = X$, the correlation of X with itself, $E(X^2)$, results; this is called the *second moment* of the random variable X. The covariance $COV(X,X)$ is called the *variance* of the random variable and is given the special notation σ_X^2. $\sigma_X = \sqrt{\sigma_X^2}$ is called the *standard deviation* of X. From the definition of covariance and (6.6),

$$\sigma_X^2 = E[(X-EX)^2] = E(X^2) - (EX)^2 .$$

By the first property of expectation, the variance is nonnegative, yielding the simple but often useful inequality

$$|EX| \le [E(X^2)]^{1/2},$$ (6.8)

a special case of the Cauchy-Schwartz inequality (see exercise 6.10 with the random variable Y set equal to the constant 1).

We turn now to correlation in the framework of random processes. The notion of an i.i.d. random process can be generalized by specifying the component random variables to be merely uncorrelated rather than independent. Although requiring the random process to be uncorrelated is a much weaker requirement, the specification is sufficient for many applications, as will be seen in several ways. In particular, in this chapter, the basic laws of large numbers require only the weaker assumption and hence are more general than they would be if independence were required. To define the class of uncorrelated processes, it is convenient to introduce the notions of autocorrelation functions and covariance functions of random processes.

Given a random process $\{X_t; t \in \mathbf{I}\}$, the *autocorrelation function* $R_X(t,s); t,s \in \mathbf{I}$ is defined by

$$R_X(t,s) = E(X_t X_s); \quad \text{all } t,s \in \mathbf{I}.$$

The *autocovariance function* or simply the *covariance function* $K_X(t,s); t,s \in \mathbf{I}$ is defined by

$$K_X(t,s) = COV(X_t, X_s).$$

Observe that (6.6) relates the two functions by

$$K_X(t,s) = R_X(t,s) - (EX_t)(EX_s).$$ (6.9)

Thus the autocorrelation and covariance functions are equal if the process has zero mean, that is, if $EX_t = 0$ for all t. The covariance function of a process $\{X_t\}$ can be viewed as the autocorrelation function of the process $\{X_t - EX_t\}$ formed by removing the mean from the given process to form a new process having zero mean.

The autocorrelation function of a random process is given by the correlation of all possible pairs of samples; the covariance function is the covariance of all possible pairs of samples. Both functions provide a measure of how dependent the samples are and will be seen to play a crucial role in laws of large numbers. Note that both definitions are valid for random processes in either discrete time or continuous time and having either a discrete alphabet or a continuous alphabet.

In terms of the correlation function, a random process $\{X_t; t \in \mathbf{I}\}$ is said to be *uncorrelated* if

$$R_X(t,s) = \begin{cases} E(X_t^2) & \text{if } t = s \\ EX_t EX_s & \text{if } t \ne s, \end{cases}$$

that is, if the random variables X_t and X_s are uncorrelated for all possible times $t \neq s$. Equivalently, in terms of the covariance function, a random process is said to be uncorrelated if

$$K_X(t,s) = \begin{cases} \sigma_{X_t}^2 & \text{if } t = s \\ 0 & \text{if } t \neq s . \end{cases}$$

The reader should not overlook the obvious fact that if a process is i.i.d. or uncorrelated, the random variables are independent or uncorrelated *only if taken at different times.* That is, X_t and X_s will not be independent or uncorrelated when $t = s$, only when $t \neq s$ (except, of course, in such trivial cases as that where $\{X_t\} = \{a_t\}$, a sequence of constants where $EX_t X_t = a_t a_t = EX_t EX_t$ and hence X_t is uncorrelated with itself).

We make two final observations before turning to use the ideas of expectation and correlation to analyze the behavior of sample averages: First, remember that the four basic properties of expectation have nothing to do with independence. In particular, whether or not the random variables involved are independent or uncorrelated, one can always interchange the expectation operation and the summation operation (property 3), because expectation is linear. On the other hand, one cannot interchange the expectation operation with the product operation (this is not a property of expectation) unless the random variables involved are uncorrelated, e.g., when they are independent. Second, an i.i.d. process is also a discrete time uncorrelated random process with identical marginal distributions. The converse statement is not true in general; that is, the notion of an uncorrelated process is more general than that of an i.i.d. process. Correlation measures only a weak pairwise degree of independence. A random process could even be pairwise independent (and hence uncorrelated) but still not be i.i.d. (exercise 6.24).

SAMPLE AVERAGES

In many applications, engineers analyze the accuracy of estimates, the probability of detector error, etc., as a function of the amount of data available. This and the next sections are a prelude to such analyses.

In this section we study the behavior of the arithmetic average of the first n values of a discrete time random process with either a discrete or a continuous alphabet. Specifically, the variance of the average is considered as a function of n.

Suppose we are given a process $\{X_n\}$. The sample average of the first n values of $\{X_n\}$ is $S_n = n^{-1} \sum_{i=0}^{n-1} X_i$. The mean of S_n is found easily using the linearity of expectation (expectation property 3) as

$$ES_n = E[n^{-1}\sum_{i=0}^{n-1} X_i] = n^{-1}\sum_{i=0}^{n-1} EX_i . \tag{6.10}$$

Hence the mean of the sample average is the same as the average of the mean of the random variables produced by the process. Suppose now that we assume that the mean of the random variables is a constant, $EX_i = \bar{X}$ independent of i. Then $ES_n = \bar{X}$. In terms of estimation theory, if one estimates an unknown random process mean, $\bar{X}$, by S_n, then the estimate is said to be *unbiased* because the expected value of the estimate is equal to the value being estimated. Obviously an unbiased estimate is not unique, so being unbiased is only one desirable characteristic of an estimate (exercise 6.19).

Next consider the variance of the sample average:

$$\sigma_{S_n}^2 = E[(S_n - E(S_n))^2]$$

$$= E[(n^{-1}\sum_{i=0}^{n-1} X_i - n^{-1}\sum_{i=0}^{n-1} EX_i)^2]$$

$$= E[(n^{-1}\sum_{i=0}^{n-1} (X_i - EX_i))^2]$$

$$= n^{-2}\sum_{i=0}^{n-1}\sum_{j=0}^{n-1} E[(X_i - EX_i)(X_j - EX_j)] .$$

The reader should be certain that the preceding operations are well understood, as they are frequently encountered in analyses. Note that expanding the square requires the use of separate dummy indices in order to get all of the cross products. Once expanded, linearity of expectation permits the interchange of expectation and summation.

Recognizing the expectation in the sum as the covariance function, the variance of the sample average becomes

$$\sigma_{S_n}^2 = n^{-2}\sum_{i=0}^{n-1}\sum_{j=0}^{n-1} K_X(i,j) . \tag{6.11}$$

Note that so far we have used none of the specific knowledge of the process, i.e., the above formula holds for general discrete time processes and does not require such assumptions as time-constant mean, time-constant variance, identical marginal distributions, independence, uncorrelated processes, etc. If we now use the assumption that the process is uncorrelated, the covariance becomes zero except when $i = j$, and expression (6.11) becomes

$$\sigma_{S_n}^2 = n^{-2}\sum_{i=0}^{n-1} \sigma_{X_i}^2 . \tag{6.12}$$

If we now also assume that the variance $\sigma^2_{X_i}$ are equal to some constant value σ^2_X for all times i, e.g., the process has identical marginal distributions as for an i.i.d. process, then the equations become

$$\sigma^2_{S_n} = n^{-1}\sigma^2_X \ . \tag{6.13}$$

Thus, for uncorrelated discrete time random processes with mean and variance not depending on time, the sample average has expectation equal to the (time-constant) mean of the process, and the variance of the sample average tends to zero as $n \rightarrow \infty$. Of course we have only specified sufficient conditions. Expression (6.11) goes to zero with n under more general circumstances, as we shall see later.

For now, however, we stick with uncorrelated processes with mean and variance independent of time and require only a definition to obtain our first law of large numbers, a result implicit in equation (6.13).

CONVERGENCE AND LAWS OF LARGE NUMBERS

The preceding section demonstrated a form of convergence for the sequence of random variables, $\{S_n\}$, the sequence of sample averages, that is different from convergence as it is seen for a nonrandom sequence. To review, a nonrandom sequence $\{x_n\}$ is said to converge to the limit x if for every $\epsilon > 0$ there exists an N such that $|x_n - x| < \epsilon$ for every $n > N$. The preceding section did not see S_n converge in this sense. Nothing was said about the convergence of the individual realizations $S_n(\omega)$ as a function of ω. Only the variance of the sequence $\sigma_{S_n}{}^2$ was shown to converge to zero in the usual ϵ, N sense. The variance calculation probabilistically averages across ω. For any particular ω, the realization S_n may, in fact, *not* converge to zero.

Thus, in order to make precise the notion of convergence of sample averages to a limit, we need to make precise the notion of convergence of a sequence of random variables. In this section we will describe four notions. Initially we will confine interest to two notions. The first is *convergence in mean square*, convergence of the type seen in the last section, which leads to a result called a *mean ergodic theorem*. The second is called *convergence in probability*, which is implied by the first and leads to a result called the weak law of large numbers. The second result will follow from the first via a simple but powerful inequality relating probabilities and expectations. Let us turn to the first definition.

A sequence of random variables Y_n; $n = 1, 2, \ldots$ is said to *converge in mean square* or *converge in quadratic mean* to a random variable Y if

$$\lim_{n \to \infty} E[(Y_n - Y)^2] = 0 \ .$$

This is also written $Y_n \to Y$ in mean square or $Y_n \to Y$ in quadratic mean.

If Y_n converges to Y in mean square, we state this convergence mathematically by writing

$$\underset{n \to \infty}{\text{l.i.m.}} Y_n = Y,$$

where l.i.m. is an acronym for "limit in the mean."

Thus a sequence of random variables converges in mean square to another random variable if the second moment of the difference converges to zero in the ordinary sense of convergence of a sequence of real numbers. Although the definition encompasses convergence to a random variable with any degree of "randomness," in most applications the random variable is a degenerate random variable, i.e., a constant. In particular, the sequence of sample averages, $\{S_n\}$, of the preceding section is seen to converge in this sense. We state the convergence in the form of a theorem for a time-constant mean and time-constant variance uncorrelated random process:

Theorem 6.2.   (A mean ergodic theorem): Let $\{X_n\}$ be a discrete time uncorrelated random process such that $EX_n = \bar{X}$ is finite and $\sigma_{X_n}^2 = \sigma_X^2 < \infty$ for all n; that is, the mean and variance are the same for all sample times. Then

$$\underset{n \to \infty}{\text{l.i.m.}} \frac{1}{n} \sum_{i=0}^{n-1} X_i = \bar{X},$$

that is, $\frac{1}{n} \sum_{i=0}^{n-1} X_i \to \bar{X}$ in mean square.

Proof. The proof follows directly from the last section with $S_n = \frac{1}{n} \sum_{i=0}^{n-1} X_i$, $ES_n = EX_i = \bar{X}$. To summarize, from (6.13),

$$\lim_{n \to \infty} E[(S_n - \bar{X})^2] = \lim_{n \to \infty} E[(S_n - ES_n)^2] = \lim_{n \to \infty} \sigma_{S_n}^2 = \lim_{n \to \infty} \frac{\sigma_X^2}{n} = 0.$$

This theorem is called *a* mean ergodic theorem because it is a special case of the more general mean ergodic theorem—it is a special case since it holds only for uncorrelated random processes. In chapter 7 we will consider more general conditions where the process may not be uncorrelated but where the correlation decreases as the distance between samples increases so that $\sigma_{S_n}^2$ still tends to zero. In other words, a form of asymptotic uncorrelation will yield the same result. For now, suffice it to say that the sample average converges in mean square to a constant if the average of the mean sequence $\{EX_i\}$ converges to the constant and (6.11) converges to zero (exercise 6.25).

We now turn to the second form of convergence. A sequence of random variables Y_n; $n = 1,2,...$ is said to *converge in probability* to a random variable Y if for every $\epsilon > 0$,

$$\lim_{n \to \infty} \Pr(|Y_n - Y| > \epsilon) = 0 .$$

Thus a sequence of random variables converges in probability if the probability that the nth member of the sequence differs from the limit by more than an arbitrarily small ϵ goes to zero as $n \to \infty$. Note that just as with convergence in mean square, convergence in probability is silent on the question of convergence of individual realizations $Y_n(\omega)$. You could, in fact, have no realizations converge individually and yet have convergence in probability. All convergence in probability states is that at each n, $\Pr(\omega: |Y_n(\omega) - Y(\omega)| > \epsilon)$ tends to zero with n. Suppose at time n a given subset of Ω satisfies the inequality, at time $n+1$ a different subset of Ω satisfies the inequality, at time $n+2$ still a different subset satisfies the inequality, etc. As long as the subsets have diminishing probability, convergence in probability can occur without convergence of the individual sequences.

Also, as in convergence in the mean square sense, convergence in probability is to a random variable. However, in most applications of the definition, we are interested in convergence to a degenerate random variable—i.e., a constant.

We next show that convergence in mean square is the stronger of the two notions; that is, if it converges in mean square, then it also converges in probability, but not necessarily vice versa. This is accomplished via an important inequality.

The Markov Inequality.   Given a nonnegative random variable U with finite expectation EU, for any $a > 0$ we have

$$\Pr(U \geq a) = P_U([a, \infty)) \leq \frac{EU}{a} .$$

Proof. We consider the discrete case, but the pdf case is essentially the same. Since U can take on only nonnegative values, its pmf is nonzero only for nonnegative letters; thus if we write

$$EU = \sum_u u p_U(u) = \sum_{u:\, u \geq a} u p_U(u) + \sum_{u:\, u < a} u p_U(u) ,$$

then the rightmost sum is nonnegative and can be deleted to yield the lower bound

$$EU \geq \sum_{u:\, u \geq a} u p_U(u) \geq \sum_{u:\, u \geq a} a p_U(u)$$

$$= a \sum_{u:\, u \geq a} p_U(u) = a \Pr(U \geq a) \, ,$$

completing the proof.

Observe that if a random variable U is nonnegative and has small expectation, say $EU \leq \epsilon$, then the Markov inequality implies that

$$\Pr(U \geq \sqrt{\epsilon}) \leq \sqrt{\epsilon} \, .$$

A useful special case of the Markov inequality is the Tchebychev inequality:

The Tchebychev Inequality.   Given a random variable V with mean $\bar{V}$ and variance σ_V^2, for any $\epsilon > 0$,

$$\Pr(|V - \bar{V}| \geq \epsilon) \leq \frac{\sigma_V^2}{\epsilon^2} \, .$$

Proof. The result follows by applying the Markov inequality to the nonnegative random variable $U = |V - \bar{V}|$ and observing that

$$\Pr(U \geq \epsilon) = \Pr(U^2 \geq \epsilon^2) \, .$$

Note that the Tchebychev inequality implies that

$$\Pr(|V - \bar{V}| \geq \gamma \sigma_v) \leq \frac{1}{\gamma^2} \, ,$$

that is, the probability that V is farther from its mean by more than γ times its standard deviation (the square root of its variance) is no greater than γ^{-2}.

Lemma 6.2.   If Y_n converges to Y in mean square, then it also converges in probability.

Proof. From the Tchebychev inequality applied to $|Y_n - Y|$, we have for any $\epsilon > 0$

$$\Pr(|Y_n - Y| > \epsilon) \leq \frac{E(|Y_n - Y|^2)}{\epsilon^2} \, .$$

The right-hand term goes to zero as $n \to \infty$ by definition of convergence in mean square.

Although convergence in mean square implies convergence in probability, the reverse statement cannot be made; i.e., they are not equivalent. This is shown by a simple counterexample. Let Y_n be a discrete random variable with pmf

$$p_{Y_n}(y) = \begin{cases} 1 - 1/n & \text{if } y = 0 \\ 1/n & \text{if } y = n \end{cases}.$$

Convergence in probability to zero without convergence in mean square is easily verified.

Combining lemma 6.2 with the mean ergodic theorem 6.2 yields the following famous result, one of the original limit theorems of probability theory:

The Weak Law of Large Numbers.   Let $\{X_n\}$ be a discrete time process with finite mean $EX_n = \bar{X}$ and variance $\sigma_{X_n}^2 = \sigma_X^2 < \infty$ for all n. If the process is uncorrelated, then the sample average $n^{-1} \sum_{i=0}^{n-1} X_i$ converges to $\bar{X}$ in probability.

The best-known (and earliest) application of the weak law of large numbers is to i.i.d. processes such as the Bernoulli process. Note that the i.i.d. specification is not needed, however. All that is used for the weak law of large numbers is constant means, constant variances, and uncorrelation. The actual distributions could be time varying and dependent within these constraints. The *weak* law is called weak because convergence in probability is one of the weaker forms of convergence. Convergence of individual realizations of the random process is not assured. This could be very annoying because in many practical engineering applications, we have only one realization to work with (i.e., only one ω), and we need to calculate averages that converge as determined by actual calculations, e.g., with a computer.

The *strong* law of large numbers considers convergence with probability one (also called convergence almost everywhere or almost surely), which states that with probability one, $\lim_{n \to \infty} S_n(\omega)$ exists in the usual sense of a limit of a sequence of real numbers and that this limit equals $\bar{X}$. Such strong theorems are much harder to prove, but fortunately are satisfied in most engineering situations. We now define the stronger form of convergence:

A sequence of random variables Y_n; $n = 1, 2, \ldots$ is said to *converge with probability one* (or *converge almost everywhere*) to a random variable Y if

$$\Pr(\{\omega: \lim_{n \to \infty} |Y_n(\omega) - Y(\omega)| = 0\}) = 1.$$

The definition means that with probability one, for every realization of the random process, for every $\epsilon > 0$ there exists an N, *possibly depending on* ω, such that $|Y_n - Y| < \epsilon$ if $n > N$. That is, unfortunately you cannot, in general, have uniform convergence—a convergence rate that does not depend on ω. Furthermore, the qualification "with probability one" is needed to be completely correct. The underlying sample space may have a collection of spurious nonconvergent sequences that have probability zero. This latter point is of no real practical significance, however.

Observe that the principal difference in the definitions of convergence in probability and convergence with probability one is that the limit is outside the probability in the former and inside the probability in the latter.

*Convergence with Probability One vs. Convergence in Probability

Convergence with probability one implies convergence in probability. Although the terms sound the same, they are really quite different. Convergence with probability one applies to the individual realizations of the random process. Convergence in probability does not. We show that probability one convergence is stronger in the following lemma:

Lemma 6.3. If Y_n converges with probability one to Y, then it also converges in probability.

Proof. Given an $\epsilon > 0$, define the sequence of sets

$$F_n(\epsilon) = \{\omega: |Y_m(\omega) - Y(\omega)| > \epsilon \text{ for some } m \geq n\} .$$

We have defined a decreasing sequence of sets, i.e., $F_n \subset F_{n-1}$, all n. Thus $\Pr(F_n)$ decreases monotonically with n. The definition of probability one convergence assures that the decrease is to zero. Therefore,

$$\lim_{n \to \infty} \Pr(F_n(\epsilon)) = \lim_{n \to \infty} \Pr(|Y_n - Y| > \epsilon) = 0 ,$$

which establishes convergence in probability.

Convergence in probability does not imply convergence with probability one; i.e., they are not equivalent. This can be proved by counterexample (exercise 6.31). There is, however, a test that can be applied to determine convergence with probability one when convergence in probability has been verified:

Lemma 6.4. Y_n converges to Y with probability one if for any $\epsilon > 0$,

$$\sum_{n=1}^{\infty} \Pr(|Y_n - Y| > \epsilon) < \infty .$$

Proof. Let $F(\epsilon)$ be the set of all ω such that the corresponding sequence, $Y_n(\omega)$, does not satisfy the convergence criterion; i.e., for some $\epsilon > 0$,

$$F(\epsilon) = \{\omega: |Y_n - Y| > \epsilon, \text{ for } some \ n \geq N, \text{ for } any \ N < \infty\} .$$

Now define a similar set at the nth time only,

$$F_n(\epsilon) = \{\omega: |Y_n - Y| > \epsilon\} .$$

Clearly the set of nonconvergent ω is a subset of the union of the latter sequence of sets; i.e., for any $N < \infty$,

$$F(\epsilon) \subset \bigcup_{n > N}^{\infty} F_n(\epsilon) = G_N(\epsilon) .$$

The probability of the set of nonconvergent ω is bounded from above by the probability of the union, which is in turn bounded by the sum of the probabilities of the F_n. The sum of the probabilities goes to zero with N by the statement of the lemma. Thus, since N is arbitrary, we must have

$$\Pr(F(\epsilon)) \leq \Pr(\lim_{n \to \infty} G_N(\epsilon)) = 0 .$$

Note that the lemma requires more information than we used to establish the weak law of large numbers. In particular, the sum in lemma 6.4 diverges if one attempts to use the slow rate of convergence to zero of the Tchebychev inequality.

Convergence with probability one does not imply—nor is it implied by—convergence in mean square. This can be shown by counterexamples (exercise 6.26).

CONVERGENCE IN DISTRIBUTION

The three forms of convergence we have considered so far are convergence in mean square, convergence in probability, and convergence with probability one. These three forms of convergence find their greatest use in proving convergence to a constant. We will now introduce a fourth and final form of convergence—a form of convergence in which we are usually interested not in convergence to a constant but in the limiting form of the distribution of the sequence.

A sequence of random variables Y_n; $n = 1,2,\ldots$ is said to *converge in distribution* to a random variable Y if the corresponding sequence of cumulative distribution functions F_{Y_n}; $n = 1,2,\ldots$ converges to the cdf of Y, F_Y, at every point of continuity of F_Y (that is, for every y for which $F_Y(y)$ is continuous).

Convergence in distribution is a very weak form of convergence and is implied by the other three forms that we have presented (exercise 6.26). Note that in no sense do any of the Y_n have to equal Y. Only the sequence of distributions has to approach a limit. Note also that it is the

cdf sequence that is converging and *not* the pmf or pdf sequence, which
need not converge.

The most important example of convergence in distribution is
provided by the central limit theorem, which will be presented in chapter
11. The central limit theorem, roughly stated, says that sums of suitably
scaled random variables approach a limiting Gaussian distribution under
very broad conditions.

To summarize the relationship among the forms of convergence, convergence in distribution is implied by convergence in probability, which is
in turn implied by either convergence in mean square or convergence with
probability one. Convergence in mean square does not imply convergence
with probability one, nor does convergence with probability one imply convergence in mean square.

A wide variety of limit results of the forms we have discussed may be
found in the literature. Such results make up a substantial portion of a
branch of mathematics called ergodic theory. Results vary in the types of
convergence considered and the assumptions on the random process necessary to ensure such convergence. We have so far described only the main
forms of convergence and shown convergence in mean square and in probability for uncorrelated discrete time processes with constant mean and
variance. More general results and the corresponding statements for continuous time are considered in the next chapter.

EXERCISES

6.1 Find the means, second moments, and variances for the random variables described by the uniform, binomial, geometric, and Poisson
pmf's. (While this problem and the next can be solved by looking up
the results somewhere, the student is strongly urged first to attempt
to find the requested quantities from scratch. This will help pinpoint
any rusty tools for the evaluation of sums and integrals.)

6.2 Find the means, second moments, and variances for the random variables described by the exponential, doubly exponential, and Gaussian
pdf's.

6.3 The *Cauchy* pdf is defined by

$$f_X(x) = \frac{1}{\pi} \frac{1}{1+x^2}; \quad x \in \mathbf{R}.$$

Find *EX*. *Hint:* This is a trick question. Check the definition of
Riemann integration over $(-\infty, \infty)$ before deciding on a final answer.

6.4 If $\{X_n\}$ is an uncorrelated process with constant first and second
moments, does it follow for an arbitrary function g that

$$n^{-1}\sum_{i=0}^{n-1} g(X_i) \underset{n\to\infty}{\longrightarrow} Eg(X)$$

in mean square? Show that it does follow if the process is i.i.d.

6.5 Apply exercise 6.4 to indicator functions to prove that relative frequencies of order n converge to pmf's in mean square and in probability for i.i.d. random processes. That is, if $r_a^{(n)}$ is defined as in the chapter, then $r_a^{(n)} \to p_X(a)$ as $n \to \infty$ in both senses for any a in the range space of X.

6.6 Does the weak law of large numbers hold for the random process of exercise 5.10, where a bias was selected uniformly on $[0,1]$ and then a coin with that bias flipped forever? In any case, is it true that S_n converges? If so, to what? Repeat for the mixture of exercise 5.14.

6.7 Define the subsets of the real line

$$F_n = \{r: |r| > \frac{1}{n}\} ,n = 1,2,...$$

and

$$F = \{0\} .$$

Show that

$$F^c = \bigcup_{n=1}^{\infty} F_n .$$

Use this fact, the Tchebychev inequality, and the continuity of probability to show that if a random variable X has variance 0, then $\Pr(|X - EX| \ge \epsilon) \le 0$ independent of ϵ and hence $\Pr(X = EX) = 1$.

6.8 Show that for a discrete random variable X,

$$|E(X)| \le E(|X|) .$$

Repeat for a continuous random variable.

6.9 This exercise considers some useful properties of autocorrelation or covariance functions.

(a)   Use the fact that $E[(X_t - X_s)^2] \ge 0$ to prove that if $EX_t = EX_0$ for all t and $E(X_t^2) = R_X(t,t) = R_X(0,0)$ for all t—that is, if the mean and variance do not depend on time—then

$$|R_X(t,s)| \le R_X(0,0)$$

and

$$|K_X(t,s)| \le K_X(0,0) .$$

Thus both functions take on their maximum value when $t = s$. This can be interpreted as saying that no random variable can be more correlated with a given random variable than it is with itself.

(b) Show that autocorrelations and covariance functions are symmetric functions, e.g., $R_X(t,s) = R_X(s,t)$.

6.10 The Cauchy-Schwarz Inequality: Given random variables X and Y, define $a = E(X^2)^{1/2}$ and $b = E(Y^2)^{1/2}$. By considering the quantity $E[(X/a \pm Y/b)^2]$ prove the following inequality:

$$|E(XY)| \le E(X^2)^{1/2}E(Y^2)^{1/2} .$$

6.11 Given two random processes $\{X_t; t \in \mathbf{I}\}$ and $\{Y_t; t \in \mathbf{I}\}$ defined on the same probability space, the *cross correlation function* $R_{XY}(t,s); t,s \in \mathbf{I}$ is defined as

$$R_{XY}(t,s) = E(X_t Y_s) .$$

since $R_X(t,s) = R_{XX}(t,s)$. Show that R_{XY} is not, in general, a symmetric function of its arguments. Use the Cauchy-Schwartz inequality of exercise 6.10 to find an upper bound to $|R_{XY}(t,s)|$ in terms of the autocorrelation functions R_X and R_Y.

6.12 Let Θ be a random variable described by a uniform pdf on $[-\pi,\pi]$ and let Y be a random variable with mean m and variance σ^2; assume that Θ and Y are independent. Define the random process $\{X(t); t \in \mathbf{R}\}$ by $X(t) = Y\cos(2\pi f_0 t + \Theta)$, where f_0 is a fixed frequency in hertz. Find the mean and autocorrelation function of this process. Find the limiting time average

$$\lim_{T \to \infty} \frac{1}{T} \int_0^T X(t)dt .$$

(Only in trivial processes such as this can one find exactly such a limiting time average.)

6.13 Let $\{X(t); t \in \mathbf{R}\}$ be a continuous time random process with zero mean and autocorrelation $R_X(t,s)$. Let Θ be a uniform random variable as in exercise 6.12 that is independent of $X(t)$ for all t. Define the random process $\{Y(t)\}$ by $Y(t) = (a_0 + b_0 X(t))\cos(2\pi f_0 t + \Theta)$, where f_0 is a fixed frequency and a_0, b_0 are fixed constants. Find the autocorrelation function R_Y. (The process $\{Y(t)\}$ is called an amplitude-modulated version of the original process—that is, a standard commercial radio AM signal.)

6.14 Let $r_a^{(n)}$ denote the relative frequency of the letter a in a sequence $x_0,...,x_{n-1}$. Show that if we define $p(a) = r_a^{(n)}$, then $p(a)$ is a valid pmf. (This pmf is called the "sample pmf," "sample distribution," or "empirical distribution.")

6.15 Given two sequences of random variables $\{X_n; n=1,2,...\}$ and $\{Y_n; n=1,2,...\}$ and a random variable X, suppose that with probability one $|X_n - X| \le Y_n$ all n and that $EY_n \to 0$ as $n \to \infty$. Prove that $EX_n \to EX$ and that X_n converges to X in probability as $n \to \infty$.

6.16 This exercise provides another example of the use of covariance functions. Say that we have a discrete time random process $\{X_n\}$ with a covariance function $K_X(t,s)$ and a mean function $m_n = EX_n$. Say that we are told the value of the past sample, say $X_{n-1} = \alpha$, and we are asked to make a good guess of the next sample on the basis of the old sample. Furthermore, we are required to make a *linear* guess or estimate, called a prediction, of the form

$$\hat{X}_n(\alpha) = a\alpha + b ,$$

for some constants a and b. Use ordinary calculus techniques to find the values of a and b that are "best" in the sense of minimizing the mean squared error

$$E[(X_n - \hat{X}_n(X_{n-1}))^2] .$$

Give your answer in terms of the mean and covariance function. Generalize to a linear prediction of the form

$$\hat{X}_n(X_{n-1}, X_{n-m}) = a_1 X_{n-1} + a_m X_{n-m} + b ,$$

where m is an arbitrary integer, $m \ge 2$. When is $a_m = 0$?

6.17 We developed the mean and variance of the sample average S_n for the special case of uncorrelated random variables. Evaluate the mean and variance of S_n for the opposite extreme, where the X_i are highly correlated in the sense that $E[X_i X_k] = E[X_i^2]$ for all i,j.

6.18 Let $\{X_n\}$, $\{Y_n\}$, $\{V_n\}$, and $\{W_n\}$ be as in exercise 5.13.

  (a)   Find EW_n.

  (b)   Find the covariance function $K_W(t,s)$.

  (c)   Evaluate the *cross-correlation* function defined by

$$R_{X,W}(t,s) = E[X_t W_s] .$$

6.19 Given n independent random variables X_i, $i = 1, 2, \ldots, n$, with variances σ_i^2 and means m_i, $i = 1, 2, \ldots, n$, find an expression for the variance of the random variable

$$Y = \sum_{i=1}^{n} a_i X_i ,$$

where the a_i are fixed real constants. Now let the mean be constant; i.e., $m_i = m$. Find the minimum variance of Y over the choice of the $\{a_i\}$ subject to the constraint that $EY = m$. The result is called the *minimum variance unbiased estimate of* m.

6.20 The random process of example [4.27] can be expressed as follows: Let Θ be a continuous random variable with a pdf

$$f_\Theta(\theta) = \frac{1}{2\pi}; \; \theta \in [-\pi, \pi]$$

and define the process $\{X(t); t \in \mathbf{R}\}$ by

$$X(t) = \cos(t + \Theta) .$$

   (a)   Find the cdf $F_{X(0)}(x)$.

   (b)   Find $EX(t)$.

   (c)   Find the covariance function $K_X(t,s)$.

6.21 Let $\{X_n\}$ be a random process with mean m and autocorrelation function $R_X(n,k)$, and let $\{W_n\}$ be an i.i.d. random process with zero mean and variance σ_W^2. Assume that the two processes are independent of each another; that is, any collection of the X_i is independent of any collection of the W_i. Form a new random process $Y_n = X_n + W_n$. *Note*: This is a common model for a communication system or measurement system with $\{X_n\}$ a "signal" process or "source," $\{W_n\}$ a "noise" process, and $\{Y_n\}$ the "received" process; see exercise 4.24, for example.

   (a)   Find the mean EY_n and covariance $K_Y(t,s)$ in terms of the given parameters.

   (b)   Find the cross-correlation function defined by

$$R_{X,Y}(k,j) = E[X_k Y_j] .$$

   (c)   As in exercise 6.16, find the minimum mean squared error estimate of X_n of the form

$$\hat{X}(Y_n) = aY_n + b.$$

       The resulting estimate is called a *filtered* value of X_n.

   (d)   Extend to a linear filtered estimate that uses Y_n and Y_{n-1}.

6.22 Let X be a random variable for which $Ee^{uX} = M_X(u)$ exists, where u is an arbitrary real number. $M_X(u)$ is called the *moment-generating function* of X and, if X is continuous and u is complex, is the two-sided Laplace transform of the pdf. Find an expression for the nth moment of X, $E(X^n)$, in terms of M_X. Find a pdf for which $M_X(u)$ does not exist. $M_X(ju)$ is called the *characteristic function* of X. When does the characteristic function exist? Find an expression for the moments of X in terms of derivatives of the characteristic function.

6.23 Find the moment-generating function and characteristic function for a Gaussian random variable and a Bernoulli random variable. Calculate the mean and variance of each from one of the functions and compare with the result obtained from direct calculations on the pdf or pmf.

6.24 Suppose that there are two independent data sources $\{W_i(n), i=1,2\}$. Each data source is modeled as a Bernoulli random process with parameter 1/2. The two sources are encoded for transmission as follows: First, three random processes $\{Y_i(n); \ i=1,2,3\}$ are formed, where $Y_1 = W_1$, $Y_2 = W_2$, $Y_3 = W_1 + W_2$, and where the last sum is taken modulo 2 and is formed to provide redundancy for noise protection in transmission. These are time-multiplexed to form a random process $\{X(3n+i) = Y_i(n)\}$. Show that $\{X(n)\}$ has identically distributed components and is pairwise independent but is not i.i.d.

6.25 Show that the convergence of the average of the means in (6.10) to a constant and the convergence of equation (6.11) to zero are sufficient for a mean ergodic theorem of the form of theorem 6.2. In what sense, if any, does $\{S_n\}$ converge?

6.26 The purpose of this exercise is to demonstrate the relationships among the four forms of convergence that we have presented. In each case, $(([0,1], B([0,1]), P)$ is the underlying probability space, with probability measure described by the uniform pdf. For each of the following sequences of random variables, determine the pmf of $\{Y_n\}$, the senses in which the sequence converges, and the random variable and pmf to which the sequence converges.

(a)

$$Y_n(\omega) = \begin{cases} 1 \text{ if } n \text{ is odd and } \omega < 1/2 \text{ or } n \text{ is even and } \omega > 1/2 \\ 0 \text{ otherwise .} \end{cases}$$

(b)

$$Y_n(\omega) = \begin{cases} 1 \text{ if } \omega < 1/n \\ 0 \text{ otherwise .} \end{cases}$$

(c)

$$Y_n(\omega) = \begin{cases} n \text{ if } \omega < 1/n \\ 0 \text{ otherwise .} \end{cases}$$

(d) Divide $[0,1]$ into a sequence of intervals $\{F_n\} = \{[0,1],[0,1/2),[1/2,1],[0,1/3],[1/3,2/3],[2/3,1],[0,1/4],...\}$. Let

$$Y_n(\omega) = \begin{cases} 1 \text{ if } \omega \in F_n \\ 0 \text{ otherwise .} \end{cases}$$

(e)

$$Y_n(\omega) = \begin{cases} 1 \text{ if } \omega < 1/2 + 1/n \\ 0 \text{ otherwise .} \end{cases}$$

6.27 Suppose that X is a random variable with mean m and variance σ^2. Let g_k be a deterministic periodic pulse train such that g_k is 1 whenever k is a multiple of a fixed positive integer N and g_k is 0 for all other k. Let U be a random variable that is independent of X such that $p_U(u) = 1/N$ for $u = 0,1,...,N-1$. Define the random process Y_n by

$$Y_n = Xg_{U+n} \,,$$

that is, Y_n looks like a periodic pulse train with a randomly selected amplitude and a randomly selected phase. Find the mean and covariance functions of the Y process. Find a random variable $\hat{Y}$ such that

$$\lim_{n \to \infty} \frac{1}{n} \sum_{i=0}^{n-1} Y_i = \hat{Y}$$

in the sense of convergence with probability one. (This is an example of a process that is simple enough for the limit to be evaluated explicitly.) Under what conditions on the distribution of X does the limit equal EY_0 (and hence the conclusion of the weak law of large numbers holds for this process with memory)?

6.28 Let $\{X_n\}$ be an i.i.d. zero-mean Gaussian random process with autocorrelation function $R_X(0)=\sigma^2$. Let $\{U_n\}$ be an i.i.d. random process with $\Pr(U_n=1) = \Pr(U_n=-1)=1/2$. Define a new random process $\{Y_n\}$ by

$$Y_n = U_n X_n \,.$$

(a) Find the autocorrelation function $R_Y(k,j)$.

(b) Find the characteristic function $M_{Y_n}(ju)$.

(c) Is $\{Y_n\}$ an i.i.d. process?

(d) Does the sample average

$$S_n = n^{-1} \sum_{i=0}^{n-1} Y_i$$

converge in mean square? If so, to what?

6.29 Use the Markov inequality to establish the *Chernoff bound*:

$$\Pr(X>x) \le E(e^{\lambda(X-x)}), \text{ for any } x \in \mathbf{R} \,,$$

where λ is an arbitrary nonnegative constant that can be used to optimize the bound. Use the Chernoff bound together with lemma 6.4 to show that the strong law of large numbers holds for an i.i.d Gaussian random process.

6.30 Assume that $\{X_n\}$ is an i.i.d. zero-mean Gaussian random process with $R_X(0)=\sigma^2$, that $\{U_n\}$ is an i.i.d. binary random process with

$\Pr(U_n=1) = 1-\epsilon$ and $\Pr(U_n=0) = \epsilon$ (in other words, $\{U_n\}$ is a Bernoulli process with parameter $1-\epsilon$), and that the processes $\{X_n\}$ and $\{U_n\}$ are mutually independent of each another. Define a new random process

$$V_n = X_n U_n .$$

(This is a model for the output of a communication channel that has the X process as an input but has "dropouts"— that is, occasionally sets an input symbol to zero.)

(a)   Find the mean EV_n and characteristic function $M_{V_n}(ju) = Ee^{juV_n}$.

(b)   Find the mean-squared error $E[(X_n - V_n)^2]$.

(c)   Find $\Pr(X_n \neq V_n)$.

(d)   Find the covariance of V_n.

(e)   Is the following true?

$$\left[\text{l.i.m.}_{n\to\infty}\frac{1}{n}\sum_{i=0}^{n-1}X_i\right]\left[\text{l.i.m.}_{n\to\infty}\frac{1}{n}\sum_{i=0}^{n-1}U_i\right] = \text{l.i.m.}_{n\to\infty}\frac{1}{n}\sum_{i=0}^{n-1}V_i$$

6.31 Show that convergence in distribution is implied by the other three forms of convergence.

6.32 Let $\{X_n\}$ be a finite-alphabet i.i.d. random process with marginal pmf p_X. The *entropy* of an i.i.d. random process is defined as

$$H(X) = -\sum_x p_X(x)\log p_X(x) = E(-\log p_X(X)) ,$$

where care must be taken to distinguish the use of the symbol X to mean the name of the random variable in $H(X)$ and p_X and its use as the random variable itself in the argument of the left-hand expression. If the logarithm is base two then the units of entropy are called *bits*. Use the weak law of large numbers to show that

$$-\frac{1}{n}\sum_{i=0}^{n-1}\log p_X(X_i) \underset{n\to\infty}{\to} H(X)$$

in the sense of convergence in probability. Show that this implies that

$$\lim_{n\to\infty} \Pr(|p_{X_0,\dots,X_{n-1}}(X_0,\dots,X_{n-1}) - 2^{-nH(X)}| > \epsilon) = 0$$

for any $\epsilon > 0$. This result was first developed by Claude Shannon and is sometimes called the *asymptotic equipartion property* of information theory. It forms one of the fundamental results of the mathematical theory of communication. Roughly stated, with high probability an i.i.d. process will produce for large n an n-dimensional sample vector $x^n = (x_0, x_1, \dots, x_{n-1})$ such that the nth-order probability mass function evaluated at x^n is approximately $2^{-nH(X)}$; that is, the process produces long vectors that appear to have an approximately uniform distribution over some collection of possible vectors.

6.33 Let $\{U_n\}$ be an i.i.d. Gaussian random process with mean 0 and variance σ^2. Suppose that Z is a random variable having a uniform distribution on $[0,1]$. Suppose Z represents the value of a measurement taken by a remote sensor and that we wish to guess the value of Z based on a noisy sequence of measurements $Y_n = Z + U_n$, $n = 0,1,2,\ldots$, that is, we observe only Y_n and wish to estimate the underlying value of Z. To do this we form a sample average and define the estimate

$$\hat{Z}_n = \frac{1}{n} \sum_{i=0}^{n-1} Y_i .$$

(a)   Find a simple upper bound to the probability

$$\Pr(|\hat{Z}_n - Z| > \epsilon)$$

that goes to zero as $n \to \infty$. (This means that our estimator is asymptotically good.)

Suppose next that we have a two-dimensional random process $\{U_n, W_n\}$ (i.e., the ouput at each time is a random pair or a two-dimensional random variable) with the following properties: Each pair (U_n, W_n) is independent of all past and future pairs (U_k, W_k), $k \neq n$. Each pair (U_n, W_n) has an identical joint cdf $F_{U,W}(u,w)$. For each n $EU_n = EW_n = 0$, $E(U_n^2) = E(W_n^2) = \sigma^2$, and $E(U_n W_n) = \rho\sigma^2$. (The quantity ρ is called the *correlation coefficient*.) Instead of just observing a noisy sequence $Y_n = Z + U_n$, we also observe a separate noisy measurement sequence $X_n = Z + W_n$ (the same Z, but different noises). Suppose further that we try to improve our estimate of Z by using both of these measurements to form an estimate

$$\tilde{Z} = a\frac{1}{n} \sum_{i=0}^{n-1} Y_i + (1-a)\frac{1}{n} \sum_{i=0}^{n-1} X_i$$

for some a in $[0,1]$.

(b)   Show that $|\rho| \leq 1$. Find a simple upper bound to the probability

$$\Pr(|\tilde{Z}_n - Z| > \epsilon)$$

that goes to zero as $n \to \infty$. What value of a gives the smallest upper bound in part (b) and what is the resulting bound? (Note as a check that the bound should be no worse than part (a) since the estimator of part (a) is a special case of that of part (b).) In the special case where $\rho = -1$, what is the best a and what is the resulting bound?

6.34 Suppose that $\{X_n\}$ are i.i.d. random variables described by a common marginal distribution F. Suppose that the random variables

$$S_n = \frac{1}{n} \sum_{i=0}^{n-1} X_i$$

also have the distribution F for all positive integers n. Find the form of the distribution F. (This is an example of what is called a *stable distribution*.) Suppose that the $1/n$ in the definition of S_n is replaced by $1/\sqrt{n}$. What must F then be?

7

STATIONARITY AND ERGODIC
PROPERTIES

In chapter 6 we defined the expectation of functions of random processes. In many applications we are interested in relating such expectations or probabilistic averages to long-term sample averages or time averages. These relations are important for primarily two reasons: First, we may wish to predict the behavior of long-term sample averages such as the average energy across a resistor, the average expenditures of customers in a given store, the average solar flux in November, and so on. If we have a model for a process and the model is such that one can equate long-term sample averages and expectations, then expectations provide a prediction of the long-term sample average that can be used in design and planning. Second, we may lack a model of a given physical process and may wish to construct a model by estimating various parameters of the observed random process, viz., the unknown expectation of a constant-mean random process. Intuitively, the sample average of a random process is a reasonable estimate of the unknown mean. This intuition can be made rigorous if we can show that limiting sample averages equal corresponding expectations.

In chapter 6 we considered some simple conditions under which sample averages converge to the expectation—laws of large numbers—and the possible senses in which convergence can occur. In this chapter several of the notions of the preceding chapter are generalized. This provides some useful classifications of random processes and some more general laws of large numbers.

STATIONARITY

The sample functions of all except the most trivial random processes are time-varying with a high probability. In many cases, however, the *probabilistic* behavior does not change with time. Consider a binary communication system. Although the bit stream itself is time-varying, the probability of the communication system's transmitting a 1 at any given instant is likely to be about the same as the probability of transmitting a 1 ten minutes later. This is an example of a *stationarity* property: a probabilistic attribute of the random process that does not change with time. For example, if a discrete time random process $\{X_n\}$ is i.i.d., then its marginal distribution does not depend on the sample time; that is, for some common distribution P_X,

$$P_{X_n} = P_X; \text{ all } n \ . \tag{7.1a}$$

Equivalently, the cdf's that describe the distributions satisfy

$$F_{X_n} = F_X; \text{ all } n \ , \tag{7.1b}$$

for some common cdf F_X.

Since the marginal distribution does not depend on the specific time of the sample, expectations of the samples and expectations of functions of the samples have the same property; that is, equation (7.1) implies that

$$EX_n = EX; \text{ all } n \ . \tag{7.2a}$$

$$\sigma^2_{X_n} = \sigma^2_X \ , \tag{7.2b}$$

and so on for the expectation of any function of a single sample X_n.

For more general random processes, e.g., the uncorrelated processes considered in chapter 6, the distributions of single samples and expectations of functions of single samples may or may not depend on time as in (7.1) and (7.2). These equations are examples of stationarity properties of a random process: probabilities or expectations that are "stationary" in the sense that they do not depend on time. We could dub them "marginal" stationary properties since they consider only marginal distributions and hence probabilistic properties involving only single samples. However, more commonly the properties of equations (7.1) and (7.2) are called "first-order" stationarity properties since they deal with only first-order expectations and distributions. Processes satisfying (7.1) and (7.2) are said to be *first-order stationary*. Clearly these equations can hold for single samples without an analogous property holding for more than one sample. For example, we could have a Gaussian random process with non-time-varying mean and variance but time-varying covariance between samples. The process is first-order stationary but it has second-order nonstationarity properties.

If a process is i.i.d., then it has considerably more stationarity properties than those considered so far. For example, if we choose two sample times t and s and consider the joint distribution, P_{X_t, X_s}, of the sample values at those times, then the joint distribution will depend only on the separation between the two samples and not on their absolute values; that is,

$$P_{X_t, X_s} = P_{X_0, X_{s-t}} \tag{7.3a}$$

or, in terms of cdf's,

$$F_{X_t, X_s} = F_{X_0, X_{s-t}}; \text{ all } s, t . \tag{7.3b}$$

This in turn implies that second-order expectations such as the autocorrelation and covariance functions depend only on the time difference between the samples, called the *lag* or *delay*. That is, we can set

$$R_X(t, s) = R_X(s - t); \text{ all } t, s$$

$$K_X(t, s) = K_X(s - t); \text{ all } t, s , \tag{7.4}$$

to emphasize the fact that the given functions depend on their two arguments only through the difference of the arguments. It must be admitted that equation (7.4) uses bad notation, using the same notation for a function with two arguments as for a function with one argument, but it is a standard abuse of notation, and we shall also adopt it.

Taken together, the stationarity properties of the distributions and expectations in (7.1) though (7.4) are called second-order stationarity properties since they show that first- and second-order moments of the given process do not depend on time, that is, on the time origin. Another way of saying this is that these functions are *time-invariant*. Processes satisfying (7.1) through (7.4) are said to be *second-order stationary* random processes. Thus i.i.d. processes are a (very) special case of second-order stationary random processes.

Just as there are first-order stationary processes that are not second-order stationary, there are also second-order stationary processes that do not have higher order stationarity properties.

For i.i.d. processes, however, it is obvious from the defining equations of (5.8) that we can look at distributions and expectations of any order and observe that all finite-dimensional distributions and all corresponding expectations are unaffected by time shifts or, equivalently, do not depend on the time origin or are time-invariant. We can say more than this. From chapter 5, the entire process distribution is specified by the family of such finite-dimensional distributions through the Kolmogorov specification. Hence it turns out that since all finite-dimensional distributions are unchanged by time shifts, the same is true of the process distribution. That is, if we are given two events F and G in sequence space where

F is formed by shifting all of the sequences in G one time unit, then the probability of the shifted event is the same as that of the unshifted event. Roughly speaking, the complete probabilistic description of the process is unchanged by time shifting or, equivalently, redefining the time origin. We will pursue this idea further in a subsequent section in a more general context than i.i.d. processes.

Thus, for a discrete time process, the i.i.d. condition is a sufficient condition for stationarity properties on any events to hold in the sense we are describing—the invariance of probabilities (and hence expectations) under time shifts. It is not, however, a necessary condition. We shall encounter many processes with memory that have the same stationarity properties. We will also encounter processes with memory that have some stationarity properties but not an all-encompassing invariance under time shifts. Hence we give names to some of the stationarity properties as a classification scheme for random processes. We focus on three forms of stationarity: (1) weak stationarity or wide-sense stationarity, wherein the first- and second-order moments of a process are time-invariant; (2) n^{th}-order stationarity, in which distributions and expectations involving up to $n < \infty$ samples at a time are time-invariant; and (3) strict stationarity or simply stationarity, in which the complete probabilistic description of the process is time-invariant. Observe that in the strictly stationary case, all possible moments and finite-dimensional distributions do not depend on the time origin; that is, they are time-invariant. Clearly stationarity implies n^{th}-order stationarity, which for $k \geq 2$ implies weak stationarity, but the reverse implications are not true. We now make these ideas precise with definitions.

Let $\{X_t;\ t \in \boldsymbol{I}\}$ be a random process (continuous or discrete time). The process is said to be *weakly stationary* (or *wide-sense stationary* or *stationary in the weak (or wide) sense*) if

$$EX_t = EX\ ;\quad \text{all } t \in \boldsymbol{I}\ ,\tag{7.5}$$

that is, the left-hand side does not depend on t, and

$$R_X(t,t+\tau) = R_X(\tau)\ ;\quad \text{all } t,\tau : t, t+\tau \in \boldsymbol{I}\ ,\tag{7.6}$$

that is, the left-hand side does not depend on t.

Thus weak stationarity is verified by checking only the mean function EX_t and the autocorrelation function $R_X(t,t+\tau)$ for time invariance; that is, they must not depend on t. Obviously, one can use the covariance function instead of the autocorrelation function in (7.6). We note in passing that a function of two arguments that depends only on their difference, as do the autocorrelation function and covariance function of a weakly stationary process, is called a *Toeplitz function*. Much of the theory of weakly stationary processes is simply an application of the theory of Toeplitz functions.

Strict stationarity is both harder to define and harder to demonstrate.

A random process $\{X_t; \ t \in \boldsymbol{I}\}$ is said to be *stationary* (or *strictly stationary* or *stationary in the strict sense*) if the probability of every sequence or waveform event is unchanged by a time shift; that is, if m is the process distribution, then for any τ such that $t + \tau \in \boldsymbol{I}$ for all $t \in \boldsymbol{I}$ we have that

$$m(\{x_t; t \in \boldsymbol{I}\}: \{x_t; t \in \boldsymbol{I}\} \in F) = m(\{x_t; t \in \boldsymbol{I}\}: \{x_{t+\tau}; t \in \boldsymbol{I}\} \in F) \qquad (7.7)$$

if the equation makes sense; that is, if F is a sequence or waveform event and if the set of sequences or waveforms obtained by shifting every member of F by τ is also an event.

Because of the equivalence of process distributions and the family of all possible finite-dimensional cdf's, it can be shown that a process is stationary if and only if for every finite dimension k, every collection of sample times $(t_0, t_1, \ldots, t_{k-1})$, and every delay τ such that $t_i, t_i + \tau \in \boldsymbol{I}$; $i = 0, 1, \ldots, k-1$ we have that

$$F_{X_{t_0 + \tau}, X_{t_1 + \tau}, \ldots, X_{t_{k-1} + \tau}}(\mathbf{x}) = F_{X_{t_0}, X_{t_1}, \ldots, X_{t_{k-1}}}(\mathbf{x}); \text{ all } \mathbf{x} \in \mathbf{R}^k \ . \qquad (7.8)$$

That is, the joint cdf's for all possible finite collections of samples of the random process are unchanged by time shifting. Observe that this will also be true if the analogous equation holds for either joint pdf's or joint pmf's.

A process is said to be n^{th}-*order stationary* if (7.8) holds for all sample times and every finite dimension $k \le n$.

EXAMPLES OF STATIONARY PROCESSES

I.i.d. Processes

We have already argued that an i.i.d. process is stationary. In fact, the i.i.d. process is one of the few processes for which a relatively easy demonstration of stationarity is possible. Since the cdf's are specified by the pmf's or pdf's, stationarity of the process follows for discrete alphabet i.i.d. processes from (7.8) and the fact that

$$p_{X_{t_0 + \tau}, \ldots, X_{t_{k-1} + \tau}}(\mathbf{x}) = \prod_{i=0}^{k-1} p_X(x_i) = p_{X_{t_0}, \ldots, X_{t_{k-1}}}(\mathbf{x}) \ .$$

Replacing the p's by f's demonstrates stationarity for continuous alphabet i.i.d. processes.

We will rarely be able to prove that a random process is stationary in this book. Only in certain special cases is the proof straightforward. For example, we have shown a straightforward proof for i.i.d. processes and will shortly show a proof for certain Gaussian random processes. It is often easy, however, to determine that a nonstationary process is *not* stationary. For example, if any moments (mean, autocorrelation, etc.) depend on the time origin, then the process is not stationary. (See, for example, exercise 7.4.)

We have already argued that both i.i.d. processes and uncorrelated processes with time-invariant mean and variance are weakly stationary. This can be directly proved by noting that in both cases (7.1) holds and that the covariance function is

$$K_X(k) = \sigma_X^2 \delta_k , \tag{7.9}$$

where δ_k is the Kronecker delta defined by

$$\delta_k = \begin{cases} 1 \text{ if } k = 0 \\ 0 \text{ if } k \neq 0 . \end{cases}$$

Equivalently, the correlation function is

$$R_X(k) = E(X^2)\delta_k + (EX)^2(1 - \delta_k) \tag{7.10}$$

so that (7.5) and (7.6) are satisfied.

Gaussian Processes

Recall from chapter 5 that a Gaussian process is completely described by its mean function m_t and its covariance function $\Lambda(t,s)$. As one might guess, the names of these functions come from the fact that they provide the first- and second-order expectations of the process; that is, if $\{X_t; t \in I\}$ is a discrete time or continuous time Gaussian random process described by these functions, then

$$EX_t = m_t, \text{ all } t \in I \tag{7.11}$$

and

$$K_X(t,s) = E[(X_t - EX_t)(X_s - EX_s)] = \Lambda(t,s); \text{ all } t,s \in I . \tag{7.12}$$

Equations (7.11) and (7.12) can be proved by resort to integral tables. For readers well versed in matrix transformations of multidimensional integrals, they can also be proved with relative ease. (See exercise 7.5.) However, the computations are tedious to perform directly using more elementary methods. Vector generalizations of the scalar moment-generating properties of characteristic functions considered in chapter 11 provide an alternative means of computing expectations and can be used to show (7.11) and (7.12). These methods are often simpler, but for the moment we shall be content just to quote the relations and use them.

We digress for a moment to give a word of explanation on the assumptions made on the matrix Λ in defining the Gaussian process.

Recall from chapter 5 that we required the function to be symmetric in its arguments and positive definite. Clearly symmetry is a necessary condition for a covariance function since

$$K_X(t,s) = E[(X_t - EX_t)(X_s - EX_s)]$$
$$= E[(X_s - EX_s)(X_t - EX_t)] = K_X(s,t) .$$

While it is not necessary for a covariance function to be positive definite, it must be at least nonnegative definite: To see this, fix a finite collection of sample times $t_0, t_1, \ldots, t_{k-1}$ and observe that for any real vector $\mathbf{a} = (a_0, \ldots, a_{k-1})$ we have that

$$\mathbf{a}^t \Lambda \mathbf{a} = \sum_{i=0}^{k-1}\sum_{j=0}^{k-1} a_i a_j K_X(t_i,t_j) = E\left\{ [\sum_{i=0}^{k-1} a_i(X_i - EX_i)]^2 \right\} \geq 0 , \quad (7.13)$$

and hence K_X is nonnegative definite. Observe that we have shown that a covariance function of any real random process must be symmetric and nonnegative definite; the process need not be Gaussian for the covariance to have these properties. If (7.13) is equal to zero for any nonzero $\mathbf{a}$, then Λ is singular so that its inverse does not exist. Thus, although we shall later see how to extend the definition of a Gaussian process to the more general case of nonnegative definite covariance functions, for the moment we need to require positive definiteness in order to ensure the existence of the inverse matrix in the vector Gaussian pdf.

If a Gaussian random process is weakly stationary, then we have from (7.5) and (7.6) that

$$m_t = m; \text{ all } t \in \mathbf{I} \qquad (7.14)$$

and

$$\Lambda(t,s) = K_X(s - t); \text{ all } t,s \in \mathbf{I} . \qquad (7.15)$$

These functions completely describe all joint pdf's and hence cdf's or distributions of the Gaussian process. Since these functions are time-invariant—that is, they do not depend on t, they depend only on the time difference or lag—the resulting joint pdf's also will all be time-invariant. Thus if a Gaussian random process is weakly stationary, it is also strictly stationary! This is one of the nice properties of this particular process.

*THE SHIFT

The definition of stationarity can be given a more compact form in terms of an underlying probability space. For this definition we need to introduce the notion of the shift transformation. We confine interest to the discrete time case and use the process sequence probability space for simplicity. Given the probability space $(\mathbf{R}^I, \mathbf{B}(\mathbf{R})^I, m)$ we define the (left) shift operator (or left shift mapping) $T: R^I \to R^I$ as follows: If $\omega \in R^I$ is any particular infinite sequence in the probability space given by $\omega = \{x_t; t \in I\}$, then $T\omega = \{x_{t+1}; t \in I\}$. For example, in the one-sided case we have

$$T\omega = T(x_0, x_1, \ldots) = (x_1, x_2, \ldots) . \qquad (7.16)$$

In the two-sided case,

$$T\omega = T(\ldots, x_{-1}, x_0, x_1, \ldots) = (\ldots, x_0, x_1, x_2, \ldots) . \qquad (7.17)$$

Thus the mapping T simply shifts a sequence to the left by one time unit. In the two-sided process case, we can also define a right shift, which shifts every coordinate one time unit to the right and is the inverse operation to the left shift. (In the two-sided case, T is a one-to-one transformation and hence is invertible.) In the one-sided case, however, such a right shift is not well defined. (In the one-sided case, T is a many-to-one operation and is not invertible.)

Now let F be any event defined on the process sequence event space. Let $T^{-1}F$ denote the inverse image of the event F under the shift operation T; that is, it is the collection of all sequences that when shifted land in F:

$$T^{-1}F = \{x: Tx \in F\} = \{\{x_t; t \in I\}: \{x_{t+1}; t \in I\} \in F\} .$$

Assuming that the operation T is measurable (which can be proved), then $T^{-1}F$ is also an event. For example, if $F = \{x: x_0 \in [0,1), x_{10} < 3\}$, then the inverse image is $T^{-1}F = \{x: x_1 \in [0,1), x_{11} < 3\}$. Note that even though we may not have a right shift operation defined on individual sequences, as in the one-sided case, we can consider T^{-1} to be a right shift operation defined on *sets* since $T^{-1}F$ has all the sequences that shift left into F. We can now rephrase the definition of stationarity as follows:

A discrete time random process with process distribution m is stationary if and only if

$$m(T^{-1}F) = m(F); \text{ all } F \in \boldsymbol{B}(\mathbf{R})^I , \qquad (7.18)$$

that is, if the probability of a sequence event is unaffected by shifting. (Measurability of T ensures that the probabilities of the inverse images of events are well defined.)

In much of the mathematics and most of the mathematical engineering literature, stationarity is often defined in the form of $m(TF) = m(F)$ for all events F. It turns out, however, that this definition is less general than ours: in particular, it is not valid for one-sided processes. The reason is that if the left shift T is measurable, then one is only assured that inverse images of the form $T^{-1}F$ are events; forward images of the form TF may not be events (see, e.g., the above example), and hence the definition involving $m(TF)$ may not make sense.

In ergodic theory literature, a probability measure satisfying this relation is also said to be *invariant* or *T*-invariant. As before, the basic idea is that shifting does not change probabilities and hence does not change expectations either.

ERGODIC THEOREMS

In this section we describe two laws of large numbers or ergodic theorems that are more general than those encountered in chapter 6. Following the proof of these theorems we define and discuss what is meant by an ergodic *process* and give a very general ergodic theorem. As we will see, an ergodic process is not the same as a process that satisfies an ergodic theorem. A process that satisfies an ergodic theorem is defined as follows:

A discrete time random process $\{X_n; n \in I\}$ is said to satisfy an ergodic theorem if there exists a random variable $\hat{X}$ such that in some sense

$$n^{-1} \sum_{i=0}^{n-1} X_i \xrightarrow[n \to \infty]{} \hat{X} .$$

The type of convergence determines the type of the ergodic theorem. For example, if the convergence is in mean square, the result is called a *mean ergodic theorem*. If the convergence is with probability one, it is called an *almost everywhere* or *pointwise ergodic theorem*.

A continuous time random process $\{X(t); t \in I\}$ is said to satisfy an ergodic theorem if there exists a random variable $\hat{X}$ such that

$$T^{-1} \int_0^T X(t)dt \xrightarrow[T \to \infty]{} \hat{X} ,$$

where again the type of convergence determines the type of ergodic theorem.

The continuous time ergodic theorem is the integral analog of the discrete time theorem. That is, the sample average or time average is still obtained as a normalized average of sample values, but the average is a sum for discrete time and an integral for continuous time. We are glossing over the whole question of existence of an integral of a random process. This is a nontrivial issue, since an integral is itself defined as a limit, and we have seen that there are several notions of limits of random variables. In particular, even for fixed T the integral $T^{-1} \int_0^T X(t)dt$ could mean a limit in the quadratic mean or a limit with probability one of a sequence of sums of random variables.

Note that in general we do not insist that the time average converge to the common expectation of the random process samples, simply that it converge to something. This added generality is important for two reasons. First, ergodic theorems often hold even for nonstationary random processes where EX_t does depend on time. For example, a process may have transients or periodicities and hence not be stationary. The long-term sample average may still converge, but not to a common expectation of the samples. This fact is rarely mentioned in texts on random processes and can lead to a belief that stationarity is a necessary condition for the convergence of time averages; it is not! Second, even if the process is stationary, sample averages may converge to something other than the expectation—the typical example being the randomly selected coin-flipping process of exercises 5.10 and 5.14. Exercise 5.14 provides an especially simple example when given the following interpretation: Suppose that nature at the beginning of time randomly selects one of two coins with equal probability, one having bias p and the other having bias $q \neq p$. After the coin is selected it is flipped once per second forever. The output random process is a one-zero sequence depending on whether or not heads occurs. Clearly the time average will converge, but it will converge to p if the first coin was selected and to q if the second coin was selected; that is, the time average will converge to a random variable. In particular, it will not converge to the expected value $p/2 + q/2$. As another viewpoint of the same phenomenon: In many applications the random process model may incorporate some lack of knowledge of the observer, e.g., the exact bias of a coin being flipped. The crucial point for applications is the knowledge that the sample average of heads will converge to *something* (and hence the observer can in principle learn it), not that it will converge to a prespecified value. The something is in general a random variable.

Ergodic theorems lie at the heart of applied random processes because they guarantee that such time averages—which model physically measurable quantities—will converge. It is important to realize that such convergence may take place under quite general assumptions and that, in particular, neither i.i.d. nor even stationarity is required.

We next give an example of two fairly general mean ergodic theorems for random processes. The first theorem provides a commonly encountered sufficient condition for a mean ergodic theorem that is an intuitive generalization of the results of chapter 6. The second theorem gives general necessary and sufficient conditions for a mean ergodic theorem and points out that even weak stationarity is not required.

TWO MEAN ERGODIC THEOREMS

The first result is quite simple, and the proof is only slightly more complicated than that for uncorrelated random process.

Theorem 7.1.   Sufficient conditions for a weakly stationary discrete time random process $\{X_n; n \in I\}$ to have a mean ergodic theorem with limiting sample average EX are that $K_X(0) < \infty$ and

$$\lim_{n \to \infty} K_X(n) = 0 . \tag{7.19}$$

Comment. The condition given in (7.19) can be interpreted as a definition of an asymptotically uncorrelated process, that is, an asymptotic form of the definition of uncorrelation. This condition occurs, for example, if the covariance is zero past some finite lag. Another example of a covariance meeting this condition is a geometric covariance $K_X(k) = \rho^{|k|}$ for $\rho < 1$.

Similar results exist for the continuous time case with the sums replaced by integrals. They require additional technical assumptions, however, to ensure that the integrals make sense. The proofs are far more difficult because of the additional details and hence will not be given.

Proof. As in the proof for the uncorrelated case that was given in chapter 6, the theorem will follow if we can show that the variance of the sample average as given in (6.11) goes to zero as $n \to \infty$. Since the process is weakly stationary we can regroup the terms in the sum to write

$$\sigma_{S_n}^2 = n^{-2} \sum_{i=0}^{n-1} \sum_{j=0}^{n-1} K_X(i-j)$$

$$= n^{-1} K_X(0) + 2n^{-1} \sum_{k=1}^{n-1} (1 - \frac{k}{n}) K_X(k) .$$

This weighted sum goes to zero as $n \to \infty$ if $K_X(k)$ goes to zero as $k \to \infty$ by the theorem of arithmetic means of convergent sequences. While this is a believable result from elementary analysis, we give a proof for completeness. Since $K_X(k) \to 0$, given ϵ we can choose an N so large that for $k > N$ we have that $K_X(k) < \epsilon/2$. From exercise 6.9 we know that $|K_X(k)|$ attains its maximum at $k = 0$; that is,

$$|K_X(k)| \le K_X(0) . \tag{7.20}$$

Thus we have for $n > N$ that

$$\sigma_{S_n}^2 \le n^{-1} K_X(0) + 2n^{-1} \sum_{k=1}^{n-1} (1 - \frac{k}{n}) |K_X(k)|$$

$$\le n^{-1} K_X(0) + 2n^{-1} \sum_{k=1}^{N} (1 - \frac{k}{n}) |K_X(0)| + n^{-1} \sum_{k=N+1}^{n-1} (1 - \frac{k}{n}) \epsilon$$

$$\le \frac{2N}{n} K_X(0) + \frac{\epsilon}{2} ,$$

where we have used the fact that

$$\sum_{k=1}^{j} k = \frac{j(j+1)}{2} .$$

Thus we have

$$\sigma_{S_n}^2 \leq \frac{2NK_X(0)}{n} + \frac{\epsilon}{2} \, .$$

The upper bound goes to $\epsilon/2$ as $n \to \infty$. Thus for arbitrarily small ϵ we can make $\sigma_{S_n^2}$ smaller than ϵ by making n large enough; thus $\sigma_{S_n^2} \to 0$ as $n \to \infty$, proving the theorem.

Our second ergodic theorem provides more general conditions than those of theorem 7.1 for convergence in mean square. The conditions are both necessary and sufficient for the sample average to exist and equal a constant.

Theorem 7.2.   Let $\{X_n\}$ be a discrete time random process with mean function EX_n and covariance function $K_X(k,j)$. (The process need not be even weakly stationary.) Necessary and sufficient conditions for the existence of a constant m such that

$$\text{l.i.m.}_{n \to \infty} n^{-1} \sum_{i=0}^{n-1} X_i = m \tag{7.21}$$

are that

$$\lim_{n \to \infty} n^{-1} \sum_{i=0}^{n-1} EX_i = m \tag{7.22}$$

and

$$\lim_{n \to \infty} n^{-2} \sum_{i=0}^{n-1} \sum_{k=0}^{n-1} K_X(i,k) = 0 \, . \tag{7.23}$$

Thus if the arithmetic average of the means converges to m and if the process is asymptotically uncorrelated in the sense of (7.23), then the sample average converges in quadratic mean.

* *Proof.* First assume that (7.21) holds. We shall show that this implies both (7.22) and (7.23) and hence that they are necessary conditions. First, the inequality $|EX| \leq E|X|$, the Cauchy-Schwartz inequality, and (7.21) imply that

$$\left| n^{-1} \sum_{i=0}^{n-1} EX_i - m \right| \leq E\left[\left| n^{-1} \sum_{i=0}^{n-1} X_i - m \right| \right]$$

$$\leq E\left[\left| n^{-1} \sum_{i=0}^{n-1} X_i - m \right|^2 \right]^{1/2} \xrightarrow[n \to \infty]{} 0 \, ,$$

which proves (7.22). Next observe that

$$n^{-2} \sum_{i=0}^{n-1} \sum_{k=0}^{n-1} K_X(i,k) = E[|n^{-1} \sum_{k=0}^{n-1} (X_k - EX_k)|^2] . \tag{7.24}$$

From the triangle inequality ($|a-b| \leq |a-c| + |c-b|$) we can bound the absolute value squared in (7.24) to obtain

$$n^{-2} \sum_{i=0}^{n-1} \sum_{k=0}^{n-1} K_X(i,k) \leq E[(|n^{-1} \sum_{k=0}^{n-1} X_i - m| + |m - n^{-1} \sum_{k=0}^{n-1} EX_i|^2]$$

$$= E[|n^{-1} \sum_{k=0}^{n-1} X_i - m|^2] +$$

$$2E[(|n^{-1} \sum_{k=0}^{n-1} X_i - m| \times |m - n^{-1} \sum_{k=0}^{n-1} EX_i|)] + E[|m - n^{-1} \sum_{k=0}^{n-1} EX_i|)^2] .$$

Of the terms in the last expression, the leftmost goes to zero by definition of l.i.m. in (7.21). The rightmost term goes to zero from (7.22), which we have already shown to hold. These two facts and the Cauchy-Schwartz inequality then imply that the middle term also goes to zero. This proves (7.14).

Conversely, assume that (7.22) and (7.23) hold. Then, using the triangle inequality again, we write

$$E[|n^{-1} \sum_{k=0}^{n-1} X_i - m|^2] \leq E[(|n^{-1} \sum_{k=0}^{n-1} (X_k - EX_k)| + |n^{-1} \sum_{k=0}^{n-1} EX_k - m|)^2]$$

$$= E[|n^{-1} \sum_{k=0}^{n-1} (X_k - EX_k)|^2] +$$

$$2E[|n^{-1} \sum_{k=0}^{n-1} (X_k - EX_k)| \times |m - n^{-1} \sum_{k=0}^{n-1} EX_i|] + E[|m - n^{-1} \sum_{k=0}^{n-1} EX_i|)^2] .$$

Consider the final expression: The leftmost term goes to zero by (7.23), the rightmost term goes to zero by (7.22), and the middle term goes to zero by the Cauchy-Schwartz inequality and plus (7.22) and (7.23).

Theorem 7.2 shows that one can expect a sample average to converge to a constant in mean square if and only if the average of the means converges and if the memory dies out asymptotically; that is, if the covariance decreases as the lag increases in the sense of (7.23).

* ERGODICITY

A fairly common source of confusion in engineering treatments of random processes is the distinction between ergodic theorems and the property of ergodicity, which we have not yet defined. Many engineering texts (most, in fact) *define* a process to be ergodic if its sample averages converge to its expected averages. This is unfortunate for several reasons: First, it is inaccurate historically and mathematically. The concept of ergodicity arose long ago in statistical mechanics and physics and has a precise definition for the random process application. The concept is well established in the physics and mathematical literature and bears no resemblance to the "engineering" definition. Second, it gives the erroneous impression that ergodicity is required for the convergence of sample averages in practice, a misconception that severely limits the range of applications of the theory. The convergence of sample averages corresponds to the satisfaction of an ergodic theorem, not necessarily to the possession of the rather abstract property of ergodicity—a much more restrictive property. For this reason we have focused on ergodic theorems and not on ergodicity. For completeness, however, we pause to give the precise definition, to make a few observations, and to state a general ergodic theorem for ergodic processes.

   As in the section on the shift transformation, suppose we have a process sequence probability space $(\mathbf{R}^I, \mathbf{\mathcal{B}}(\mathbf{R})^I, m)$, that is, a directly given random process $\{X_t; t \in \mathbf{I}\}$ with process distribution m. The shift transformation is as defined in (7.16) through (7.18). As before, we confine our attention to discrete time random processes for simplicity. A process event F is said to be *invariant* with respect to the shift transformation if and only if $T^{-1}F = F$. This means that the set of sequences in F and the set of sequences that when left-shifted are in F are the same. Roughly speaking, if we apply the operation T^{-1}, which shifts an entire event to the right one unit, to the set F, then we simply get the set F again. We can iterate on this operation to conclude that $T^{-2}F = T^{-1}(T^{-1}F) = T^{-1}F = F$ and so on, so that if F is invariant, then $T^{-n}F = F$; no matter how many times we shift an invariant event to the right, we get the same thing.

   At first exposure this abstract concept seems hard to grasp. However, we will provide examples to show that in any application, there are many invariant events that are natural to consider. Obviously $\mathbf{R}^I$ is invariant. So also is the null event. So is the event that consists of a sequence of all ones (or any other constant) for all time. We will consider less obvious examples shortly. But first we define ergodicity.

   A random process is said to be *ergodic* if for *any* invariant event, F, either $m(F) = 0$ or $m(F) = 1$. Thus, if an event is "closed" under time shifts, then it must have all of the probability or none of it.

   This definition is not very intuitive, but some interpretations and examples may shed a little light. First observe, however, that the definition has nothing to do with stationarity or, on the face of it, with convergent sample averages. It simply states that events that are unaffected by shifting must have probability either zero or one. The sample space, $\mathbf{R}^I$, is invariant and has probability one. The null event is invariant and has probability zero. We give a less trivial example. Suppose that a two-sided process has distribution m as follows:

$$m(...,x_{-1}=1,x_0=0,x_1=1,x_2=0,x_3=1,...) = p$$

$$m(...,x_{-1}=0,x_0=1,x_1=0,x_2=1,x_3=0,...) = 1-p \ ,$$

that is, the two possible alternating one-zero sequences have probability one. Note that F ={all alternating one-zero sequences} is an invariant event and has probability one. Any other invariant event that does not include F has probability zero. Thus the random process is ergodic. One of the folk theorems in much of the engineering trade is that ergodicity implies stationarity. Our simple example provides a counterexample. Note that the process is *not* stationary unless we also have $p=1/2$.

The example just given is somewhat artificial. It was chosen to illustrate concisely the principles involved. There are many important nonartificial examples in engineering practice. We will illustrate one of these in terms of a coin-flipping model. Given a binary sequence space modeling the output of a Bernoulli process of coin flips, the set of all sequences such that the relative frequency of heads converges to a fixed value p is an invariant set. This set is invariant because a single shift does not affect the relative frequency in the limit—perhaps one less head is counted, depending on the time origin. Another invariant event is the set of all sequences for which the relative frequency of heads converges to q for some fixed q. If a process is ergodic, then these two events, which are disjoint, must have zero probability, or at most one of them must have probability 1; that is, either the relative frequency must converge with probability 1 to p or it must converge with probability 1 to q, or neither. In particular, it cannot converge with probability 1/2 to p and with probability 1/2 to q. Thus, if nature flips a fair coin at the beginning of time and then flips another coin of bias p forever if a head occurs and a coin of bias q forever if a tail occurs, then the resulting random process is *not* ergodic. Sample averages still converge, however, either to p or q, depending on which coin is selected. Note that this process is, however, stationary. Note also that the definition requires that *any* invariant event have probability zero or one. Obviously, the invariant event which consists of all sequences for which the relative frequency converges to either p or q has probability one. This is not enough.

As a final example of an ergodic process that is often encountered in modeling physical processes, suppose that m is a process distribution for a stationary process. The process is said to be *strongly mixing* if for all events F and G we have that

$$\lim_{n \to \infty} m(F \cap T^{-n}G) = m(F)m(G) \ ,$$

that is, the probability of the intersection of one event with the shift of another looks more and more like the product of the probabilities as we shift the second event farther away. Strong mixing can be viewed as a form of asymptotic independence. Note that for the special case of a one-dimensional rectangle, the equation becomes

$$\Pr(X_0 \epsilon F \text{ and } X_n \epsilon G) \underset{n \to \infty}{\longrightarrow} \Pr(X_0 \epsilon F)\Pr(X_n \epsilon G) = P_X(F)P_X(G) \ ,$$

so that individual samples are *asymptotically independent* as they become more and more separated. (Compare this with the notion of asymptotic uncorrelation of theorem 7.1.) It is easy to see that a stationary strong mixing process is ergodic: Simply let $F = G$ be an invariant set, and the definition of strong mixing implies that

$$\lim_{n \to \infty} m(F \cap T^{-n}F) = \lim_{n \to \infty} m(F \cap F) = m(F) = m(F)m(F) \ .$$

The only way for $m(F)$ to equal $m(F)^2$, however, is for $m(F)$ to be 0 or 1; hence the process is ergodic.

As a final observation, ergodicity of a process means that there can be no disjoint sequence events having nonzero probability and having the property that one cannot get from one event to the other by shifting the sequence. Ergodicity requires that if any event is a trap in the sense that it can never be left by shifting, then that event must have all or none of the probability!

The importance of ergodicity derives from the fact that if one wishes to verify that *all* convergent sample averages converge to a constant and not to a random variable, then a necessary condition is that the process be ergodic. (It is not necessary for the process to be stationary!) As the example shows, this may eliminate models in which the parameters of the process may themselves have been selected in a random manner, e.g., may not be known *a priori*. We observe without proof that i.i.d. processes are ergodic and stationary Gaussian processes are ergodic if they have strictly positive definite covariance functions K_X such that $K_X(k)$ is finite for all k and goes to 0 as $k \to \infty$. (The standard proof that i.i.d. processes are ergodic first shows that they are strongly mixing.)

We now state without proof the best-known ergodic theorem for stationary and ergodic random processes.

Theorem 7.3.   (Birkhoff-Khinchin): Let $\{X_t ; t \epsilon I\}$ be a stationary random process on $(\mathbf{R}^I, \mathbf{B}(\mathbf{R})^I, m)$. We denote the sample points ω and the shift transformation T as before in (7.16) and (7.17). Let $f(\omega)$ be an arbitrary function on $\mathbf{R}^I$ with finite expectation Ef. Then the sample average of f satisfies

$$n^{-1} \sum_{k=0}^{n-1} f(T^k \omega) \underset{n \to \infty}{\to} \hat{f} \;,$$

where $\hat{f}$ is a random variable and the convergence is with probability one. If in addition the second moment of f, Ef^2, is finite, convergence is in quadratic mean. If, in addition to being stationary, the random process is ergodic, then $\hat{f} = Ef$ with probability one.

Comment. Note that if we define a new random process $\{Y_n\}$ by

$$Y_n(\omega) = f(T^n \omega) \;,$$

then the theorem states that the sample average of the Y_n, $\dfrac{1}{n} \sum_{i=0}^{n-1} Y_i$, converges. An important generality of this result is that if the X process meets the given conditions, then *all* sample averages of processes formed in this way from the X process converge. Intuitively, all sequences of measurements on a stationary process have convergent sample averages.

To make the theorem more concrete, we cite three examples. If $f(\omega) = x_0$, then the sample average as in theorem 7.1 results. If $f(\omega) = x_0^2$, then the sample average second moment results. If $f(\omega) = x_0 x_k$, then the sample average autocorrelation of lag k results.

Theorem 7.3 is very powerful and reasonably general. The main difficulties lie in the verification of the ergodic property and stationarity. Indeed, it is probably this theorem that led to the misconception that ergodic processes have to be stationary. As we have shown, this is not true. Random processes need only be stationary if theorem 7.3 is to be applied. Finally, we remark that for a *stationary* random process, ergodicity as we have defined it is equivalent to the requirement that *all* sample averages with finite expectation converge to a constant. Obviously this definition is more limited because of the stationarity requirement.

EXERCISES

7.1 Show that if $\{X_t; t \in I\}$ is a stationary random process and $f: \mathbf{R} \to \mathbf{R}$ is a (measurable) function, then the random process $\{f(X_t); t \in I\}$ is also stationary.

7.2 Is the process of exercise 6.12 weakly stationary? Strictly stationary?

7.3 Is the amplitude-modulated waveform of exercise 6.13 weakly stationary if the original process is? Is the same conclusion true if the uniform random variable Θ is removed from inside the cosine?

7.4 Say that X and Y are independent random variables, both with a pmf $p(x) = 2/3$ for $x = -1$ and $p(x) = 1/3$ for $x = 2$. Define the random process $Z(t) = Y\cos t + X\sin t$. Show that the process is weakly stationary but not stationary.

7.5   Prove equations (7.11) and (7.12) for the mean and covariance functions of a Gaussian random process, $\{X_t\}$, from the defining integrals by introducing the transformation $\mathbf{y} = \mathbf{x} - \mathbf{m}$ in evaluating the vector expectation $E\mathbf{X}$ and the matrix expectation $E[(\mathbf{X} - E\mathbf{X})(\mathbf{X} - E\mathbf{X})^t]$.

7.6   Let $\{X_n; n \in \mathbf{Z}\}$ be a two-sided discrete time random process. A general causal nonlinear filter can be modeled as a mapping f taking input sequences into an output random process defined by

$$Y_n = f(X_n, X_{n-1}, X_{n-2}, \ldots) .$$

Suppose that the input X process is stationary. Show that the output Y process is also stationary.

7.7   Suppose that you have a random process on $(\mathbf{R}^I, \mathbf{B}(\mathbf{R})^I, m)$ with elementary events $\omega = (\ldots, x_{-1}, x_0, x_1, \ldots)$. The process distribution is defined in terms of three parameters as follows:

$$m(\omega = (\ldots 0,0,0,0,\ldots)) = p$$

$$m(\omega = (\ldots 1,1,1,1,\ldots)) = q$$

$$m(\omega = (\ldots 0,1,0,1,\ldots)) = r$$

$$m(\omega = (\ldots 1,0,1,0,\ldots)) = 1 - p - q - r .$$

Find values of the parameters such that the process is neither stationary nor ergodic, stationary but not ergodic, ergodic but not stationary, and both stationary and ergodic.

7.8   Suppose that a discrete time process $\{X_n\}$ is stationary and strongly mixing. Let F be an arbitrary one-dimensional Borel set and let 1_F denote its indicator function. Prove that

$$\underset{n \to \infty}{\text{l.i.m.}} \frac{1}{n} \sum_{i=0}^{n-1} 1_F(X_i) = \Pr(X_0 \epsilon F) .$$

Hint: Consider theorem 7.1 and the random process $\{Y_n\}$ defined by $Y_n = 1_F(X_n)$.

7.9   Show that limits in the quadratic mean are linear; that is, if $\underset{n \to \infty}{\text{l.i.m.}} X_n = \hat{X}$ and $\underset{n \to \infty}{\text{l.i.m.}} Y_n = \hat{Y}$, then for any real constants a and b,

$$\underset{n \to \infty}{\text{l.i.m.}} (aX_n + bY_n) = a\hat{X} + b\hat{Y} .$$

Use this result together with the mean ergodic theorem for uncorrelated processes to prove the following mean ergodic theorem: If $\{X_n\}$ is uncorrelated and has mean m and if Y_n is a random process obtained by passing $\{X_n\}$ through a linear filter with pulse response h_k with $h_0=1$ and $h_1=r$ and $h_k=0$ for $k>1$, then there is a constant c such that

$$\text{l.i.m.}_{n \to \infty} \frac{1}{n} \sum_{i=0}^{n-1} Y_n = c .$$

Evaluate c. As a check, obtain the same answer using theorem 7.1.

8

SECOND-ORDER MOMENTS
AND LINEAR SYSTEMS

In chapters 6 and 7 we have seen that the second-order moments of a random process—the mean and covariance—play a fundamental role in describing the relation of limiting sample averages and expectations. We have seen, e.g., in exercise 6.16, and we shall see again that these moments also play a key role in other aspects of random processes, linear least squares estimation in particular. Because of the fundamental importance of these particular moments, this chapter is devoted to their computation for several important examples. The primary focus is on a second-order moment analog of a derived distribution problem: Suppose that we are given the second-order moments of one random process and that this process is then used as an input to a linear system; what are the resulting second-order moments of the output random process? These results are collectively known as second-order moment input/output or I/O relations for linear systems.

Linear systems may seem to be a very special case. As we will see, their most obvious attribute is that they are easier to handle analytically, which leads to more complete, useful, and stronger results than can be obtained for the class of all systems. This special case, however, plays a central role and is by far the most important class of systems. The design of engineering systems frequently involves the determination of an optimum system—perhaps the optimum signal detector for a signal in noise, the filter that provides the highest signal-to-noise ratio, the optimum receiver, etc. Surprisingly enough, the optimum is frequently a linear system. Even when the optimum is not linear, often a linear system is a good enough approximation to the optimal system so that a linear system is used for the sake of economical design. For these reasons it is of interest to study the properties of the output random process from a linear system that is driven by a specified input random process. In this chapter we consider only second-order moments; in the next chapter we consider examples in which one can develop a more complete probabilistic description of the output process. As one might suspect, the less complete second-order descriptions are possible under far more general conditions.

With the knowledge of the second-order properties of the output process when a linear system is driven by a given random process, one will have the fundamental tools for the analysis and optimization of such linear systems. As an example of such analysis, the chapter closes with an application of second-order moment theory to the design of systems for linear least squares estimation.

Because the primary engineering application of these systems is to noise discrimination, we will group them together under the name "linear filters." This designation denotes the suppression or "filtering out" of noise from the combination of signal and noise. The methods of analysis are not limited to this application, of course.

As usual, we emphasize discrete time in the development, with the obvious extensions to continuous time provided by integrals. Furthermore, we restrict attention in the basic development to linear time-invariant filters. The extension to time-varying systems is obvious but cluttered with obfuscating notation. Time-varying systems will be encountered briefly in the final section, when considering recursive estimation.

LINEAR FILTERING OF RANDOM PROCESSES

Suppose that a random process $\{X(t); t \in \boldsymbol{I}\}$ (or $\{X_t; t \in \boldsymbol{I}\}$) is used as an input to a linear time-invariant system described by a pulse response h. Hence the output process, say $\{Y(t)\}$ or $\{Y_t\}$, is described by the convolution integral of (2.22) in the continuous time case or the convolution sum of (2.29) in the discrete time case. To be precise, we have to be careful about how the integral or sum is defined; that is, integrals and infinite sums of random processes are really limits of random variables, and those limits can converge in a variety of ways, such as quadratic mean or with probability one. For the moment we will assume that the convergence is pointwise (that is, with probability one), i.e., that each realization or sample function of the output is related to the corresponding realization of the input via (2.22) or (2.29). That is, we take

$$Y(t) = \int\limits_{s:\, t-s \in I} X(t-s)h(s)ds \tag{8.1}$$

or

$$Y_n = \sum_{k:\, n-k \in I} X_{n-k}h_k \tag{8.2}$$

to mean actual equality for all elementary events ω on the underlying probability space Ω. More precisely, with probability one,

$$Y(t,\omega) = \int\limits_{s:\, t-s \in I} X(t-s,\omega)h(s)ds$$

or

$$Y_n(\omega) = \sum_{k:\, n-k \in I} X_{n-k}(\omega)h_k \;,$$

respectively. Rigorous consideration of conditions under which the various limits exist is straightforward for the discrete time case. It is obvious that the limits exist for the so-called finite impulse response (FIR) discrete time filters where only a finite number of the h_k are nonzero and hence the sum has only a finite number of terms. It is also possible to show mean square convergence for the general discrete time convolution if the input process has finite mean and variance and if the filter is stable in the sense of (2.30). In particular, for a two-sided input process, (8.2) converges in quadratic mean; i.e.,

$$\underset{N \to \infty}{\text{l.i.m.}} \sum_{k=0}^{N} X_{n-k}h_k$$

exists for all n. Then the application of lemma 6.4 can be used to establish probability one convergence if convergence of (2.30) for the pulse response is fast enough. The theory is far more complicated in the continuous time case. As usual, we will by and large ignore these problems and just assume that the convolutions are well defined.

Unfortunately, (2.24) and (2.30) are not satisfied for most sample functions of most random processes and hence in general one cannot take Fourier transforms of both sides of (8.1) and (8.2) and obtain a useful spectral relation. Even if one could, the Fourier transform of a random process would be a random variable for each value of frequency! Because of this, the frequency domain theory for random processes is quite different from that for deterministic processes. Relations such as (2.26) may on occasion be useful for intuition, but they must be used with extreme care.

With the foregoing notation and preliminary considerations, we now turn to the analysis of discrete time linear filters with random process inputs.

DISCRETE TIME SECOND-ORDER I/O RELATIONS

Ideally one would like to have a complete specification of the output of a linear system as a function of the specification of the input random process. Usually this is a difficult proposition because of the complexity of the computations required. However, it is a relatively easy task to determine the mean and covariance function at the output. As we will show, the output mean and covariance function depend only on the input mean and covariance function and on no other properties of the input random process. Furthermore, in many, if not most, applications, the mean and covariance functions of the output are all that are needed to solve the problem at hand.

Linear filter input/output (I/O) relations are most easily developed using the convolution representation of a linear system. Let $\{X_n\}$ be a discrete time random process with mean function $m_n = EX_n$ and covariance function $K_X(n,k) = E[(X_n - m_n)(X_k - m_k)]$. Let $\{h_k\}$ be the unit pulse response of a discrete time linear filter. For notational convenience we assume that the pulse response is causal. The noncausal case simply involves a change of the limits of summation. Next we will find the mean and covariance functions for the output process $\{Y_n\}$ that is given in the convolution equation of (8.2).

From (8.2) the mean of the output process is found using the linearity of expectation as

$$EY_n = \sum_k h_k EX_{n-k} = \sum_k h_k m_{n-k} , \qquad (8.3)$$

assuming, of course, that the sum exists. The sum does exist if the filter is stable and the input mean is bounded. That is, if there is a constant, $m < \infty$, such that $|m_n| \le |m|$ for all n and if the filter is stable in the sense of equation (2.30), then

$$|EY_n| = |\sum_k h_k m_{n-k}| \le \max_k |m_{n-k}| \sum_k |h_k|$$

$$\le |m| \sum_k |h_k| < \infty .$$

If the input process $\{X_n\}$ is weakly stationary, then the input mean function equals the constant, m, and

$$EY_n = m \sum_k h_k , \qquad (8.4a)$$

which is the dc response of the filter times the mean. For reference we specify the precise limits for the two-sided random process where $I = Z$ and for the one-sided input random process where $I = Z_+$:

$$EY_n = m \sum_{k=0}^{\infty} h_k \ , \ \boldsymbol{I} = \boldsymbol{Z} \tag{8.4b}$$

$$EY_n = m \sum_{k=0}^{n} h_k \ , \ \boldsymbol{I} = \boldsymbol{Z}_+ \ . \tag{8.4c}$$

Thus, for weakly stationary input random processes, the output mean exists if the input mean is finite and the filter is stable. In addition, it can be seen that for two-sided weakly stationary random processes, the expected value of the output process does not depend on the time index n since then the limits of the summation do not depend on n. For one-sided weakly stationary random processes, however, the output mean is not constant with time but approaches a constant value as $n \to \infty$ if the filter is stable. Note that this means that if a one-sided stationary process is put into a linear filter, the output is in general not stationary!

If the filter is not stable, the magnitude of the output mean is unbounded with time. For example, if we set $h_k = 1$ for all k in (8.4c), then $EY_n = (n+1)m$, which very strongly depends on the time index n and which is unbounded.

Turning to the calculation of the output covariance function, we use equations (8.2) and (8.3) to evaluate the covariance with some bookkeeping as

$$
\begin{aligned}
K_Y(k,j) &= E[(Y_k - EY_k)(Y_j - EY_j)] \\
&= E[\left(\sum_n h_n(X_{k-n} - m_{k-n})\right)\left(\sum_m h_m(X_{j-m} - m_{j-m})\right)] \\
&= \sum_n \sum_m h_n h_m E[(X_{k-n} - m_{k-n})(X_{j-m} - m_{j-m})] \\
&= \sum_n \sum_m h_n h_m K_X(k-n, j-m) \ .
\end{aligned}
\tag{8.5}
$$

As before, the range of the sums depends on the index set used. Since we have specified causal filters, the sums run from 0 to ∞ for two-sided processes and from 0 to k and 0 to j for one-sided random processes.

It can be shown that the sum of (8.5) converges if the filter is stable in the sense of (2.30) and if the input process has bounded variance; i.e., there is a constant $\sigma^2 < \infty$ such that $|K_X(n,n)| < \sigma^2$ for all n (exercise 8.17).

If the input process is weakly stationary, then K_X depends only on the difference of its arguments. This is made explicit by replacing $K_X(m,n)$ by $K_X(m-n)$. Then (8.5) becomes

$$K_Y(k,j) = \sum_n \sum_m h_n h_m K_X((k-j) - (n-m)) \ . \tag{8.6a}$$

Specifying the limits of the summation for the one-sided and two-sided cases, we have that

$$K_Y(k,j) = \sum_{n=0}^{\infty} \sum_{m=0}^{\infty} h_n h_m K_X((k-j)-(n-m)); \; \boldsymbol{I} = \mathbf{Z} \qquad (8.6b)$$

and

$$K_Y(k,j) = \sum_{n=0}^{k} \sum_{m=0}^{j} h_n h_m K_X((k-j)-(n-m)); \; \boldsymbol{I} = \mathbf{Z}_+ . \qquad (8.6c)$$

If the sum of (8.6b) converges (e.g., if the filter is stable and $K_X(n,n)$ = $K_X(0) < \infty$), then two interesting facts follow: First, if the input random process is weakly stationary and if the processes are two-sided, then the covariance of the output process depends only on the time lag; i.e., $K_Y(k,j)$ can be replaced by $K_Y(k-j)$. Note that this is not the case for a one-sided process, even if the input process is stationary and the filter stable! This fact, together with our earlier result regarding the mean, can be summarized as follows:

Given a two-sided random process as input to a linear filter, if the input process is weakly stationary and the filter is stable, the output random process is also weakly stationary. The output mean and covariance functions are given by

$$EY_n = m \sum_{k=0}^{\infty} h_k \qquad (8.7a)$$

$$K_Y(k) = \sum_{n=0}^{\infty} \sum_{m=0}^{\infty} h_n h_m K_X(k-(n-m)) . \qquad (8.7b)$$

The second observation is that (8.6) or (8.7b) is a double discrete convolution! The direct evaluation of (8.6), while straightforward in concept, can be an exceedingly involved computation in practice. As we will see, the computations can be simplified by using transform techniques. Before turning to transform techniques, however, we consider a special (and very important) case where the evaluation is not too involved.

LINEARLY FILTERED UNCORRELATED PROCESSES

If the input process $\{X_n\}$ to a discrete time linear filter with pulse response $\{h_k\}$ is an uncorrelated process with mean m and variance σ^2 (for example, if it is i.i.d.), then $K_X(k) = \sigma^2 \delta_k$. For the two-sided case we have easily from (8.7b) that

$$K_Y(k) = \sigma^2 \sum_{n=0}^{\infty} h_n h_{n-k} = \sigma^2 \sum_{n=k}^{\infty} h_n h_{n-k}; \; \boldsymbol{I} = \mathbf{Z} , \qquad (8.8a)$$

where the lower limit of the sum follows from the causality of the filter. For a one-sided process, (8.6c) yields

$$K_Y(k,j) = \sigma^2 \sum_{n=0}^{k} h_n h_{n-(k-j)}; \quad I = Z_+ . \tag{8.8b}$$

Note that if $k > j$, then the sum can be taken over the limits $n = k - j$ to k since causality of the filter implies that the first few terms are 0. If $k < j$, then all of the terms in the sum may be needed. The covariance for the one-sided case appears to be asymmetric, but recalling that h_l is 0 for negative l, we can write the terms of the sum of (8.8b) in descending order to obtain

$$\sigma^2(h_k h_j + h_{k-1} h_{j-1} + \dots + h_0 h_{j-k})$$

if $j \geq k$ and

$$\sigma^2(h_k h_j + h_{k-1} h_{j-1} + \dots + h_{k-j} h_0)$$

if $j \leq k$. By defining the function $\min(k,j)$ to be the smaller of k and j, we can rewrite (8.8b) in two symmetric forms:

$$K_Y(k,j) = \sigma^2 \sum_{n=0}^{\min(k,j)} h_{k-n} h_{j-n}; \quad I = Z_+ \tag{8.8c}$$

and

$$K_Y(k,j) = \sigma^2 \sum_{n=0}^{\min(k,j)} h_n h_{n+|k-j|} . \tag{8.8d}$$

In the two-sided case, the expression (8.8a) for the output covariance function is the convolution of the unit pulse response with its *reflection* h_{-k}. Such a convolution between a waveform or sequence and its own reflection is also called a sample autocorrelation.

We next consider specific examples of this computation. These examples point out that two processes—one one-sided and the other two-sided—that are apparently similar can have quite different properties.

[8.1]

Suppose that an uncorrelated discrete time two-sided random process $\{X_n\}$ with mean m and variance σ^2 is put into a linear filter with causal pulse response $h_k = r^k$, $k \geq 0$, with $|r| < 1$. Let $\{Y_n\}$ denote the output process. Find the output mean and covariance.

From the geometric series summation formula,

$$\sum_{k=0}^{\infty} |r|^k = \frac{1}{1-|r|} ,$$

and hence the filter is stable. From (8.4),

$$EY_n = m \sum_{k=0}^{\infty} r^k = \frac{m}{1-r}; \quad n \in Z .$$

From (8.8a), the output covariance for nonnegative k is

$$K_Y(k) = \sigma^2 \sum_{n=k}^{\infty} r^n r^{n-k}$$

$$= \sigma^2 r^{-k} \sum_{n=k}^{\infty} (r^2)^n = \sigma^2 \frac{r^k}{1-r^2}$$

using the geometric series formula. Repeating the development for negative k (or appealing to symmetry) we find that in general the covariance function is

$$K_Y(k) = \sigma^2 \frac{r^{|k|}}{1-r^2} ; \; k \in \mathbf{Z} .$$

Observe in particular that the output variance is

$$\sigma_Y^2 = K_Y(0) = \frac{\sigma^2}{1-r^2} .$$

As $|r| \to 1$ the output variance grows without bound. However, as long as $|r| < 1$, the variance is defined and the process is clearly weakly stationary.

[8.2]

Suppose that a one-sided uncorrelated process $\{X_n\}$ with mean m and variance σ^2 is put into a one-sided filter with pulse response as in example [8.1]. Let $\{Y_n\}$ be the resulting one-sided output process. Find the output mean and covariance.

This time (8.4) yields

$$EY_n = m \sum_{k=0}^{n} r^k = m \frac{1-r^{n+1}}{1-r}$$

from the geometric series formula. From (8.8), the covariance is

$$K_Y(k,j) = \sigma^2 \sum_{n=0}^{\min(k,j)} r^{2n+|k-j|} = \sigma^2 r^{|k-j|} \frac{1-r^{2(\min(k,j)+1)}}{1-r^2} .$$

Observe that since $|r| < 1$, if we let $n \to \infty$, then the mean of this example goes to the mean of the preceding example in the limit. Similarly, if one fixes the lag $|k-j|$ and lets k (and hence j) go to ∞, then in the limit the one-sided covariance looks like the two-sided example. This simple example points out a typical form of nonstationarity: A linearly filtered uncorrelated process is not stationary by any definition, but as one gets farther and farther from the origin, the parameters look more and more stationary. This can be considered as a form of asymptotic stationarity. In fact, a process is defined as being asymptotically weakly stationary if the mean and covariance converge in the sense just given. One can view such processes as having transients that die out with time. It is not difficult to show that if a process is asymptotically weakly stationary and if the limiting mean and covariance meet the conditions of the ergodic theorem, then the process itself will satisfy the ergodic theorem. Intuitively stated, transients do not affect the long-term sample averages.

[8.3]

Next consider the one-sided process of example [8.2], but now choose the pulse response with $r = 1$; that is, $h_k = 1$ for all $k \geq 0$. Find the output mean and covariance. (Note that this filter is *not* stable.)

Applying (8.4) and (8.8) yields

$$EY_n = m \sum_{k=0}^{n} h_k = m(n+1)$$

and

$$K_Y(k,j) = \sigma^2(\min(k,j)+1) = \sigma^2 \min(k+1, j+1) .$$

Observe that like example [8.2], the process of example [8.3] is not weakly stationary. Unlike [8.2], however, it does not behave asymptotically like a weakly stationary process—even for large time, the moments very much depend on the time origin. Thus the nonstationarities of this process are not only transients—they last forever! In a sense, this process is much more nonstationary than the previous one and, in fact, does not have a mean ergodic theorem. If the input process is Gaussian with zero mean, then we will see in chapter 9 that the output process $\{Y_n\}$ is also Gaussian. Such a Gaussian process with zero mean and with the covariance function of this example is called the discrete time *Wiener process*. We will see more of Wiener processes in chapter 9.

CONTINUOUS TIME LINEAR FILTERS

For each of the discrete time filter results there is an analogous continuous time result. We now develop some of these analogs. For simplicity, however, we consider only the simpler case of two-sided processes. Let $\{X(t)\}$ be a two-sided continuous time input random process to a linear time-invariant filter with impulse response $h(t)$.

We can evaluate the mean and covariance functions of the output process in terms of the mean and covariance functions of the input random process by using the same development as was used for discrete random processes. This time we will have integrals instead of sums. Let $m(t)$ and $K_X(t,s)$ be the respective mean and covariance functions of the input process. Then the mean function of the output process is

$$EY(t) = \int E[X(t-s)]h(s)ds = \int m(t-s)h(s)ds . \qquad (8.9)$$

The covariance function of the output random process is obtained by computations analogous to (8.5) as

$$K_Y(t,s) = \int d\alpha \int d\beta K_X(t-\alpha,s-\beta)h(\alpha)h(\beta) . \qquad (8.10)$$

Thus if $\{X(t)\}$ is weakly stationary with mean $m = m(t)$ and covariance function $K_X(\tau)$, then

$$EY(t) = m \int h(t)dt \qquad (8.11)$$

and

$$K_Y(t,s) = \int d\alpha \int d\beta K_X((t-s)-(\alpha-\beta))h(\alpha)h(\beta) . \qquad (8.12)$$

In analogy to the discrete time result, the output mean is constant for a two-sided random process, and the covariance function depends on only the time difference. Thus a weakly stationary two-sided process into a stable linear time-invariant filter yields a weakly stationary output process in both discrete and continuous time. We leave it to the reader to develop conclusions that are parallel to the discrete time results for one-sided processes.

LINEAR MODULATION

In this section we consider a different form of linear system: a linear modulator. Unlike the filters considered thus far, these systems are generally time-varying and contain random parameters. They are simpler than the general linear filters, however, in that the output depends on the input in an instantaneous fashion; that is, the output at time t depends only on the input at time t and not on previous inputs.

In general, the word *modulation* means the methodical altering of one waveform by another. The waveform being altered is often called a carrier, and the waveform or sequence doing the altering, which we will model as a random process, is called the signal. Physically, such modulation is usually done to transform an information-bearing signal into a process suitable for communication over a particular medium; e.g., simple amplitude modulation of a carrier sinusoid by a signal in order to take advantage of the fact that the resulting high-frequency signals will better propagate through the atmosphere than will audio frequencies.

The emphasis will be on continuous time random processes since most communication systems involve at some point such a continuous time link. Several of the techniques, however, work virtually without change in a discrete environment.

The prime example of linear modulation is the ubiquitous amplitude modulation or AM used for much of commercial broadcasting. If $\{X(t)\}$ is a continuous time weakly stationary random process with zero mean and covariance function $K_X(\tau)$, then the output process

$$Y(t) = (a_0 + a_1 X(t))\cos(2\pi f t + \theta) \qquad (8.13)$$

is called amplitude modulation of the cosine by the original process. The parameters a_0 and a_1 are called modulation constants. Observe that linear modulation is not a linear operation in the normal linear systems sense unless the constant a_0 is 0. (It is, however, an *affine* operation—linear in the sense that straight lines in the two-dimensional $x - y$ space are said to be linear.) Nonetheless, as is commonly done, we will refer to this operation as linear modulation.

The phase term θ may be a fixed constant or a random variable, say Θ. (We point out a subtle source of confusion here: If Θ is a random variable, then the system is affine or linear for the input process only when the actual sample value, say θ, of Θ is known.) We usually assume for convenience that Θ is a random variable, independent of the X process and uniformly distributed on $[0, 2\pi]$—one complete rotation of the carrier phasor in the complex plane. This is a mathematical convenience, that, as we will see, makes $Y(t)$ weakly stationary. Physically it corresponds to the simple notion that we are modeling the modulated waveform as seen by a receiver. Such a receiver will not know *a priori* the phase of the transmitter oscillator producing the sinusoid. Furthermore, although the transmitted phase could be monitored and related to the signal as part of the transmission process, this is never done with AM. Hence, so far as the receiver is concerned, the phase is equally likely to be anything; that is, it has a uniform distribution independent of the signal.

If $a_0 = 0$, the modulated process is called double sideband suppressed carrier (DSB or DSB-SC). The a_0 term clearly wastes power, but it makes for easier and cheaper recovery or demodulation of the original process, as explained in any text on elementary communication theory. Our goal here is only to look at the second-order properties of the AM process.

Observe that for any fixed phase angle, say $\Theta = 0$ for convenience, a system taking a waveform and producing the DSB modulated waveform is indeed linear in the usual linear systems sense. It is actually simpler than the output of a general linear filter since the output at a given time depends only on the input at that time.

Since Θ and the X process are independent, we have that the mean of the output is

$$EY(t) = (a_0 + a_1 EX(t)) E\cos(2\pi f t + \Theta) .$$

But Θ is uniformly distributed. Thus for any fixed time and frequency,

$$E\cos(2\pi ft + \Theta) = \int_0^{2\pi} \cos(2\pi ft + \theta)\frac{d\theta}{2\pi}$$

$$= \frac{1}{2\pi} \int_0^{2\pi} \cos(2\pi ft + \theta)d\theta = 0 \qquad (8.14)$$

since the integral of a sinusoid over a period is zero; hence $EY(t) = 0$ whether or not the original signal has zero mean.

The covariance function of the output is given by the following expansion of the product $Y(t)Y(s)$ using (8.13):

$$K_Y(t,s) = a_0^2 E[\cos(2\pi ft + \Theta)\cos(2\pi fs + \Theta)] +$$

$$a_0 a_1 (EX(t)E[\cos(2\pi ft + \Theta)\cos(2\pi fs + \Theta)] +$$

$$a_0 a_1 EX(s)E[\cos(2\pi ft + \Theta)\cos(2\pi fs + \Theta)])$$

$$+ a_1^2 K_X(t,s)E[(\cos 2\pi ft + \Theta)(\cos 2\pi fs + \Theta)] .$$

Using the fact that the original process has zero mean eliminates the middle lines in the preceding. Combining the remaining two terms and using the cosine identity,

$$\cos x \cos y = \frac{1}{2}\cos(x + y) + \frac{1}{2}\cos(x - y) ,$$

$$K_Y(t,s) = (a_0^2 + a_1^2 K_X(t,s))$$

$$\left[\frac{1}{2}E\cos(2\pi f(t + s) + 2\Theta) + \frac{1}{2}E\cos(2\pi f(t - s)) \right] .$$

Exactly as in the mean computation of (8.14), the expectation of the term with the Θ in it is zero, leaving

$$K_Y(\tau) = \frac{1}{2}(a_0^2 + a_1^2 K_X(\tau))\cos(2\pi f \tau) .$$

Thus we have demonstrated that amplitude modulation of a carrier by a weakly stationary random process results in an output that is weakly stationary. In the next section we will study the frequency characteristics of linear filters, including AM.

POWER SPECTRAL DENSITIES

We now return to an issue raised earlier: We wish to avoid the complications of computing the multiple convolutions required in (8.7b) and (8.12) to find the second-order properties of the two-sided output process of a linear filter driven by a random process. Throughout this section we consider only two-sided processes since it is in this case that Fourier techniques are most useful. In both discrete and continuous time, the covariance function of the output can be found by first convolving the input autocorrelation with the pulse response h_k or $h(t)$ and then convolving the result with the reflected pulse response h_{-k} or $h(-t)$. A way of avoiding the double convolution is found in Fourier transforms. Taking the Fourier transform (continuous or discrete time) of the double convolution yields the transform of the covariance function, which can be used to arrive at the output covariance function— essentially the same result with (in many cases) less overall work. We will show the development for discrete time. Using (8.7b),

$$\mathbf{F}(\{K_Y\}) = \sum_k (\sum_n \sum_m h_n h_m K_X(k-(n-m)))e^{-j2\pi fk}$$

$$= \sum_n \sum_m h_n h_m (\sum_k K_X(k-(n-m))e^{-j2\pi f(k-(n-m))})e^{-j2\pi f(n-m)}$$

$$= (\sum_n h_n e^{-j2\pi fn})(\sum_m h_m e^{+j2\pi fm})\mathbf{F}(K_X)$$

$$= \mathbf{F}(\{K_X\})\mathbf{F}(\{h\})\mathbf{F}(\{h\})^* ,$$

where the brackets are used to emphasize that the Fourier transform depends on the argument for all values of time; that is, on the entire function, and the asterix denotes complex conjugate. In this development we have shown implicitly that convolution in the time domain corresponds to multiplication in the frequency domain and that the transform of a reflected time function is the complex conjugate of the transform of the unreflected function. Motivated by this formula, we give a symbol and a name to the transform of a covariance function:

Given a weakly stationary random process $\{X_t\}$ with covariance function $K_X(\tau)$, the *power spectral density* $S_X(f)$ of the process is defined as the Fourier transform of the covariance function; that is,

$$S_X(f) = \mathbf{F}(\{K_X\}) = \begin{cases} \sum_k K_X(k)e^{-j2\pi fk}, & \text{discrete time} \\ \int K_X(\tau)e^{-j2\pi f\tau}d\tau, & \text{continuous time} . \end{cases} \qquad (8.15)$$

Combining these observations and definition we have the following:

Let $\{X_t\}$ be the input random process to a linear time-invariant filter with transfer function $H(f)$. If $\{X_t\}$ has spectral density $S_X(f)$, then the output random process $\{Y_t\}$ has spectral density

$$S_Y(f) = |H(f)|^2 S_X(f) . \tag{8.16}$$

Under suitable technical conditions the Fourier transform can be inverted to obtain the covariance function from the power spectral density. Since frequency is always a continuous dummy variable, the inversion is an integral transform in both continuous and discrete time. The limits of integration are different, however. Note in (8.15) that the discrete time spectral density is periodic in f with period equal to unity. Therefore the power spectral density can be inverted by integration over a single period. Thus the reader can verify from the definitions of (8.15) that

$$K_X(\tau) = \begin{cases} \int_{-1/2}^{1/2} S_X(f)e^{j2\pi f\tau}df, & \text{discrete time, integer } \tau \\ \int_{-\infty}^{\infty} S_X(f)e^{j2\pi f\tau}df, & \text{continuous time, continuous } \tau . \end{cases} \tag{8.17}$$

The limits of -1/2 to +1/2 for the discrete time integral correspond to the fact that time is measured in units; e.g., adjacent outputs are one second or one minute or one year apart. Sometimes, however, the discrete time process is formed by sampling a continuous time process at every, say, T seconds, and it is desired to retain seconds as the unit of measurement. Then it is more convenient to incorporate the scale factor T into the time units and scale (8.15) and the limits of (8.17) accordingly—i.e., kT replaces k in (8.15), and the limits become $-1/2T$ to $1/2T$.

The result just given is equivalent to, but in many applications much simpler than, the double convolution formula for obtaining the output covariance function from the input covariance function and pulse response. The result holds for both discrete and continuous time.

As a first trivial example, suppose that a discrete time random process $\{X_n\}$ is uncorrelated so that $K_X(k) = \sigma^2 \delta_k$ and hence

$$S_X(f) = \sum_k \sigma^2 \delta_k e^{-j2\pi fk} = \sigma^2; \text{ all } f ,$$

since the only nonzero term in the sum is the $k=0$ term. Because the power spectral density is flat, in analogy to the flat electromagnetic spectrum of white light, such a process is said to be *white* or *white noise*. The inverse Fourier transform of the white noise spectral density is easily found from (8.17) (or simply by uniqueness) to be $K_X(k) = \sigma^2 \delta_k$. Thus a discrete time random process is white if and only if it is uncorrelated.

As a more interesting example, consider the power spectral density of example [8.1]. The output spectral density can be found directly by taking the Fourier transform of the output covariance as

$$S_Y(f) = \sum_{k=-\infty}^{\infty} \frac{\sigma^2 r^{|k|}}{1-r^2} e^{-j2\pi fk} ,$$

a summation that can be evaluated using the geometric series formula—first from 1 to ∞ and then from 0 to $-\infty$—and then summing the two complex terms. The reader should perform this calculation as an exercise. It is easier, however, to find the output spectral density through the linear system I/O relation. The transfer function of the filter is evaluated by a single application of the geometric series formula as

$$H(f) = \sum_{k=0}^{\infty} r^k e^{-j2\pi fk} = \frac{1}{1-re^{-j2\pi f}} .$$

Therefore the output spectral density from (8.16) is

$$S_Y(f) = \frac{\sigma^2}{|1-re^{-2\pi f}|^2} = \frac{\sigma^2}{1+r^2-2r\cos(2\pi f)} .$$

By a quick table lookup the reader can verify that the inverse transform of the output spectral density agrees with the covariance function previously found.

Power spectral densities inherit the property of symmetry from covariance functions. As seen from the definition in chapter 6, covariance functions are symmetric $(K_X(t,s)=K_X(s,t))$ Therefore $K_X(\tau)$ is an even function. From (8.15) it can be seen with a little juggling that $S_X(f)$ is also symmetric; that is, $S_X(-f) = S_X(f)$ for all f.

Another property that spectral densities inherit from covariance functions is the property of nonnegativity. The fact that covariance functions are nonnegative definite implies that power spectral densities are nonnegative functions; that is, $S_X(f) \geq 0$. The general result is a deep and difficult result of Fourier theory known as Bochner's theorem. However, that the result is true can be established by the following simple considerations. If $EX_t = 0$ then

$$\int S_X(f)df = K_X(0) = E[X_t^2] , \qquad (8.18)$$

the average power across a unit resistor if the process corresponds to a voltage. This integral is clearly nonnegative. Now suppose that we wish to find the power of a process, say $\{X_t\}$ in some frequency band $f \in F$. Then a physically natural way to accomplish this would be to pass the given process through a bandpass filter with transfer function $H(f)$ equal to 1 for $f \in F$ and 0 otherwise and then to measure the output power. This is depicted in Figure 8.1 for the special case of a frequency interval $F = \{f: f_0 \leq |f| < f_0 + \Delta f\}$. Calling the output process $\{Y_t\}$, we have from (8.18) that the output power is

$$K_Y(0) = \int S_Y(f)df = \int |H(f)|^2 S_X(f)df = \int_F S_X(f)df . \qquad (8.19)$$

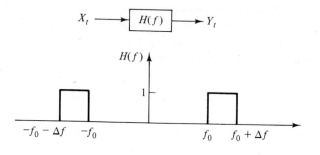

$$E(Y_t^2) = \int S_x(f) \, df$$
$$f: \; f_0 \le |f| \le f_0 + \Delta f$$

$$= \text{power in } X_t \text{ in frequency}$$
$$\text{range } f_0 < |f| < f_0 + \Delta f.$$

Figure 8.1   Power spectral density.

Thus to find the average power contained in any frequency band we integrate the power spectral density over the frequency band. Because the average power must be nonnegative, it follows that any power spectral density must be nonnegative. To elaborate further, suppose that this is not true; i.e., suppose that $S_X(f)$ is negative for some range of frequencies. If we put $\{X_t\}$ through a filter that passes only those frequencies, the filter output power would have to be negative—clearly an impossibility.

From the foregoing considerations it can be deduced that the name *power spectral density* derives from the fact that $S_X(f)$ is a nonnegative function that is integrated to get power; that is, a "spectral" (meaning frequency content) density of power.

The power spectral density of the AM process that we considered in the section on linear modulation can be found directly by transforming the covariance function or by using standard Fourier techniques: The transform of a covariance function times a cosine is the convolution of the original power spectral density with the generalized Fourier transform of the cosine—that is, a pair of impulses. This yields a pair of replicas of the original power spectral densities, centered at plus and minus the carrier frequency f and symmetric about f, as depicted in Figure 8.2. If further filtering is desired, e.g., to remove one of the symmetric halves of the power spectral density to form single sideband modulation, then the usual linear techniques can be applied, as indicated by (8.16).

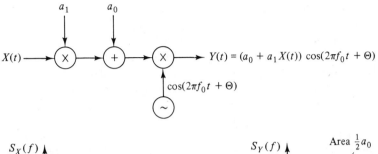

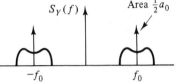

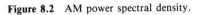

Figure 8.2   AM power spectral density.

WHITE NOISE

Let $\{X_n\}$ be an uncorrelated weakly stationary discrete time random process with zero mean. We saw in chapter 6 that for such a process the covariance function is a pulse at the origin; that is,

$$K_X(\tau) = \sigma_X^2 \delta_\tau \, ,$$

where δ_τ is a Kronecker delta function. As noted in the preceding section, taking the Fourier transform results in the spectral density

$$S_X(f) = \sigma_X^2; \text{ all } f \, ,$$

that is, the power spectral density of such a process is flat over the entire frequency range. We remarked that a process with such a flat spectrum is said to be white. We now make this definition formally for both discrete and continuous time processes:

A random process $\{X_t\}$ is said to be *white* if its power spectral density is a constant for all f. (A white process is also almost always assumed to have a zero mean, an assumption that we will make unless explicitly stated otherwise.)

The concept of white noise is clearly well defined and free of analytical difficulties in the discrete time case. In the continuous time case, however, there is a problem. Recall from (8.18) that the average power in a process is the integral of the power spectral density. In the discrete time case, integrating a constant over a finite range causes no problem. In the continuous time case, we find from (8.18) that a white noise process has infinite average power. In other words, if such a process existed, it would blow up the universe! A quick perusal of the stochastic systems literature shows, however, that this problem has not prevented continuous time white noise process models from being popular and useful. The resolution of the apparent paradox is fairly simple: Indeed, white noise is a physically impossible process. But there do exist noise sources that have a flat power spectral density over a range of frequencies that is much larger than the bandwidths of subsequent filters or measurement devices. In fact, this is exactly the case with the thermal noise process that we will treat in chapter 11. There a physics-based derivation shows that the random process has a covariance function of the form $K_X(\tau) = kTR\alpha e^{-\alpha|\tau|}$, where k is Boltzman's constant, T is the absolute temperature, and R and α are parameters of the physical medium. The application of (8.15) results in the power spectral density

$$S_X(f) = kTR\frac{2\alpha^2}{\alpha^2+(2\pi f)^2}.$$

As $\alpha \to \infty$, the power spectral density tends toward the value $2kTR$ for all f; that is, the process looks like white noise over a large bandwidth. Thus, for example, the total noise power in a bandwidth $(-B,B)$ is approximately $2kTR \times 2B$, a fact that has been verified closely by experiment.

If such a process is put into a filter having a transfer function whose magnitude becomes negligible long before the power spectral density of the input process decreases much, then the output process power spectral density $S_Y(f) = |H(f)|^2 S_X(f)$ will be approximately the same at the output as it would have been if $S_X(f)$ were flat forever since $S_X(f)$ is flat for all values of f where $|H(f)|$ is nonnegligible. Thus, so far as the output process is concerned, the input process can be either the physically impossible white noise model or a more realistic model with finite power. However, since the input white noise model is much simpler to work with analytically, it is usually adopted.

In summary, continuous time white noise is often a useful model for the input to a filter when we are trying to study the output. Commonly the input random process is represented as being white with flat spectral density equal to $N_0/2$. The factor of 2 is included because of the "two-sided" nature of filter transfer functions; viz. a low pass filter with cutoff frequency B applied to the white noise input will have output power equal to N_0B in accordance with (8.19). Such a white noise process makes mathematical sense, however, only if seen through a filter. The process itself is not rigorously defined. Its covariance function, however, can be represented in terms of a Dirac delta function for the purposes of analytical manipulations. Note that in (8.17) the generalized Fourier transform of the flat spectrum results in a Dirac delta function or unit impulse. In particular, if the continuous time white noise random process has power spectral density

$$S_X(f) = \frac{N_0}{2} ,$$

then it will have a covariance function

$$K_X(\tau) = \frac{N_0}{2}\delta(\tau) .$$

Thus adjacent samples of the random process are uncorrelated (and hence also independent if the process is Gaussian) *no matter how close together in time the samples are!* At the same time, the variance of a single sample is infinite. Clearly such behavior is physically impossible. It is reasonable, however, to state qualitatively that adjacent samples are uncorrelated at all times greater than the shortest time delays in subsequent filtering.

Perhaps the nicest attribute of white noise processes is the simple form of the output power spectral density of a linear filter driven by white noise. If a discrete or continuous time random process has power spectral density $S_X(f) = N_0/2$ for all f and it is put into a linear filter with transfer function $H(f)$, then from (8.16) the output process $\{Y_t\}$ has power spectral density

$$S_Y(f) = |H(f)|^2\frac{N_0}{2} . \tag{8.20}$$

The result given in (8.20) is of more importance than first appearances indicate. A basic result of the theory of weakly stationary random processes, called the spectral factorization theorem, states that if a random process $\{Y_t\}$ has a spectral density $S_Y(f)$ such that

$$\int \ln S_Y(f)df > -\infty \quad \text{(discrete time)}$$

or

$$\int \frac{\ln S_Y(f)}{1+f^2}df > -\infty \quad \text{(continuous time)} ,$$

then the power spectral density has the form of (8.20) for some *causal* linear stable time-invariant filter. That is, the second-order properties of any random process satisfying these conditions can be *modeled* as the output of a causal linear filter driven by white noise. Such random processes are said to be *physically realizable* and comprise all random processes seen in practice. This result is of extreme importance in estimation, detection, prediction, and system identification. We note in passing that in such models the white noise driving process is called the *innovations* process of the output process if the filter has a causal and stable inverse.

*DIFFERENTIATING RANDOM PROCESSES

We have said that linear systems can often be described by means other than convolution integrals, e.g., difference equations for discrete time and differential equations in continuous time. In this section we explore the I/O relations for a simple continuous time differentiator in order to demonstrate some of the techniques involved for handling such systems. In addition, the results developed will provide another interpretation of white noise.

Suppose now that we have a continuous time random process $\{X(t)\}$ and we form a new random process $\{Y(t)\}$ by differentiating; that is,

$$Y(t) = \frac{d}{dt}X(t) .$$

In this section we will take $\{X(t)\}$ to be a zero-mean random process for simplicity. Results for nonzero-mean processes are found by noting that $X(t)$ can be written as the sum of a zero-mean random process plus the mean function $m(t)$. That is, we can write $X(t) = X_0(t)+m(t)$ where $X_0(t) = X(t)-m(t)$. Then, the derivative of $X(t)$ is the derivative of a zero-mean random process plus the derivative of the mean function. The derivative of the mean function is a derivative in the usual sense and hence provides no special problems.

To be strictly correct, there is a problem in interpreting what the derivative means when the thing being differentiated is a random process. A derivative is defined as a limit, and as we found in chapter 6, there are several notions of limits of sequences of random variables. Care is required because the limit may exist in one sense but not necessarily in another. In particular, two natural definitions for the derivative of a random process correspond to convergence with probability one and convergence in mean square. As a first possibility we could assume that each sample function of $Y(t)$ is obtained by differentiating each sample function of $X(t)$; that is, we could use ordinary differentiation on the sample functions. This gives us a definition of the form

$$Y(t,\omega) = \frac{d}{dt}X(t,\omega)$$

$$= \lim_{\Delta t \to 0} \frac{X(t+\Delta t,\omega)-X(t,\omega)}{\Delta t} .$$

If $P(\{\omega\text{:the limit exists}\})=1$, then the definition of differentiation corresponds to convergence with probability one. Alternatively, we could define $Y(t)$ as a limit in quadratic mean of the random variables $\frac{X(t+\Delta t)-X(t)}{\Delta t}$ as Δt goes to zero (which does not require that the derivative exist with probability one on sample functions). With this definition we obtain

$$Y(t) \;=\; \underset{\Delta t \to 0}{\mathrm{l.i.m.}} \frac{X(t+\Delta t)-X(t)}{\Delta t} \; .$$

Clearly a choice of definition of derivative must be made in order to develop a theory for this simple problem and, more generally, for linear systems described by differential equations. We will completely avoid the issue here by sketching a development with the assumption that all of the derivatives exist as required. We will blithely ignore careful specification of conditions under which the formulas make sense. (Mathematicians sometimes refer to such derivations as formal developments: Techniques are used as if they are applicable and to see what happens. This often provides the answer to a problem, which, once known, can then be proved rigorously to be correct.)

Although we make no attempt to prove it, the result we will obtain can be shown to hold under sufficient regularity conditions on the process. In engineering applications these regularity conditions are almost always either satisfied, or if they are not satisfied, the answers that we obtain can be applied anyway, with care.

Formally define a process $\{Y_{\Delta t}(t)\}$ for a fixed Δt as the following difference, which approximates the derivative of $X(t)$:

$$Y_{\Delta t}(t) \;=\; \frac{X(t+\Delta t)-X(t)}{\Delta t} \; .$$

This difference process is perfectly well defined for any fixed $\Delta t > 0$ and in some sense it should converge to the desired $Y(t)$ as $\Delta t \to 0$. We can easily find the following correlation:

$$E[Y_{\Delta t}(t)Y_{\Delta s}(s)] \;=\; E[\frac{X(t+\Delta t)-X(t)}{\Delta t}\,\frac{X(s+\Delta s)-X(s)}{\Delta s}] \;=\;$$
$$\frac{R_X(t+\Delta t, s+\Delta s)-R_X(t+\Delta t, s)-R_X(t, s+\Delta s)+R_X(t,s)}{\Delta t\,\Delta s} \; .$$

If we now (formally) let Δt and Δs go to zero, then, if the various limits exist, this formula becomes

$$R_Y(t,s) \;=\; \frac{\partial}{\partial t\,\partial s} R_X(t,s) \; . \tag{8.21}$$

As previously remarked, we will not try to specify complete conditions under which this sleight of hand can be made rigorous. Suffice it to say that if the conditions on the X process are sufficiently strong, the formula is valid. Intuitively, since differentiation and expectation are linear operations, the formula follows from the assumption that the linear operations commute, as they usually do. There are, however, serious issues of existence involved in making the proof precise.

One obvious regularity condition to apply is that the double derivative of (8.21) exists. If it does and the processes are weakly stationary, then we can transform (8.21) by using the property of Fourier transforms that differentiation in the time domain corresponds to multiplication by f in the frequency domain. Then for the double derivative to exist we obtain the requirement that the spectral density of $\{X(t)\}$ have finite second moment; i.e., if $S_Y(f) = f^2 S_X(f)$, then

$$\int_{-\infty}^{\infty} S_Y(f)df < \infty .$$

(8.22)

As a rather paradoxical application of (8.21), suppose that we have a one-sided continuous time Gaussian random process $\{X(t); t \geq 0\}$ that has zero mean and a covariance function that is the continuous time analog of example [8.3]; that is, $K_X(t,s) = \sigma^2 \min(t,s)$. (The Kolmogorov construction guarantees that there is such a random process; that is, that it is well defined.) This process is known as the continuous time Wiener process, a process that we will encounter again in the next chapter. Strictly speaking, the double derivative of this function does not exist because of the discontinuity of the function at $t = s$. From engineering intuition, however, the derivative of such a step discontinuity is an impulse, suggesting that

$$R_Y(t,s) = \sigma^2 \delta(t-s) ,$$

the covariance function for Gaussian white noise! Because of this formal relation, Gaussian white noise is sometimes described as the formal derivative of a Wiener process. We have to play loose with mathematics to find this result, and the sloppiness cannot be removed in a straightforward manner. In fact, it is known from the theory of Wiener processes that they have the strange attribute of producing with probability one sample waveforms that are continuous but *nowhere differentiable!* Thus we are considering white noise as the derivative of a process that is not differentiable. In a sense, however, this is a useful intuition that is consistent with the extremely pathological behavior of sample waveforms of white noise—an idealized concept of a process that can not really exist anyway.

*LINEAR ESTIMATION

In this section we give another application of second-order moments in linear systems by showing how they arise in one of the basic problems of communication: estimating the outcomes of one random process based on observations of another process using a linear filter. We will obtain the classical orthogonality principle and the Wiener-Hopf equation and consider solutions for various simple cases. The main purpose of this section is not to teach estimation theory but rather to get additional practice manipulating second-order moments and to provide more evidence for their importance.

We will focus on discrete time for the usual reasons, but the continuous time analogs are found, also as usual, by replacing the sums by integrals.

Suppose that we are given a record of observations of values of one random process; e.g., we are told the values of $\{Y_i; N < i < M\}$, and we are asked to form the best estimate of a particular sample, say X_n of another, related, random process $\{X_k; k \in I\}$. We refer to the collection of indices of observed samples by $\boldsymbol{K} = (N, M)$. We permit N and M to take on infinite values. For convenience we assume throughout this section that both processes have zero means for all time. We place the strong constraint on the estimate that it must be linear; that is, the estimate $\hat{X}_n$ of X_n must have the form

$$\hat{X}_n = \sum_{k:\, n-k \in \boldsymbol{K}} h_k Y_{n-k} = \sum_{k \in \boldsymbol{K}} h_{n-k} Y_k$$

for some pulse response h. We wish to find the "best" possible filter h, perhaps under additional constraints such as causality. One possible notion of best is to define the error

$$\epsilon_n = X_n - \hat{X}_n$$

and define that filter to be best within some class if it minimizes the mean squared error $E(\epsilon_n^2)$; that is, a filter satisfying some constraints will be considered optimum if no other filter yields a smaller expected squared error. The filter accomplishing this goal is often called a linear least squared error (LLSE) filter.

Many constraints on the filter or observation times are possible. Typical constraints on the filter and on the observations are the following:

1. We have a noncausal filter that can "see" into the infinite future and a two-sided infinite observation $\{Y_k; k \in \boldsymbol{Z}\}$. Here we consider $N = -\infty$ and $M = \infty$. This is clearly not completely possible, but it may be a reasonable approximation for a system using a very long observation record to estimate a sample of a related process in the middle of the record.

2. The filter is causal ($h_k = 0$ for $k < 0$), a constraint that can be incorporated by assuming that $n \geq M$; that is, that samples occurring after the one we wish to estimate are not observed. When $n > M$ the estimator is sometimes called a predictor since it estimates the value of the desired process at a time later than the last observation. Here we assume that we observe the entire past of the Y process; that is, we take $N = -\infty$ and observe $\{Y_k; k < M\}$. If, for example, the X process and the Y process are the same and $M = n$, then this case is called the one-step predictor (based on the semi-infinite past).

3.   The filter is causal, and we have a finite record of T seconds; that is, we observe $\{Y_k; M - T \leq k < M\}$.

As one might suspect, the fewer the constraints, the easier the solution but the less practical the resulting filter. We will develop a general characterization for the optimum filters, but we will provide specific solutions only for certain special cases. We formally state the basic result as a theorem and then prove it.

Theorem 8.1.   Suppose that we are given a set of observations $\{Y_k; k \in \boldsymbol{K}\}$ of a zero-mean random process $\{Y_k\}$ and that we wish to find a linear estimate $\hat{X}_n$ of a sample X_n of a zero-mean random process $\{X_n\}$ of the form

$$\hat{X}_n = \sum_{k: n-k \in \boldsymbol{K}} h_k Y_{n-k} \ . \tag{8.23}$$

If the estimation error is defined as

$$\epsilon_n = X_n - \hat{X}_n \ ,$$

then for a fixed n no linear filter can yield a smaller expected squared error $E(\epsilon_n^2)$ than a filter h (if it exists) that satisfies the relation

$$E(\epsilon_n Y_k) = 0; \text{ all } k \in \boldsymbol{K} \ , \tag{8.24}$$

or, equivalently,

$$E(X_n Y_k) = \sum_{i: n-i \in \boldsymbol{K}} h_i E(Y_{n-i} Y_k); \text{ all } k \in \boldsymbol{K} \ . \tag{8.25}$$

If $R_Y(k,j) = E(Y_k Y_j)$ is the autocorrelation function of the Y process and $R_{X,Y}(k,j) = E(X_k Y_j)$ is the cross correlation function of the two processes, then (8.25) can be written as

$$R_{X,Y}(n,k) = \sum_{i: n-i \in \boldsymbol{K}} h_i R_Y(n-i,k) \ ; \text{ all } k \in \boldsymbol{K} \ . \tag{8.26}$$

If the processes are jointly weakly stationary in the sense that both are individually weakly stationary with a cross-correlation function that depends only on the difference between the arguments, then, with the replacement of k by $n - k$, the condition becomes

$$R_{X,Y}(k) = \sum_{i: n-i \in \boldsymbol{K}} h_i R_Y(k-i); \text{ all } k: n-k \in \boldsymbol{K} \ . \tag{8.27}$$

Comments. Two random variables U and V are said to be orthogonal if $E(UV) = 0$. Therefore equation (8.24) is known as the *orthogonality principle* because it states that the optimal filter causes the estimation error to be orthogonal to the observations. Note that (8.24) implies not only that the estimation error is orthogonal to the observations, but that it is also orthogonal to all linear combinations of the observations. Relation (8.27) with $K = (-\infty, n)$ is known as the Wiener-Hopf equation. To be useful in practice, we must be able to find a pulse response that solves one of these equations. We shall later find solutions for some simple cases. A more general treatment is beyond the intended scope of this book. Our emphasis here is to demonstrate an example in which determination of an optimal filter for a reasonably general problem requires the solution of an equation given in terms of second-order moments.

Proof. Suppose that we have a filter h that satisfies the given conditions. Let g be any other linear filter with the same input observations and let $\tilde{X}_n$ be the resulting estimate. We will show that the given conditions imply that g can yield an expected squared error no better than that of h. Let $\tilde{\epsilon}_n = X_n - \tilde{X}_n$ be the estimation error using g so that

$$E(\tilde{\epsilon}_n^2) = E((X_n - \sum_{i:\, n-i\in K} g_i Y_{n-i})^2) \;.$$

Add and subtract the estimate using h satisfying the conditions of the theorem and expand the square to obtain

$$E(\tilde{\epsilon}_n^2) = E((X_n - \sum_{i:\, n-i\in K} h_i Y_{n-i} + \sum_{i:\, n-i\in K} h_i Y_{n-i} - \sum_{i:\, n-i\in K} g_i Y_{n-i})^2) =$$

$$E((X_n - \sum_{i:\, n-i\in K} h_i Y_{n-i})^2) + 2E((X_n - \sum_{i:\, n-i\in K} h_i Y_{n-i})(\sum_{i:\, n-i\in K} (h_i - g_i) Y_{n-i}))$$

$$+ E((\sum_{i:\, n-i\in K} (h_i - g_i) Y_{n-i})^2) \;.$$

The first term on the right is the expected squared error using the filter h, say $E(\epsilon_n^2)$. The last term on the right is the expectation of something squared and is hence nonnegative. Thus we have the lower bound

$$E(\tilde{\epsilon}_n^2) \geq$$

$$E(\epsilon_n^2) + 2 \sum_{i:\, n-i\in K} (h_i - g_i) \left\{ E(X_n Y_{n-i}) - \sum_{j:\, n-j\in K} h_j E(Y_{n-j} Y_{n-i}) \right\} \;,$$

where we have brought one of the sums out, used different dummy variables for the two sums, and interchanged some expectations and sums. From (8.25), however, the bracketed term is zero for each i in the index set being summed over, and hence the entire sum is zero, proving that

$$E(\tilde{\epsilon}_n^2) \geq E(\epsilon_n^2) \;,$$

which completes the proof of the theorem.

Note that from (8.23) through (8.27) we can write the mean square error for the optimum linear filter as

$$E(\epsilon_n^2) = E(\epsilon_n(X_n - \sum_{k:\, n-k \in \boldsymbol{K}} h_k Y_{n-k}) = E(\epsilon_n X_n)$$

$$= R_X(n,n) - \sum_{k:\, n-k \in \boldsymbol{K}} h_k R_{X,Y}(n, n-k)$$

in general and

$$E(\epsilon_n^2) = R_X(0) - \sum_{k:\, n-k \in \boldsymbol{K}} h_k R_{X,Y}(k)$$

for weakly stationary processes.

Older proofs of the result just given use the calculus of variations, that is, calculus minimization techniques. The method we have used, however, is simple and intuitive and shows that a filter satisfying the given equations actually yields a global minimum to the mean squared error and not only a local minimum as usually obtained by calculus methods. A popular proof of the basic orthogonality principle is based on Hilbert space methods and the projection theorem, the generalization of the standard geometric result that the shortest line from a point to a plane is the projection of the point on the plane—the line passing through the point which meets the plane at a right angle (is orthogonal to the plane). The projection method also proves that the filter of (8.24) yields a global minimum.

We consider four examples in which the theorem can be applied to construct an estimate. The first two are fairly simple and suffice for a brief reading.

[8.4]

Suppose that the processes are jointly weakly stationary, that we are given the entire two-sided realization of the random process $\{Y_n\}$, and that there are no restrictions on h. Equation (8.27) then becomes

$$R_{X,Y}(k) = \sum_{i \in \boldsymbol{Z}} h_i R_Y(k-i); \text{ all } k \in \boldsymbol{Z} .$$

This equation is a simple convolution and can be solved by standard Fourier techniques. Take the Fourier transform of both sides and define the transform of the cross-correlation function $R_{X,Y}$ to be the cross spectral density $S_{X,Y}(f)$. We obtain $S_{X,Y}(f) = H(f)S_Y(f)$ or

$$H(f) = \frac{S_{X,Y}(f)}{S_Y(f)} ,$$

which can be inverted to find the optimal pulse response h:

$$h(k) = \int_{-1/2}^{1/2} \frac{S_{X,Y}(f)}{S_Y(f)} e^{j2\pi kf} df \; ,$$

which yields an optimum estimate

$$\hat{X}_n = \sum_{i=-\infty}^{\infty} h_i Y_{n-i} \; .$$

Thus we have an explicit solution for the optimal linear estimator for this case in terms of the second-order properties of the given processes. Note, however, that the resulting filter is *not* causal in general. Another important observation is that the filter itself does not depend on the sample time n at which we wish to estimate the X process; e.g., if we want to estimate X_{n+1}, we apply the same filter to the shifted observations; that is,

$$\hat{X}_{n+1} = \sum_{i=-\infty}^{\infty} h_i Y_{n+1-i} \; .$$

Thus in this example not only have we found a means of estimating X_n for a fixed n, but the same filter also works for *any* n. When one filter works for all estimate sample times by simply shifting the observations, we say that it is a time-invariant or stationary estimator. As one might guess, such time invariance is a consequence of the weak stationarity of the processes.

The most important application of example [8.4] is to "infinite smoothing," where $Y_n = X_n + V_n$. $\{V_n\}$ is a noise process that is uncorrelated with the signal process $\{X_n\}$, i.e., $R_{X,V}(k) = 0$ for all k. Then $R_{X,Y} = R_X$ and hence $R_Y = R_X + R_V$, so that

$$H(f) = \frac{S_X(f)}{S_X(f) + S_V(f)} \; .$$

[8.5]

Again assume that the processes are jointly weakly stationary. Assume that we require a causal linear filter h but that we observe the infinite past of the observation process. Assume further that the observation process is white noise; that is, $R_Y(k) = \frac{N_0}{2}\delta_k$. Then $\mathbf{K} = \{n, n-1, n-2,...\}$, and equation (8.27) becomes the Wiener-Hopf equation

$$R_{X,Y}(k) = \sum_{i: n-i \in \mathbf{K}} h_i \frac{N_0}{2}\delta_{k-i} = h_k \frac{N_0}{2}; \; k \in \mathbf{Z}_+ \; .$$

This equation easily reduces because of the delta function to

$$h_k = \frac{2}{N_0} R_{X,Y}(k), \; k \in \mathbf{Z}_+ \; .$$

Thus we have for this example the optimal estimator

$$\hat{X}_n = \sum_{k=0}^{\infty} \frac{2}{N_0} R_{X,Y}(k) Y_{n-k} \ .$$

As with the previous example, the filter does not depend on n, and hence the estimator is time-invariant.

The case of a white observation process is indeed special, but it suggests a general approach to solving the Wiener-Hopf equation, which we sketch next.

[8.6]

Assume joint weak stationarity and a causal filter on a semi-infinite observation sequence as in example [8.5], but do not assume that the observation process is white. In addition, assume that the observation process is physically realizable so that a spectral factorization of the form of (8.20) exists; that is,

$$S_Y(f) = |G(f)|^2$$

for some causal stable filter with transfer function $G(f)$. As previously discussed, for practical purposes, all random processes have spectral densities of this form. We also assume that the inverse filter, the filter with transfer function $1/G(f)$, is causal and stable. Again, this holds under quite general conditions. Observe in particular that you can't run into trouble with $G(f)$ being zero on a frequency interval of nonzero length because the condition in the spectral factorization theorem would be violated.

Unlike the earlier examples, this example does not have a trivial solution. We sketch a solution as a modification to the solution of example [8.5]. The given observation process may not be white, but suppose that we pass it through a linear filter r with transfer function $R(f) = 1/G(f)$ to obtain a new random process, say $\{W_n\}$. Since the inverse filter $1/G(f)$ is assumed stable, then the W process has power spectral density $S_W(f) = S_Y(f)|R(f)|^2 = 1$ for all f; that is, $\{W_n\}$ is white. One says that the W process is a *whitened* version of the Y process, sometimes called the *innovations process* of $\{Y_n\}$. Intuitively, the W process contains the same information as the Y process since we can recover the Y process from it by passing it through the filter $G(f)$ (at least in principle). Thus we can get an estimate of X_n from the W process that is just as good as (and no better than) that obtainable from the Y process. Furthermore, if we now filter the W process to estimate X_n, then the overall operation of the whitening filter followed by the estimating linear filter is also a linear filter, producing the estimate from the original observations. Since the inverse filter is causal, a causal estimate based on the W process is also a causal estimate based on the Y process.

Because W is white, the estimate of X_n from $\{W_n\}$ is given immediately by the solution to example [8.5]; that is, the filter h with the W process as input is given by $h_k = R_{X,W}(k)$ for $k \geq 0$. The cross-correlation of the X and W processes can be calculated using the standard linear filter I/O techniques. It turns out that the required cross-correlation is the inverse Fourier transform of a cross-spectral density given by

$$S_{X,W}(f) = \frac{S_{X,Y}(f)}{G(f)^*} = H(f) , \qquad (8.28)$$

where the asterisk denotes the complex conjugate. (See exercise 8.19.) Thus the optimal linear estimator given the whitened process is

$$h_k = \begin{cases} \displaystyle\int_{-1/2}^{1/2} \frac{S_{X,Y}(f)}{G(f)^*} e^{2\pi jkf}; & k \geq 0, \\ 0 & \text{otherwise}, \end{cases}$$

and the overall optimal linear estimate has the form shown in Figure 8.3. Although more complicated, we again have a filter that does not depend on n.

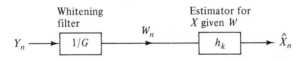

Figure 8.3 Prewhitening method.

This approach to solving the Wiener-Hopf equation is called the "prewhitening" (or "innovations" or "shaping filter") approach and it can be made rigorous under quite general conditions. That is, for all practical purposes, the optimal filter can be written in this cascade form as a whitening filter followed by a LLSE filter given the whitened observations as long as the processes are jointly weakly stationary and the observations are observed for the infinite past. The difficult part of the approach for applications is actually finding the factorization of the observation process's power spectral density.

When the observation interval is finite or when the processes are not jointly weakly stationary, the spectral factorization approach becomes quite complicated and cumbersome, and alternative methods, usually in the time domain, are required. The final example considers such an estimator.

[8.7]

Suppose that the random process we wish to estimate satisfies a difference equation of the form

$$X_{n+1} = \Phi_n X_n + U_n, \, n > 0 ,$$

where X_0 is an initial random variable and the process $\{U_n\}$ is a zero-mean process that is uncorrelated with a possibly time-varying second moment $E(U_n^2) = \Gamma_n$. $\{\Phi_n\}$ is a known sequence of constants. In other words, we know that the random process is defined by a time-varying linear system driven by noise that is uncorrelated but not necessarily stationary. Assume that the observation process has the form

$$Y_n = H_n X_n + V_n , \tag{8.30}$$

a scaled version of the X process plus observation noise, where H_n is a known sequence of constants. We also assume that the observation noise has zero mean and is uncorrelated but not necessarily stationary, say $E(V_n^2) = \Psi_n$. We further assume that the U and V processes are uncorrelated: $E(U_n V_k) = 0$ for all n and k. Intuitively, the random processes are such that new values are obtained by scaling old values and adding some perturbations. Additional noise influences our observations. Suppose that we observe $Y_0, Y_1, \ldots, Y_{n-1}$, what is the best linear estimate of X_n?

In a sense this problem is more restrictive than the Wiener-Hopf formulation of example [8.6] because we have assumed a particular structure for the process to be estimated and for the observations. On the other hand, it is not a special case of the previous model because the time-varying parameters make it nonstationary and because we restrict the observations to a finite time window (not including the current observation), often a better approximation to reality. Because of these differences, the spectral techniques of the standard Wiener-Hopf solution of example [8.6] do not apply without significant generalization and modification. Hence we consider another approach, called recursive estimation or *Kalman − Bucy filtering*, whose history may be traced to Gauss's formulas for plotting the trajectory of heavenly bodies. The basic idea is the following: Instead of considering how to operate on a complete observation record in order to estimate something at one time, suppose that we already have a good estimate $\hat{X}_n$ for X_n and that we make a single new observation Y_n. How can we use this new information to update our old estimate in a linear fashion to form a new estimate $\hat{X}_{n+1}$ of X_{n+1}? For example, can we find sequences of numbers a_n and b_n so that

$$\hat{X}_{n+1} = a_n \hat{X}_n + b_n Y_n$$

is a good estimate? One way to view this is that instead of constructing a filter h described by a convolution that operates on the past to produce an estimate for each time, we wish a possibly time-varying filter with feedback that observes its own past outputs or estimates and operates on this and a new observation to produce a new estimate. This is the basic idea of recursive filtering, which is applicable to more general models than that considered here. In particular, the standard developments in the literature consider vector generalizations of the above difference equations. We sketch a derivation for the simpler scalar case.

We begin by trying to apply directly the orthogonality principle of (8.24) through (8.26). If we fix a time n and try to estimate X_n by a linear filter as

$$\hat{X}_n = \sum_{i=1}^{n} h_i Y_{n-i} , \qquad (8.31)$$

then the LLSE filter is described by the time-dependent pulse response, say $h^{(n-1)}$, which, from (8.26), solves the equations

$$R_{X,Y}(n,l) = \sum_{i=1}^{n} h_i^{(n-1)} R_Y(n-i,l); \ l = 0,1,\ldots,n-1 , \qquad (8.32)$$

where the superscript reflects the fact that the estimate is based on observations through time $n-1$ and the fact that for this very nonstationary problem, the filter will likely depend very much on n. To demonstrate this, consider the estimate for X_{n+1}. In this case we will have a filter of the form

$$\hat{X}_{n+1} = \sum_{i=1}^{n+1} h_i^{(n)} Y_{n+1-i} , \qquad (8.33)$$

where the LLSE filter satisfies

$$R_{X,Y}(n+1,l) = \sum_{i=1}^{n+1} h_i^{(n)} R_Y(n+1-i,l); \ l = 0,1,\ldots,n . \qquad (8.34)$$

Note that (8.34) is different from (8.32), and hence the pulse responses satisfying the respective equations will also differ. In principle these equations can be solved to obtain the desired filters. Since they will in general depend on n, however, we are faced with the alarming possibility of having to apply for each time n a completely different filter $h^{(n)}$ to the entire record of observations $Y_0,\ldots,Y_n$ up to the current time, clearly an impractical system design. We shall see, however, that a more efficient means of recursively computing the estimate can be found. It will still be based on linear operations, but now they will be time-varying.

We begin by comparing (8.32) and (8.34) more carefully to find a relation between the two filters $h^{(n)}$ and $h^{(n-1)}$. If we consider $l<n$, then (8.29) implies that

$$R_{X,Y}(n+1,l) = E(X_{n+1}Y_l) = E((\Phi_n X_n + U_n)Y_l)$$

$$= \Phi_n E(X_n Y_l) + E(U_n Y_l) = \Phi_n E(X_n Y_l) = \Phi_n R_{X,Y}(n,l) .$$

Reindexing the sum of (8.34) using this relation and restricting ourselves to $l<n$ then yields

$$\Phi_n R_{X,Y}(n,l) = \sum_{i=0}^{n} h_{i+1}^{(n)} R_Y(n-i,l) =$$

$$h_1^{(n)} R_Y(n,l) + \sum_{i=1}^{n} h_{i+1}^{(n)} R_Y(n-i,l) \ ; \ l=0,1,\ldots,n-1 \ .$$

But for $l < n$ we also have that

$$R_{X,Y}(n,l) = E(X_n,Y_l) = E(\frac{Y_n - V_n}{H_n} Y_l)$$

$$= \frac{1}{H_n} E(Y_n Y_l) = \frac{1}{H_n} R_Y(n,l)$$

or

$$R_Y(n,l) = H_n R_{X,Y}(n,l) \ .$$

Substituting this result, we have with some algebra that

$$R_{X,Y}(n,l) = \sum_{i=1}^{n} \frac{h_{i+1}^{(n)}}{\Phi_n - h_1^{(n)} H_n} R_Y(n-i,l); \ l=0,1,\ldots,n-1 \ ,$$

which is the same as (8.32) if one identifies

$$h_i^{(n-1)} = \frac{h_{i+1}^{(n)}}{\Phi_n - h_1^{(n)} H_n}; \ i=1,\ldots,n \ .$$

From (8.33) the estimate for X_n is

$$\hat{X}_n = \sum_{i=1}^{n} h_i^{(n-1)} Y_{n-i} = \frac{1}{\Phi_n - h_1^{(n)} H_n} \sum_{i=2}^{n+1} h_i^{(n)} Y_{n+1-i} \ .$$

Comparing this with (8.33) yields

$$\hat{X}_{n+1} = h_1^{(n)} Y_n + (\Phi_n - h_1^{(n)} H_n) \hat{X}_n \ ,$$

which has the desired form. It remains, however, to find a means of computing the numbers $h_1^{(n)}$. Since this really depends on only one argument n, we now change notation for brevity and henceforth denote this term by $\varkappa_n$; that is,

$$\varkappa_n = h_1^{(n)} \ .$$

To describe the estimator completely we need to find a means of computing $\varkappa_n$ and an initial estimate. The initial estimate does not depend on any observations. The LLSE estimate of a random variable without observations is the mean of the random variable (see, e.g., exercise 6.16). Since by assumption the processes all have zero mean, $\hat{X}_0 = 0$ is the initial estimate.

Before computing $\varkappa_n$, we make several remarks on the estimator and its properties. First, we can rewrite the estimator as

$$\hat{X}_{n+1} = \varkappa_n(Y_n - H_n\hat{X}_n) + \Phi_n\hat{X}_n \ .$$

It is easily seen from the orthogonality principle that if $\hat{X}_n$ is a LLSE estimate of X_n given $Y_0, Y_1, \ldots, Y_{n-1}$, then $\hat{Y}_n = H_n\hat{X}_n$ is a LLSE estimate of Y_n given the same observations (exercise 8.21). Thus $\nu_n = Y_n - H_n\hat{X}_n = Y_n - \hat{Y}_n$ can be interpreted as the "new" information in the observation Y_n in the sense that our best prediction of Y_n based on previously known samples has been removed from Y_n. We can now write

$$\hat{X}_{n+1} = \varkappa_n\nu_n + \Phi_n\hat{X}_n \ .$$

This can be interpreted as saying that the new estimate is formed from the old estimate by using the same transformation Φ_n used on the actual samples and then by adding a term depending only on the new information.

It also follows from the orthogonality principle that the sequence ν_n is uncorrelated: Since $H_n\hat{X}_n$ is the LLSE estimate for k_n based on $Y_0, \ldots, Y_{n-1}$, the error ν_n must be orthogonal to past Y_l from the orthogonality principle. Hence ν_n must also be orthogonal to linear combinations of past Y_l and hence also to past ν_l. It is straightforward to show that $E\nu_n = 0$ for all n and hence orthogonality of the sequence implies that it is also uncorrelated (exercise 8.22). Because of the various properties, the sequence $\{\nu_n\}$ is called the innovations sequence of the observations process. Note the analog with example [8.6], where the observations were first whitened to form innovations and then the estimate was formed based on the whitened version.

Observe next that the innovations and the estimation error are simply related by the formula

$$\nu_n = Y_n - H_n\hat{X}_n = H_nX_n + V_n - H_n\hat{X}_n = H_n(X_n - \hat{X}_n) + V_n$$

or

$$\nu_n = H_n\epsilon_n + V_n \ ,$$

a useful formula for deriving some of the properties of the filter. For example, we can use this formula to find a recursion for the estimate error:

$$\epsilon_0 = X_0$$

$$\begin{aligned}
\epsilon_{n+1} &= X_{n+1} - \hat{X}_{n+1} \\
&= \Phi_nX_n + U_n - \varkappa_n\nu_n - \Phi_n\hat{X}_n \\
&= (\Phi_n - \varkappa_nH_n)\epsilon_n + U_n - \varkappa_nV_n; \quad n = 0,1,\ldots \ .
\end{aligned} \tag{8.35}$$

This formula immediately implies that

$$E(\epsilon_n) = 0; \; n = 0,1,\dots \; ,$$

and hence the estimate is unbiased (i.e., an estimate having an error which has zero mean is defined to be unbiased). It also provides a recursion for finding the expected squared estimation error:

$$E(\epsilon_{n+1}^2) = (\Phi_n - \varkappa_n H_n)^2 E(\epsilon_n^2) + E(U_n^2) + \varkappa_n^2 E(V_n^2) \; ,$$

where we have made use of the assumptions of the problem statement, viz., the uncorrelation of U_n and V_n sequences with each other, with $Y_0,\dots,Y_{n-1}$, $X_0,\dots,X_n$, and hence also with ϵ_n. Rearranging terms for later use, we have that

$$E(\epsilon_{n+1}^2) =$$

$$\Phi_n^2 E(\epsilon_n^2) - 2\varkappa_n H_n \Phi_n E(\epsilon_n^2) + \varkappa_n^2 (H_n^2 E(\epsilon_n^2) + \Psi_n) + \Gamma_n \; . \qquad (8.36)$$

Since we know that $E(\epsilon_0^2) = E(X_0^2)$, if we knew the $\varkappa_n$ we could recursively evaluate the expected squared errors from the formula and the given problem parameters. We now complete the system design by developing a formula for the $\varkappa_n$. This is most easily done by using the orthogonality relation and (8.35):

$$0 = E(\epsilon_{n+1} Y_n) = (\Phi_n - \varkappa_n H_n) E(\epsilon_n Y_n) + E(U_n Y_n) - \varkappa_n E(V_n Y_n) \; .$$

Consider the terms on the right. Proceeding from left to right, the first term involves

$$E(\epsilon_n Y_n) = E(\epsilon_n (H_n X_n + V_n)) = H_n E(\epsilon_n (\epsilon_n + \hat{X}_n)) = H_n E(\epsilon_n^2) \; ,$$

where we have used the fact that ϵ_n is orthogonal to V_n, to $Y_0,\dots,Y_{n-1}$ and hence to $\hat{X}_n$. The second term is zero by the assumptions of the problem. The third term requires the evaluation

$$E(V_n Y_n) = E(V_n (H_n X_n + V_n)) = E(V_n^2) \; .$$

Thus we have that

$$0 = (\Phi_n - \varkappa_n H_n) H_n E(\epsilon_n^2) - \varkappa_n E(V_n^2)$$

or

$$\varkappa_n = \frac{\Phi_n H_n E(\epsilon_n^2)}{E(V_n^2) + H_n^2 E(\epsilon_n^2)} \; .$$

Thus for each n we can solve the recursion for $E(\epsilon_n^2)$ and for the required $\varkappa_n$ to form the next estimate.

We can now combine all of the foregoing mess to produce the final answer. A recursive estimator for the given model is

$$\hat{X}_0 = 0; \; E(\epsilon_0^2) = E(X_0^2) \; , \qquad (8.37)$$

and for $n = 0,1,2,...,$

$$\hat{X}_{n+1} = \varkappa_n(Y_n - H_n\hat{X}_n) + \Phi_n\hat{X}_n; \; n = 0,1,... \, , \qquad (8.38)$$

where

$$\varkappa_n = \frac{\Phi_n H_n E(\epsilon_n^2)}{\Psi_n + H_n^2 E(\epsilon_n^2)} \qquad (8.39)$$

and where, from (8.36) and (8.39),

$$E(\epsilon_{n+1}^2) = \Phi_n E(\epsilon_n^2) + \Gamma_n - \frac{(\Phi_n H_n E(\epsilon_n^2))^2}{\Psi_n + H_n^2 E(\epsilon_n^2)} \, . \qquad (8.40)$$

Although these equations seem messy, they can be implemented numerically or in hardware in a straightforward manner. Variations of their matrix generalizations are also well suited to fast implementation. Such algorithms in greater generality are a prime focus of the areas of estimation, detection, signal identification, and signal processing.

EXERCISES

8.1 Suppose that X_n is an i.i.d. Gaussian process with mean m and variance σ^2. Let h be the pulse response $h_0 = 1$, $h_1 = r$, and $h_k = 0$ for all other k. Let $\{W_n\}$ be the output process when the X process is put into the filter described by h; that is,

$$W_n = X_n + rX_{n-1} \, .$$

Assuming that the processes are two-sided—that is, that they are defined for $n \in \mathbf{Z}$—find EW_n and $R_W(k,j)$. Is $\{W_n\}$ strictly stationary? Next assume that the processes are one-sided; that is, defined for $n \in \mathbf{Z}_+$. Find EW_n and $R_W(k,j)$. For the one-sided case, evaluate the limits of EW_n and $R_W(n,n+k)$ as $n \to \infty$.

8.2 We define the following two-sided random processes. Let $\{X_n\}$ be an i.i.d. random process with marginal pdf $f_X(x) = e^{-x}$, $x \geq 0$. Let $\{Y_n\}$ be another i.i.d. random process, independent of the X process, having marginal pdf $f_Y(y) = 2e^{-2y}$, $y \geq 0$. Define a random process $\{U_n\}$ by the difference equation

$$U_n = X_n + X_{n-1} + Y_n \, .$$

The process U_n can be thought of as the result of passing X_n through a first order moving average filter and then adding noise. Find EU_0 and $R_U(k)$.

8.3   Let $\{X(t)\}$ be a stationary continuous time random process with zero mean and autocorrelation function $R_X(\tau)$. The process $X(t)$ is put into a linear time-invariant stable filter with impulse response $h(t)$ to form a random process $Y(t)$. A random process $U(t)$ is then defined as $U(t) = Y(t)X(t-T)$, where T is a fixed delay. Find $EU(t)$ in terms of R_X, h, and T. Simplify your answer for the case where $S_X(f) = N_0/2$, all f.

8.4   Find the output power spectral densities in exercises 8.1 and 8.2.

8.5   Let $\{X_n\}$ be an i.i.d. Gaussian random process with zero mean and variance $R_X(0) = \sigma^2$. Let $\{U_n\}$ be an i.i.d. binary random process with $\Pr(U_n=1) = \Pr(U_n=-1) = 1/2$. (All processes are assumed to be two-sided in this problem.) Define the random processes $Z_n = X_n U_n$, $Y_n = U_n + X_n$, and $W_n = U_0 + X_n$, all n. Find the mean, covariance, and power spectral density of each of these processes. Find the cross-covariance functions between the processes.

8.6   Let $\{U_n\}$, $\{X_n\}$, and $\{Y_n\}$ be the same as in exercise 8.5. The process $\{X_n\}$ can be viewed as a binary signal corrupted by additive Gaussian noise. One possible method of trying to remove the noise at a receiver is to quantize the received Y_n to form an estimate $\hat{U}_n = q(Y_n)$ of the original binary sample, where

$$q(r) = \begin{cases} +1 & \text{if } r \geq 0 \\ -1 & \text{if } r < 0 . \end{cases}$$

Write an integral expression for the error probability $P_e = \Pr(\hat{U}_n \neq U_n)$. Find the mean, covariance, and power spectral density of the $\hat{U}_n$ process. Are the processes $\{U_n\}$ and $\{\hat{U}_n\}$ equivalent—that is, do they have the same process distributions? Define an error process ϵ_n by $\epsilon_n = 0$ if $\hat{U}_n = U_n$ and $\epsilon_n = 1$ if $\hat{U}_n \neq U_n$. Find the marginal pmf, mean, covariance, and power spectral density of the error process.

8.7   *Cascade filters.* Let $\{g_k\}$ and $\{a_k\}$ be the pulse responses of two discrete time causal linear filters ($g_k = a_k = 0$ for $k < 0$) and let $G(f)$ and $A(f)$ be the corresponding transfer functions, e.g.,

$$G(f) = \sum_{k=0}^{\infty} g_k e^{-j2\pi kf} .$$

Assume that $g_0 = a_0 = 1$. Let $\{Z_n\}$ be a weakly stationary uncorrelated random process with variance σ^2 and zero mean. Consider the cascade of two filters formed by first putting an input Z_n into the filter g to form the process X_n, which is in turn put into the filter a to form the output process Y_n.

(a) Let $\{d_k\}$ denote the pulse response of the overall cascade filter; that is,

$$Y_n = \sum_{k=0}^{\infty} d_k Z_{n-k} .$$

Find an expression for d_k in terms of $\{g_k\}$ and $\{a_k\}$. (*Note:* As a check on your answer you should have $d_0 = g_0 a_0 = 1$.)

(b) Let $D(f)$ be the transfer function of the cascade filter. Find $D(f)$ in terms of $G(f)$ and $A(f)$.

(c) Find the power spectral density $S_Y(f)$ in terms of σ^2, G, and A.

(d) Prove that

$$E(Y_n^2) = \int_{-1/2}^{1/2} S_Y(f)df \geq \sigma^2 .$$

Hint: Show that if $d_0 = 1$ (from part (a)), then

$$\int_{-1/2}^{1/2} |D(f)|^2 df = 1 + \int_{-1/2}^{1/2} |1-D(f)|^2 df \geq 1 .$$

8.8 *One-step prediction.* This exercise develops a basic result of estimation theory. No prior knowledge of estimation theory is required. Results from exercise 8.7 may be quoted without proof (even if you did not complete it). Let $\{X_n\}$ be as in exercise 8.7; that is, $\{X_n\}$ is a discrete time zero-mean random process with power spectral density $\sigma^2|G(f)|^2$, where $G(f)$ is the transfer function of a causal filter with pulse response $\{g_k\}$ with $g_0 = 1$. Form the process $\{\hat{X}_n\}$ by putting X_n into a causal linear time-invariant filter with pulse response $\{h_k\}$ with $h_0 = 0$; that is,

$$\hat{X}_n = \sum_{k=1}^{\infty} h_k X_{n-k} .$$

Suppose that the linear filter tries to estimate the value of X_n based on the values of X_i for all $i<n$ by choosing the pulse response $\{h_k\}$ optimally. That is, the filter estimates the next sample based on the present value and the entire past. Such a filter is called a *one-step predictor*. Define the error process $\{\epsilon_n\}$ by

$$\epsilon_n = X_n - \hat{X}_n .$$

(a) Find expressions for the power spectral density $S_\epsilon(f)$ in terms of $S_X(f)$ and $H(f)$. Use this result to evaluate $E\epsilon_n^2$.

(b)   Evaluate $S_\epsilon(f)$ and $E(\epsilon_n^2)$ for   the   case   where
        $1-H(f) = 1/G(f)$.

(c)   Use part (d) of exercise 8.7 to show that the prediction filter
        $H(f)$ of (b) in this exercise yields the smallest possible value of
        $E(\epsilon_n^2)$ for any prediction filter.

        You have just developed the optimal one-step prediction filter
        for the case of a process that can be modeled as a weakly sta-
        tionary uncorrelated sequence passed through a linear filter. As
        discussed in the text, most discrete time random processes can
        be modeled in such a fashion, at least through second-order
        properties.

(d)   *Spectral factorization.* Suppose that $\{X_n\}$ has a power spectral
        density $S_X(f)$ that satisfies

$$\int_{-1/2}^{1/2} \ln S_X(f)df < \infty .$$

        Expand $\ln S_X(f)$ in a Fourier series and write the expression for
        $\exp(\ln S_X(f))$ in terms of the series to find $G(f)$. Find the pulse
        response of the optimum prediction filter in terms of your
        result. Find the mean square error. (*Hint*: You will need to
        know what evenness of $S_X(f)$ implies for the coefficients in the
        requested series and what the Taylor series of an exponential is.)

8.9 *Binary filters.* All of the linear filters considered so far were linear in
        the sense of real arithmetic. It is sometimes useful to consider filters
        that are linear in other algebraic systems, e.g., in binary or mod 2
        arithmetic. Such systems are more appropriate, for example, when
        considering communications systems involving only binary arithmetic,
        such as binary codes for noise immunity on digital communication
        links. In this exercise we explore some of the properties of such
        filters.

        Binary arithmetic (or mod 2 arithmetic or GF (2), for "Galois field of
        2 elements" arithmetic) consists of an operation $\oplus$ defined on the
        binary alphabet $\{0,1\}$ as follows:

$$0 \oplus 0 = 1 \oplus 1 = 0$$

$$1 \oplus 0 = 0 \oplus 1 = 1$$

        The operation $\oplus$ corresponds to an "exclusive or" in logic; that is, it
        produces a 1 if one or the other but not both of its arguments is 1.
        A binary first-order autoregressive filter with input process $\{X_n\}$ and
        output process $\{Y_n\}$ is defined by the difference equation

$$Y_n = Y_{n-1} \oplus X_n , \text{ all } n .$$

Assume that the $\{X_n\}$ is a Bernoulli process with parameter p. In this case the process $\{Y_n\}$ is called a binary first-order autoregressive source.

(a)   Show that for nonnegative integers k, the autocorrelation function of the process $\{Y_n\}$ satisfies

$$R_Y(k) = E(Y_j Y_{j+k}) = \Pr(\sum_{i=1}^{k} X_i = \text{an even number}).$$

(b)   Use the result of (a) to evaluate R_Y and K_Y.
    Hint: This is most easily done using a trick. Define the random variable

$$W_k = \sum_{i=1}^{k} X_i.$$

W_k is a binomial random variable. Use this fact and the binomial theorem to show that

$$\Pr(W_k \text{ is odd}) - \Pr(W_k \text{ is even}) = (1-2p)^k.$$

Alternatively, find a linear recursion relation for $p_k = \Pr(W_k$ is odd) using conditional probability (i.e., find a formula giving p_k in terms of p_{k-1}) and then solve for p_k.

(c)   Find the power spectral density of the process $\{Y_n\}$.

8.10 Let $\{X_n\}$ be a Bernoulli random process with parameter p and let $\oplus$ denote mod 2 addition as defined in exercise 8.9. Define the first-order binary moving average process $\{W_n\}$ by the difference equation

$$W_n = X_n \oplus X_{n-1}.$$

This is a mod 2 convolution and an example of what is called a *convolutional code* in communication and information theory. Find $p_{W_n}(w)$ and $R_W(k,j)$. Find the power spectral density of the process $\{W_n\}$.

8.11 Let $\{X(t)\}$ be a continuous time zero-mean Gaussian random process with spectral density $S_X(f) = N_0/2$, all f. Let $H(f)$ and $G(f)$ be the transfer functions of two linear time-invariant filters with impulse responses $h(t)$ and $g(t)$, respectively. The process $\{X(t)\}$ is passed through the filter $h(t)$ to obtain a process $\{Y(t)\}$ and is also passed through the filter $g(t)$ to obtain a process $\{V(t)\}$; that is,

$$Y(t) = \int_{0}^{\infty} h(\tau)X(t-\tau)d\tau$$

$$V(t) = \int_0^\infty g(\tau)X(t-\tau)d\tau \ .$$

(a)   Find the cross-correlation function $R_{Y,V}(t,s) = E(Y_t V_s)$.

(b)   Under what assumptions on H and G are Y_t and V_t independent random variables?

8.12 Let $\{X(t)\}$ and $\{Y(t)\}$ be two continuous time zero-mean stationary Gaussian processes with a common autocorrelation function $R(\tau)$ and common power spectral densities $S(f)$. Assume also that $E[X(t)Y(s)] = 0$ all t,s and that $\sigma^2 = R(0)$. For a fixed frequency f_0, define the random process

$$W(t) = X(t)\cos(2\pi f_0 t) + Y(t)\sin(2\pi f_0 t) \ .$$

Find the mean $E(W(t))$ and autocorrelation $R_W(t,s)$. Is $\{W(t)\}$ weakly stationary?

8.13 Say that we are given an i.i.d. binary random process $\{X_n\}$ with equal marginal probabilities for a -1 or a +1. We form a continuous time random process $\{X(t)\}$ by assigning

$$X(t) = X_n; \ t \in [(n-1)T, nT) \ ,$$

for a fixed time T. This process can also be described as follows: Let $p(t)$ be a pulse that is 1 for $t \in [0,T)$ and 0 elsewhere. Define

$$X(t) = \sum_k X_k p(t-kT) \ .$$

This is an example of pulse amplitude modulation (PAM). If the process $X(t)$ is then used to phase-modulate a carrier, the resulting process is called a phase-shift-keyed modulation of the carrier by the process $\{X(t)\}$ (PSK). PSK is a popular technique for digital communications. Define the PSK process

$$U(t) = a_0\cos(2\pi f_0 t + \delta X(t)) \ .$$

Observe that neither of these processes is stationary, but we can force them to be at least weakly stationary by the trick of inserting uniform random variables in appropriate places. Let Z be a random variable, uniformly distributed on $[0,T]$ and independent of the original i.i.d. process. Define the random process

$$Y(t) = X(t+Z) \ .$$

Let Θ be a random variable uniformly distributed on $[0,1/f_0]$ and independent of Z and of the original i.i.d. random process. Define the process

$$V(t) = U(t + \Theta) .$$

Find the mean and autocorrelation functions of the processes $Y(t)$ and $V(t)$.

8.14 Let $\{X(t)\}$ be a Gaussian random process with zero mean and autocorrelation function

$$R_X(\tau) = \frac{N_0}{2} e^{-|\tau|} .$$

Find the power spectral density of the process. Let $Y(t)$ be the process formed by DSB-SC modulation of $X(t)$. Letting Θ be uniformly distributed in equation (8.13), sketch the power spectral density of the modulated process.

8.15 A continuous time two-sided weakly stationary Gaussian random process $\{S(t)\}$ with zero mean and power spectral density $S_S(f)$ is put into a noisy communication channel. First, white Gaussian noise $\{W(t)\}$ with power spectral density $N_0/2$ is added, where the two random processes are assumed to be independent of one another, and then the sum $S(t) + W(t)$ is passed through a linear filter with impulse response $h(t)$ and transfer function $H(f)$ to form a received process $\{Y(t)\}$. Find an expression for the power spectral density $S_Y(f)$. Find an expression for the expected squared error $E[(S(t) - Y(t))^2]$ and the so-called signal-to-noise ratio (SNR)

$$\frac{E(S(t)^2)}{E[(S(t) - Y(t))^2]} .$$

Suppose that you know that $S_Y(f)$ can be factored into the form $|G(f)|^2$, where $G(f)$ is a stable causal filter with a stable causal inverse. What is the best choice of $H(f)$ in the sense of maximizing the signal to noise ratio? What is the best causal $H(f)$?

8.16 Show that equation (8.2) converges in mean square if the filter is stable and the input process has finitely bounded mean and variance. Show that convergence with probability one is achieved if the convergence of equation (2.30) is fast enough for the pulse response.

8.17 Show that the sum of equation (8.5) converges for the two-sided weakly stationary case if the filter is stable and the input process has finitely bounded variance.

8.18 Provide a formal argument for the integration counterpart of equation (8.22); that is, if $\{X(t)\}$ is a stationary two-sided continuous time random process and $Y(t) = \int_{-\infty}^{t} X(s)ds$, then, subject to suitable technical conditions, $S_Y(f) = S_X(f)/f^2$.

8.19 Prove that equation (8.28) holds under the conditions given.

8.20 Suppose that $\{Y_n\}$ is as in example [8.1] and that $W_n = Y_n + U_n$, where U_n is a zero-mean white noise process with second moment $E(U^2) = N_0/2$. Solve the Wiener-Hopf equation to obtain a LLSE of Y_{n+m} given $\{W_i; i \leq n\}$ for $m > 0$. Evaluate the resulting mean squared error.

8.21 Prove the claim that if $\{X_n\}$ and $\{Y_n\}$ are described by equations (8.29) and (8.30) and if $\hat{X}_n$ is a LLSE estimate of X_n given $Y_0, Y_1, ..., Y_{n-1}$, then $\hat{Y}_n = H_n \hat{X}_n$ is a LLSE estimate of Y_n given the same observations.

8.22 Prove the claim that the innovations sequence $\{\nu_n\}$ of example [8.7] is uncorrelated and has zero mean. (Fill in the details of the arguments used in the text.)

8.23 Let $\{Y_n\}$ be as in example [8.2]. Find the LLSE for Y_{n+m} given $\{Y_0, Y_1, ..., Y_n\}$ for an arbitrary positive integer m. Evaluate the mean square error. Repeat for the process of example [8.3] (the same process with $r = 1$).

8.24 Specialize the recursive estimator formulas of equations (8.37) through (8.40) to the case where $\{X_n\}$ is the $\{Y_n\}$ process of example [8.2], where H_n is a constant, say a, and where $\Psi_n = N_0/2$, all n. Describe the behavior of the estimator as $n \to \infty$.

8.25 Find an expression for the mean square error in example [8.4]. Specialize to infinite smoothing.

8.26 In the section on linear estimation we assumed that all processes had zero-mean functions. In this exercise we remove this assumption. Let $\{X_n\}$ and $\{Y_n\}$ be random processes with mean functions $\{m_X(n)\}$ and $\{m_Y(n)\}$, respectively. We estimate X_n by adding a constant to equation (8.23); i.e,

$$\hat{X}_n = a_n + \sum_{k:n-k\in K} h_k Y_{n-k} .$$

   (a) Show that the minimum mean square estimate of X_n is $\hat{X}_n = m_X(n)$ if no observations are used.

   (b) Modify and prove theorem 8.1 to allow for the nonzero means.

8.27 Suppose that $\{X_n\}$ and $\{Z_n\}$ are mutually independent i.i.d. two-sided Gaussian random processes with correlations

$$R_x(k) = \sigma_x^2 \delta_k; \ R_Z(k) = \sigma_Z^2 \delta_k .$$

These processes are used to construct new processes as follows:

$$Y_n = Z_n + r\,Y_{n-1}$$
$$U_n = X_n + Y_n$$
$$W_n = U_n - r\,U_{n-1}\,.$$

Find the covariances and power spectral densities of $\{U_n\}$ and $\{W_n\}$. Find $E\left[(X_n - W_n)^2\right]$.

9

USEFUL RANDOM PROCESSES

This chapter begins with a continuation of the study of the output processes of linear systems with random process inputs. The goal is to develop the detailed structure of such processes and of other processes with similar behavior that cannot be described by a linear system model. In chapter 8, we confined interest to second-order properties of the output random process, properties that can be found under quite general assumptions on the input process and filter. In order to get more detailed probabilistic descriptions of the output process, we next further restrict the input process for the discrete time case to be an i.i.d. random process and study the resulting output process and the continuous time analog to such a process. By restricting the structure of the output process in this manner, we shall see that in some cases we can find complete descriptions of the process and not just the first and second moments. The random processes obtained in this way provide many important and useful models, including moving-average, autoregressive, autoregressive moving-average (ARMA), independent increment, counting, random walk, Markov, Wiener, Poisson, and Gaussian processes. Similar techniques are used for the development of a variety of random processes with markedly different behavior, the key tools being the characteristic function or transform of a probability distribution and the concept of conditional probability distributions. This chapter contains extensive practice in derived distributions and in specifying random processes.

DISCRETE TIME LINEAR MODELS

We begin by describing several classes of discrete time random processes that will be considered in this chapter. This provides an opportunity for collecting some useful terminology. Recall that if we have a random process $\{X_n; n \in I\}$ as input to a linear system described by a convolution, then as in equation (8.2) there is a pulse response h_k such that the output process $\{Y_n\}$ is given by

$$Y_n = \sum_{k:n-k \in I} X_{n-k} h_k . \qquad (9.1)$$

A linear filter with such a description—that is, one that can be defined as a convolution—is sometimes called a *moving-average filter* since the output is a weighted running average of the inputs. If only a finite number of the h_k are not zero, then the filter is called a finite-order moving-average filter (or an FIR filter, for "finite impulse response," in contrast to an IIR or "infinite impulse response" filter). The order of the filter is equal to the maximum minus the minimum value of k for which the h_k are nonzero. For example, if $Y_n = X_n + X_{n-1}$, we have a first-order moving-average filter. Although some authors reserve the term *moving-average filter* for a finite-order filter, we will use the broader definition we have given. A block diagram for such a filter is given in Figure 9.1.

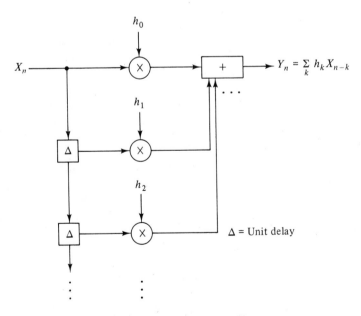

Figure 9.1 Moving average filter.

Several other names are used to describe finite-order moving-average filters. Since the output is determined by the inputs without any feedback from past or future outputs, the filter is sometimes called a feedforward or tapped delay line or transversal filter. If the filter has a well-defined transfer function $H(f)$ (e.g., it is stable) and if the transfer function is analytically continued to the complex plane by making the substitution $z = e^{j2\pi f}$, then the resulting complex function contains only zeroes and no poles in the complex plane. For this reason such a filter is sometimes called an all-zeroes filter.

In chapter 8 we considered only linear systems involving moving-average filters, that is, systems that could be represented as a convolution. This was because the convolution representation is well suited to second-order I/O relations. In this chapter, however, we will find that other representations are often more useful. Recall that a convolution is simply one example of a difference equation. Another form of difference equation describing a linear system is obtained by convolving the outputs to get the inputs instead of vice versa. For example, the output process may satisfy a difference equation of the form

$$X_n = \sum_k a_k Y_{n-k} . \qquad (9.2)$$

As in the moving-average case, the limits of the sum depend on the index set; e.g., the sum could be from $k = -\infty$ to ∞ in the two-sided case with $I = Z$ or from $k = -\infty$ to n in the one-sided case with $I = Z_+$.

The numbers $\{a_k\}$ are called regression coefficients, and the corresponding filter is called an autoregressive filter. If $a_k \neq 0$ for only a finite number of k, the filter is said to be finite-order autoregressive. The order is equal to the maximum minus the minimum value of k for which a_k is nonzero. For example, if $X_n = Y_n + Y_{n-1}$, we have a first-order autoregressive filter. As with moving-average filters, for some authors the "finite" is implicit, but we will use the more general definition. A block diagram for such a filter is given in Figure 9.2. Note that, in contrast with a finite-order moving-average filter, a finite-order autoregressive filter contains only feedback terms and no feedforward terms—the new output can be found solely from the current input and past or future outputs. Hence it is sometimes called a feedback filter. If we consider a deterministic input and transform both sides of (9.2), then we find that the transfer function of an autoregressive filter has the form

$$H(f) = \frac{1}{\sum_k a_k e^{-j2\pi kf}} .$$

Note that the analytic continuation of the transfer function into the complex plane with the substitution $z = e^{j2\pi f}$ for a finite-order autoregressive filter has poles but no zeroes in the complex plane. Hence a finite-order

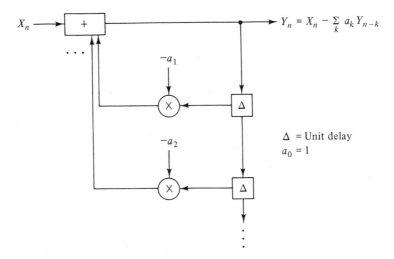

$X_n \longrightarrow$

$Y_n = X_n - \sum\limits_{k} a_k Y_{n-k}$

$-a_1$

$-a_2$

Δ = Unit delay
$a_0 = 1$

Figure 9.2   Autoregressive filter.

autoregressive filter is sometimes called an all-poles filter. An autoregressive filter may or may not be stable, depending on the location of the poles.

More generally, one can describe a linear system by a general difference equation combining the two forms—moving-average and autoregressive—as in (2.34):

$$\sum_{k} a_k y_{n-k} = \sum_{i} b_i x_{n-i} .$$

Filters with this description are called ARMA (for "autoregressive moving-average") filters. ARMA filters are said to be finite-order if only a finite number of the a_k's and b_k's are not zero. A finite-order ARMA filter is depicted in Figure 9.3. Once again, it should be noted that some authors use finite-order implicitly, a convention that we will not adopt. Applying a deterministic input and using (2.32), we find that the transfer function of an ARMA filter has the form

$$H(f) = \frac{\sum\limits_{i} b_i e^{-j2\pi i f}}{\sum\limits_{k} a_k e^{-j2\pi k f}} . \qquad (9.3)$$

As we shall see by example, one can often describe a linear system by any of these filters, and hence one often chooses the simplest model for the desired application. For example, an ARMA filter representation with only three nonzero a_k and two nonzero b_k would be simpler than either a pure autoregressive or pure moving-average representation, which would in

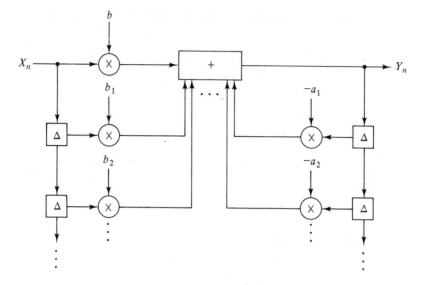

Figure 9.3   ARMA filter.

general require an infinite number of parameters. The general development of representations of one type of filter or process from another is an area of complex analysis that is outside the scope of this book. We shall, however, see some simple examples where different representations are easily found.

We are now ready to introduce three classes of random processes that are collectively called linear models since they are formed by putting an i.i.d. process into a linear system.

A discrete time random process $\{Y_n\}$ is called an *autoregressive random process* if it is formed by putting an i.i.d. random process into an autoregressive filter. Similarly, the process is said to be a *moving-average random process* or *ARMA random process* if it is formed by putting an i.i.d. process into a moving-average or ARMA filter, respectively. If a finite-order filter is used, the *order* of the process is the same as the order of the filter.

Since i.i.d. processes are uncorrelated, the techniques of chapter 8 immediately yield the power spectral densities of these processes in the two-sided weakly stationary case and yield in general the second-order moments of moving-average processes. We shall see that we can also find marginal probability distributions for moving-average processes. Perhaps surprisingly, however, the autoregressive models will prove much more useful for finding more complete specifications, that is, joint probability distributions for the output process.

PROCESSES WITH INDEPENDENT AND STATIONARY INCREMENTS

The change in value of a random process in moving forward in any given time interval is called a *jump* or *increment* of the process. The first more specific class of processes that we consider consists of random processes whose jumps or increments in nonoverlapping time intervals are independent random variables whose probability distributions depend only on the time differences over which the jumps occur. In the discrete time case, the nth output of such processes can be regarded as the sum of the first n random variables produced by an i.i.d. random process. Because the jumps in nonoverlapping time intervals then consist of sums of different i.i.d. random variables, the jumps are obviously independent. This general class of processes is of interest for three reasons: First, the class contains two of the most important examples of random processes that we will describe subsequently: the Wiener process and the Poisson counting process. Second, members of the class form building blocks for many other random process models. For example, in chapter 8 we presented an intuitive derivation of the properties of continuous time Gaussian white noise. A rigorous development would be based on the Wiener process, which we can treat rigorously with elementary tools. Third, these processes provide a useful vehicle for the introduction of several important and useful tools of probability theory: characteristic functions or transforms, conditional pmf's, conditional pdf's, and nonelementary conditional probability. In addition, independent increment processes provide specific examples of several general classes of processes: Markov processes, autoregressive processes, counting processes, and random walks. Additional examples of some of these classes of processes will be considered in later sections.

Independent and stationary increment processes are generally not stationary since, as will be seen, their probabilistic description changes with time. They possess, however, some stationarity properties. In particular, the distributions of the jumps or increments taken over fixed-length time intervals are stationary even though the distributions of the process are not. For example, if we form a process whose n^{th} output is given as the sum of n outputs of a Bernoulli process, then the resulting process is nonstationary. Specifically, if the underlying Bernoulli process has parameter p, then the mean of the nth sample of the sum process is np, and hence the process is not even weakly stationary. If, however, we form increments or differences such as $X_n = Y_n - Y_{n-1}$, then the increment process $\{X_n\}$ is stationary.

Summing successive samples is a linear operation, and hence the discrete time processes studied in this chapter can be viewed as a special case of linearly filtered memoryless processes. The special case is of particular interest, however, because of the unusual nature of the processes produced. We found in examples [8.1] and [8.2] in chapter 8 that when a filter is stable and hence the pulse response dies out with time, the effect of past input samples diminishes with time. As a result the output process is weakly stationary (two-sided) or asymptotically weakly stationary (one-sided) if the input process is weakly stationary. If we simply sum all of the samples with equal weight instead of diminishing weight, that is, if h_k is 1 for all nonnegative k, then an independent increment process is formed, and samples in the distant past affect the output just as much as do recent samples. In this case, the filter is not stable, and the output process is not even weakly stationary.

We begin the development with the class of discrete time random processes formed by summing the outputs of i.i.d. processes. This model forms the basis of much of this chapter and will be exemplified by two particular examples: the binomial counting process and the discrete time Wiener process.

Let $\{X_n; n=1,2,...\}$ be an i.i.d. process (with discrete or continuous alphabet). Let m denote its mean and σ^2 its variance. Define a new random process $\{Y_n; n=1,2,...\}$ as in example [8.3] as the output of a linear filter with pulse response $h_k = 1$ all nonnegative k and input process $\{X_n\}$; that is,

$$Y_n = \sum_{i=1}^{n} X_i; \ n=1,2,... \ . \tag{9.4a}$$

We have changed notation slightly from example [8.3] since we have (so far) defined the Y process for $n=1,2,3,...$ instead of for $n=0,1,2,3,...$, as we have done previously. This is because it will prove convenient to consider the Y process as having an initial value of 0, that is, to complete the definition of the process by defining

$$Y_0 = 0 \ . \tag{9.4b}$$

As you will see, this requirement will make the discrete time and continuous time independent increment processes more closely resemble each other. Observe that if we further let $X_0 = 0$, then by definition $\{Y_n; n\in\mathbf{Z}_+\}$ is a moving-average random process by construction with the moving-average filter $h_k = 1$ for all nonnegative k.

Since an i.i.d. input process is also uncorrelated, we can apply example [8.3] (with a slight change due to the different indexing) and evaluate the first and second moments of the Y process as

$$EY_n = mn; \ n=1,2,...$$

and

$$K_Y(k,j) = \sigma^2\min(k,j); \ k,j=1,2,... \ .$$

For later use we state these results in a slightly different notation: Since $EY_1 = m$, since $K_Y(1,1) = \sigma_{Y_1}^2 = \sigma^2$, and since the formulas also hold for $n = 0$ and for $k = j = 0$, we have that

$$EY_t = tEY_1; \quad t \geq 0 \tag{9.5a}$$

and

$$K_Y(t,s) = \sigma_{Y_1}^2 \min(t,s); \quad t,s \geq 0 . \tag{9.5b}$$

Here we explicitly consider only those values of t and s that are in the appropriate index set, here the nonnegative integers.

We now seek to use the stronger i.i.d. assumption to find out more about the Y process. It will prove useful, however, to consider an alternative representation to the linear system representation defined by (9.1) and (9.4).

An alternative model for the same process is obtained by rewriting the sum of (9.4) as a linear difference equation with initial conditions:

$$Y_0 = 0 \tag{9.6a}$$

$$Y_n = Y_{n-1} + X_n; \quad n = 1,2,3,\ldots . \tag{9.6b}$$

Observe that in this guise, $\{Y_n\}$ is a finite-order autoregressive process [see (9.2)] since it is obtained by passing the i.i.d. X process through a first-order autoregressive filter with $a_0 = 1$ and $a_1 = -1$. This description is depicted in Figure 9.4. Observe again that this filter is not stable, but it does have a transfer function, $H(f) = 1/(1 - e^{-j2\pi f})$ (which, with the substitution $z = e^{j2\pi f}$, has a pole on the unit circle).

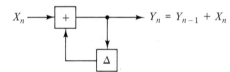

Figure 9.4   First-order autoregressive source, $a_0 = 1$, $a_1 = -1$.

As a final descriptive classification of the general process before considering a specific example, we consider the increments of the process. The increments or jumps or differences of a random process are obtained by picking a collection of ordered sample times and forming the pairwise differences of the samples of the process taken at these times. For example, given a discrete time or continuous time random process $\{Y_t; t \in \boldsymbol{I}\}$, one can choose a collection of sample times $t_0, t_1, \ldots, t_k$, $t_i \in \boldsymbol{I}$ all i, where we assume that the sample times are *ordered* in the sense that

$$t_0 < t_1 < t_2 < \ldots < t_k .$$

Given this collection of sample times, the corresponding *increments* of the process $\{Y_t\}$ are the differences

$$Y_{t_i} - Y_{t_{i-1}}; \ i = 1, 2, \ldots, k \ .$$

Note that the increments very much depend on the choice of the sample times; one would expect quite different behavior when the samples are widely separated than when they are nearby. We can now define the general class of processes with independent increments for both the discrete and continuous time cases.

A random process $\{Y_t; t \in \mathbf{I}\}$ is said to have *independent increments* or to be an *independent increment random process* if for all choices of $k = 1, 2, \ldots$ and all choices of $k + 1$ ordered sample times, the k increments $Y_{t_i} - Y_{t_{i-1}}; \ i = 1, 2, \ldots, k$ are independent random variables. An independent increment process is said to have *stationary increments* if the distribution of the increment $Y_{t+\delta} - Y_{s+\delta}$ does not depend on δ for all allowed values of $t > s$ and δ. (Observe that this is really only a first-order stationarity requirement on the increments, not by definition a strict stationarity requirement, but the jargon is standard. In any case, if the increments are independent and stationary in this sense, then they are also strictly stationary.)

We shall call a random process an *independent stationary increment* or *i.s.i.* process if it has independent and stationary increments.

We shall always make the additional assumption that 0 is the smallest possible time index; that is, that $t \geq 0$ for all $t \in \mathbf{I}$, and that $Y_0 = 0$ as in the discrete time case. We shall see that such processes are not stationary and that they must "start" somewhere or, equivalently, be one-sided random processes. We simply define the starting time as 0 for convenience and fix the starting value of the random process as 0, again for convenience. If these initial conditions are changed, the following development changes only in notational details.

A discrete time random process is an i.s.i. process if and only if it can be represented as a sum of i.i.d. random variables. To see this, observe that if $\{Y_n\}$ has independent and stationary increments, then by choosing sample times $t_i = i$ and defining $X_n = Y_n - Y_{n-1}$ for $n = 1, 2, \ldots$, then the X_n must be independent from the independent increment assumption, and they must be identically distributed from the stationary increment assumption. Thus we have that

$$Y_n = \sum_{k=1}^{n} (Y_k - Y_{k-1}) = \sum_{k=1}^{n} X_k,$$

and hence Y_n has the form of (9.4). Conversely, if Y_n is the sum of i.i.d. random variables, then increments will always have the form

$$Y_t - Y_s = \sum_{i=s+1}^{t} X_i; \ t > s,$$

that is, the form of sums of disjoint collections of i.i.d. random variables, and hence they will be independent. Furthermore, the increments will clearly be stationary since they are sums of i.i.d. random variables; in particular, the distribution of the increment will depend only on the number of samples added and not on the starting time. Unfortunately, there is no such nice construction of continuous time independent increment processes. The natural continuous time analog would be to integrate a memoryless process, but as with white noise, such memoryless processes are not well-defined. One can do formal derivations analogous to the discrete time case and sometimes (but not always) arrive at correct answers. We will use alternative and more rigorous tools when dealing with the continuous time processes. We do note, however, that while we cannot express a continuous time process with independent increments as the output of a linear system driven by a continuous time memoryless process, for any collection of sample times $t_0 = 0$, $t_1, t_2, \ldots, t_k$ we can write

$$Y_{t_n} = \sum_{i=1}^{n} (Y_{t_i} - Y_{t_{i-1}}) \tag{9.7}$$

and that the increments in the parentheses are independent—that is, we can write Y_{t_n} as a sum of independent increments (in many ways, in fact)—and the increments are identically distributed if the time intervals are identical for all increments.

Since discrete time i.s.i. processes can always be expressed as the sum of i.i.d. random variables, their first and second moments always have the form of (9.5). We shall later see that (9.5) also holds for continuous time processes with stationary and independent increments!

We again emphasize that an independent increment process may have stationary increments, but we already know from the moment calculations of (9.5) that the process itself cannot be stationary. Since the mean and covariance grow with time, independent increment processes clearly only make sense as one-sided processes.

A COUNTING PROCESS

As an example of an independent increment process we construct a process by summing the outputs of a Bernoulli process. We shall thereby obtain a quite complicated process by performing a simple operation on a simple process. The development provides practice in numerous old and new techniques.

Let $\{X_n; n = 1, 2, \ldots\}$ be a Bernoulli process with parameter p, that is, a binary i.i.d. process with alphabet $\{0, 1\}$ and a marginal pmf described by $p_X(1) = p$. Define the random process $\{Y_n; n = 1, 2, \ldots\}$ as in (9.4) and (9.6).

From (9.4), each random variable Y_n provides a count of the number of 1's appearing in the X_n process through time n. Because of this counting structure we have that either

$$Y_n = Y_{n-1} \text{ or } Y_n = Y_{n-1}+1; \; n=2,3,\dots. \tag{9.8}$$

In general, a discrete time process that satisfies (9.8) is called a *counting process* since it is nondecreasing, and when it jumps, it is always with an increment of 1. A continuous alphabet counting process is similarly defined as a process with a nondecreasing output that increases in steps of 1. The jumps in a continuous time process can, however, occur quite close to one another.

This example is a special case of an independent increment process, an autoregressive process, and a counting process. We now proceed to look at its probabilistic properties.

Observe that the second-order moments are immediately given by (9.5) as $EY_n = np$ and $K_Y(k,j) = p(1-p)\min(k,j)$. We now find the marginal pmf for Y_n, $p_{Y_n}(y) = \Pr(Y_n = y)$. This is easily found from the fundamental derived distribution formula and combinatoric arguments. Let $\mathbf{X} = (X_1,\dots,X_n)$ and define the *Hamming weight* $h(\mathbf{x})$ of a vector $\mathbf{x} = (x_1,\dots,x_n)$ as the number of coordinates of $\mathbf{x}$ that have nonzero entries; e.g., $h((0,1,1,0,0,0))=2$. The number of $\mathbf{x}$ vectors with a given Hamming weight k is given by the binomial coefficient

$$\binom{n}{k} = \frac{n!}{k!(n-k)!}.$$

Therefore

$$\begin{aligned}
p_{Y_n}(k) &= \Pr(\text{there are exactly } k \text{ ones in } \mathbf{X}) \\
&= \sum_{\mathbf{x}:h(\mathbf{x})=k} p_{\mathbf{X}}(\mathbf{x}) \\
&= \sum_{\mathbf{x}:h(\mathbf{x})=k} p^{h(\mathbf{x})}(1-p)^{n-h(\mathbf{x})} \\
&= \sum_{\mathbf{x}:h(\mathbf{x})=k} p^k (1-p)^{n-k} \\
&= \binom{n}{k} p^k (1-p)^{n-k}; k=0,1,\dots,n,
\end{aligned}$$

the binomial pmf! Because of this marginal, the given process is called the *binomial counting process*.

Observe that the process cannot be stationary since the marginal pmf for a single sample depends on the time index, n.

This derivation is fairly simple, but it is very much aimed at the specific example considered of a Bernoulli i.i.d. process being summed. The specific formulas do not readily generalize. For example, if we make the simple change of assuming that the underlying i.i.d. process has alphabet $\{+1, -1\}$ instead of $\{0,1\}$, then the nature of the resulting independent increment process changes drastically, and the previous derivation no longer applies. The new process can wander through positive and negative numbers, and is no longer a counting process. It is, in fact, called a *discrete random walk process*, which provides a model for the classical one-dimensional random walk or drunkard's walk of an object, which at each time randomly selects a direction and then takes a step in that direction. Although the random walk can be handled by a simple linear transformation from the binomial counting process, more complicated independent increment processes cannot. In order to derive properties for this and other independent increment processes, it is instructive to consider an alternate derivation that does not involve combinatorics but deals directly with the sum of independent random variables. The derivation will remain valid in more general sums of i.i.d. random variables.

DERIVED DISTRIBUTIONS OF SUMS OF INDEPENDENT RANDOM VARIABLES

For simplicity we begin with the case of only two random variables and rename them to avoid fooling with subscripts. In addition we focus first on discrete random variables.

Say we have two random variables W and Z and we form a new random variable $Y = W + Z$. Using the derived distribution technique we can find a formula for the pmf of Y given the joint pmf for W and Z as follows:

$$
\begin{aligned}
p_Y(y) &= \Pr(W + Z = y) \\
&= \sum_{w,z : w+z=y} p_{W,Z}(w,z) \\
&= \sum_w p_{W,Z}(w, y-w) ,
\end{aligned}
$$

where we have used the fact that given one of the two dummy variables w and z, the other is fixed by the requirement that $w + z = y$. In a general context the formula is as far as we can proceed. If the random variables W and Z are independent, as in our current application, however, then the pmf factors, and we have that

$$
p_Y(y) = \sum_w p_W(w) p_Z(y-w) ,
$$

a discrete convolution! We have therefore that the pmf of the sum of two independent random variables is the convolution of the two marginal pmf's. A similar result holds for continuous random variables; that is, if the random variables W and Z are described by pdf's f_W and f_Z and if they are independent, then the pdf for the sum Y is given by the ordinary convolution integral

$$f_Y(y) = \int_{-\infty}^{\infty} f_W(w)f_Z(y-w)dw \ .$$

This integral result can be proved by deriving the cdf for Y and then obtaining the pdf using the differentiation of an integral formula (4.16).

The convolution formula can be used in concept to find the pmf of Y_n recursively. Letting $W = X_1$ and $Z = Y_0$, we find the pmf of Y_1. Then letting $W = X_2$ and $Z = Y_1$, we find the pmf of Y_2. Continuing in this fashion, we find the pmf of Y_n as an n-fold convolution. The convolution result is interesting, but it has some serious drawbacks. In particular, convolutions are often complicated even with only two functions. Trying to find the pmf for Y_n for large n through the n-fold convolution is an intractable computation if attempted directly. Here, however, we can take advantage of a parallel with linear systems theory: When one needs to find the output waveform of a cascade or series of linear filters, one does not perform a sequence of convolutions. Instead one takes the Fourier or Laplace transform of the impulse responses and uses the fact that convolution in the time domain corresponds to multiplication in the frequency domain. Thus a difficult or impossible problem in one domain becomes a simple multiplication in the transform domain. The same trick works here, although we will have none of the usual engineering intuition of "frequency domain" since here we are transforming probability functions and not time waveforms.

Since the same trick will work both for the current discrete alphabet case and for the continuous alphabet case, we consider the two together for the moment. Recall from example [6.6] the following definition:

Let X be a random variable with cdf F_X. Define the *characteristic function M_X* of the cdf by

$$M_X(ju) = E(e^{juX}) = \int e^{jux} dF_X(x)$$

$$= \begin{cases} \sum_x e^{jux} p_X(x) \\ \text{or} \\ \int e^{jux} f_X(x) dx \ . \end{cases}$$

The characteristic function of a cdf or, equivalently, of a pmf or pdf is simply an exponential transform. In the form here it is called the Fourier-Stieltjes transform. We will, as usual, not be overly concerned with demonstrating the existence of such a transform. We shall usually simply assume that it exists and evaluate it when needed. Obviously, if the evaluation fails, more care will have to be exercised. We shall also assume that the student is familiar with the basic properties of Fourier or Laplace transforms and shall simply quote them as needed. Details may be found in the standard literature on the subject. In particular, we shall use the fact that such transforms have an essentially unique inverse, but we shall invert such transforms only by inspection or table lookup—we shall not concern ourselves with more general inversion techniques requiring contour integration and such.

Before considering an example of a characteristic function, we demonstrate its principal application, which we state formally.

Lemma 9.1.   Given a sequence of mutually independent random variables $X_1, X_2, \ldots, X_n$, define

$$Y_n = \sum_{i=1}^{n} X_i .$$

Then

$$M_{Y_n}(ju) = \prod_{i=1}^{n} M_{X_i}(ju).$$

Thus, for example, if the X_i are also identically distributed and hence have identical transforms, say M_X, then

$$M_{Y_n}(ju) = M_X(ju)^n .$$

Proof.   Successive application of theorem 6.1, which states that functions of independent random variables are uncorrelated, yields

$$E(e^{juY_n}) = E\left(e^{ju\sum_{i=1}^{n} X_i}\right) = E(\prod_{i=1}^{n} e^{juX_i})$$

$$= \prod_{i=1}^{n} E(e^{juX_i}) = \prod_{i=1}^{n} M_{X_i}(ju) .$$

THE BINOMIAL COUNTING PROCESS

Returning to the binomial counting process, the transform of the samples of the original Bernoulli process is easily found to be

$$M_X(ju) = \sum_{k=0}^{1} e^{juk} p_X(k) = (1-p)+pe^{ju},$$

and hence from the lemma we have that

$$M_{Y_n}(ju) = \sum_{k=0}^{n} p_{Y_n}(k)e^{juk}$$

$$= M_X(ju)^n$$

$$= ((1-p)+pe^{ju})^n$$

$$= \sum_{k=0}^{n} [\binom{n}{k}(1-p)^{n-k}p^k]e^{juk},$$

where we have invoked the binomial theorem in the last step. For the equality to hold, however, we have from the uniqueness of transforms (see example [6.6]) that $p_{Y_n}(k)$ must be the bracketed term, that is, the binomial pmf.

While the combinatoric approach provides a quick solution to this particular problem, the transform approach is useful far more generally. The following maxim should be kept in mind whenever faced with sums of independent random variables:

When given a derived distribution problem involving the sum of independent random variables, first find the characteristic function of the sum by taking the product of the characteristic functions of the individual random variables. Then find the corresponding probability function by inverting the transform. This technique is valid if and only if the random variables are independent—they do not have to be identically distributed.

*A DISCRETE RANDOM WALK

As a second example of the preceding development, consider the random walk where the i.i.d. process used has alphabet $\{+1,-1\}$ and, say, $\Pr(X_n = -1) = p$. In this case the transform of the i.i.d. random variables is

$$M_X(ju) = (1-p)e^{ju} + pe^{-ju},$$

and hence using the binomial theorem we have that

$$M_{Y_n}(ju) = ((1-p)e^{ju} + pe^{-ju})^n$$

$$= \sum_{k=0}^{n} \left[\binom{n}{k} (1-p)^{n-k} p^k \right] e^{ju(n-2k)}$$

$$= \sum_{k=-n,-n+2,\ldots,n-2,n} \left[\binom{n}{\frac{n-k}{2}} (1-p)^{\frac{n+k}{2}} p^{\frac{n-k}{2}} \right] e^{juk} \ .$$

Comparison of this formula with the definition of the characteristic function reveals that the pmf for Y_n is given by

$$p_{Y_n}(k) = \binom{n}{\frac{n-k}{2}} (1-p)^{\frac{n+k}{2}} p^{\frac{n-k}{2}} \ , \quad k=-n,-n+2,\ldots,n-2,n \ .$$

Note that Y_n must be even or odd depending on whether n is even or odd. This follows from the nature of the increments.

THE DISCRETE TIME WIENER PROCESS

We next provide another example of a discrete time independent increment process that demonstrates the application of the transform technique.

Let $\{X_n\}$ be an i.i.d. process with zero-mean Gaussian marginal pdf's and variance σ^2. Then the process $\{Y_n\}$ defined by (9.4) or (9.6) is called the *discrete time Wiener process.*

Other names for this process are the discrete time diffusion process and discrete time Brownian motion. As with the binomial counting process and all other independent increment random processes, the Wiener process is a first-order autoregressive process through the representation of (9.6). Since it is formed as sums of an i.i.d. process, it has independent and stationary increments. Its second-order moments are found from (9.5) to be $EY_n = 0$ and $K_Y(k,j) = \sigma^2 \min(k,j)$.

From lemma 9.1 the characteristic function of the pdf for Y_n can be found from the transform $M_X(ju)$ of the original i.i.d. process using the moving-average representation (9.4). Hence we evaluate this transform. For future use we find the characteristic function of a Gaussian pdf with a possibly nonzero mean as well for a zero mean. This is accomplished as follows using the "complete the square" technique (or published tables): Assume that X is a Gaussian random variable with mean m and variance σ^2. Then

$$M_X(ju) = E(e^{juX}) = \int_{-\infty}^{\infty} \frac{1}{(2\pi\sigma^2)^{1/2}} e^{-\frac{(x-m)^2}{2\sigma^2}} e^{jux} dx$$

$$= \int_{-\infty}^{\infty} \frac{1}{(2\pi\sigma^2)^{1/2}} e^{-\frac{(x^2 - 2mx - 2\sigma^2 jux + m^2)}{2\sigma^2}} dx$$

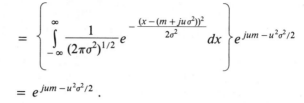

$$= e^{jum - u^2\sigma^2/2} .$$

Thus the characteristic function of a Gaussian random variable with mean m and variance σ_X^2 is

$$M_X(ju) = e^{jum - u^2\sigma^2/2} ,$$

and hence from lemma 9.1 for the identically distributed case, the characteristic function of Y_n is

$$M_{Y_n}(ju) = e^{junm - nu^2\sigma_X^2/2} . \qquad (9.9)$$

From the uniqueness property of transforms, this is the transform of a Gaussian random variable with mean nm and variance $n\sigma_X^2$! Thus the discrete time Wiener process has Gaussian marginal pdf's, and we have seen that the sum of several independent Gaussian random variables yields another Gaussian random variable. We shall later see that more general linear combinations of Gaussian random variables yield other Gaussian random vectors.

CONDITIONAL PMF'S AND SPECIFICATION

We have thus far obtained only the second-order moments and the marginal distribution of discrete time processes with independent and stationary increments. To specify such processes completely, however, we must in principle give the probability functions (e.g., the cdf's) for all possible finite collections of samples of the process; that is, we require a consistent family of joint cdf's or pmf's or pdf's as in chapter 5. Actually, here we need not worry about consistency since we have defined the process $\{Y_n\}$ on an underlying probability space (the output space of the $\{X_n\}$ process), and hence the joint distributions must be consistent if they are correctly computed from the underlying probability measure—the process distribution for the i.i.d. process. In this section we develop the specification for the simpler case of discrete alphabet processes.

To specify a discrete time discrete alphabet random process requires in principle the joint pmf's for all possible finite collections of sample times. For such a single-sided discrete time process, however, it suffices to find all pmf's of the form $p_{Y_1, Y_2, ..., Y_n}(y_1, y_2, ..., y_n)$ for $n = 1, 2, ...$ since all other pmf's may be found from this family by summing out the unwanted dummy variables. To simplify the notation somewhat, we introduce a different notation for a vector when we wish to make the dimension explicit:

Define $Y^n = (Y_1, Y_2, \ldots, Y_n)$ and $y^n = (y_1, y_2, \ldots, y_n)$; that is, we use the superscript to denote a vector of the given dimension instead of boldface when we wish to make explicit the dimension of the vector.

Thus the problem of specifying the counting process is to find expressions for $p_{Y^n}(y^n)$. Directly tackling this problem can lead to complications. In the specific case of independent increment processes and in many more general examples, it is much simpler to describe probabilistically the next output Y_n of the process if we are given the previous $n-1$ outputs $Y_1, \ldots, Y_{n-1}$, that is, to find conditional probabilities rather than joint probabilities. For example, in the binomial counting process, the next output is formed simply by adding a binary random variable to the old sum. Thus all of the conditional probability mass is concentrated on two values— the last value and the last value plus 1. In the general discrete alphabet case, a simple application of the definition of elementary conditional probability allows us to form conditional pmf's. We first give the basic definition and then apply it to the specification problem.

Let U and V denote two discrete random variables or random vectors with joint pmf $p_{U,V}$. Define the *conditional pmf* $p_{U|V}(u|v)$ of U given V by

$$p_{U|V}(u|v) = \Pr(U=u \mid V=v) = \frac{p_{U,V}(u,v)}{p_V(v)}, \qquad (9.10)$$

where, as always,

$$p_V(v) = \sum_u p_{U,V}(u,v)$$

and where the definition is only valid for those v for which $p_V(v) \neq 0$. Thus a conditional pmf is simply an elementary conditional probability.

Thus, for example,

$$p_{X_n|X_{n-1},X_{n-2},\ldots,X_1}(x_n|x_{n-1},x_{n-2},\ldots,x_1) = \Pr(X_n=x_n \mid X_i=x_i; i=1,2,\ldots,n-1).$$

We can iterate on the definition of conditional probability to obtain an expression for a joint pmf in terms of the conditional pmf's:

$$p_{Y_1,\ldots,Y_n}(y_1,\ldots,y_n) =$$

$$p_{Y_n|Y_{n-1},\ldots,Y_1}(y_n|y_{n-1},\ldots,y_1)p_{Y_1,\ldots,Y_{n-1}}(y_1,\ldots,y_{n-1}) =$$

$$= p_{Y_1}(y_1)\prod_{i=2}^{n}p_{Y_i| Y_{i-1},\ldots,Y_1}(y_i|y_{i-1},\ldots,y_1) \ . \tag{9.11}$$

Thus the conditional pmf's imply the joint pmf, and vice versa. Observe that this formula provides a generalization of the joint pmf of an i.i.d. process: If the process is i.i.d., the conditional pmf's become marginal pmf's, and we have the usual product form. In general the formula expresses the joint pmf as a product, but each term in the product depends on the values of the previous samples! The formula can be viewed as a chain rule for probabilities.

SPECIFICATION OF DISCRETE INDEPENDENT INCREMENT PROCESSES

We return to the original problem, specifying a discrete time discrete alphabet process $\{Y_n\}$ with independent and stationary increments defined by (9.6). As just discussed, we can accomplish this by finding the conditional pmf's

$$p_{Y_n| Y_{n-1},\ldots,Y_1}(y_n|y_{n-1},\ldots,y_1) = p_{Y_n| Y^{n-1}}(y_n|y^{n-1})$$

$$= \mathrm{Pr}(Y_n=y_n| Y^{n-1}=y^{n-1}) \ .$$

From the definition of the Y_n process, however, we have that the conditioning event $\{Y_i=y_i; \ i=1,2,\ldots,n-1\}$ is identical to the event $\{X_1=y_1, \ X_i=y_i-y_{i-1}; \ i=2,3,\ldots,n-1\}$. In words, the Y_n will assume the given values if and only if the X_n assume the corresponding differences since the Y_n are defined as the sum of the X_n. Furthermore, given this event, Y_n will equal y_n if and only if $X_n=y_n-y_{n-1}$, i.e., if the current sample of X_n provides the appropriate difference. Thus the conditional probability becomes

$$p_{Y_n| Y_{n-1},\ldots,Y_1}(y_n|y_{n-1},\ldots,y_1) =$$

$$\mathrm{Pr}(Y_n=y_n| X_1=y_1,X_i=y_i-y_{i-1}; \ i=2,\ldots,n-1)$$

$$= \mathrm{Pr}(X_n=y_n-y_{n-1}| X_i=y_i-y_{i-1}; \ i=1,2,\ldots,n-1) \ , \tag{9.12}$$

where we have shortened the notation of the conditioning event by defining $y_0=0$. Now, however, we have a probability entirely in terms of the given X_i variables, in particular,

$$p_{Y_n| Y_{n-1},\ldots,Y_1}(y_n|y_{n-1},\ldots,y_1) =$$

$$p_{X_n| X_{n-1},\ldots,X_2,X_1}(y_n-y_{n-1}|y_{n-1}-y_{n-2},\ldots,y_2-y_1,y_1) \ . \tag{9.13}$$

So far we have not used the fact that the X_n are i.i.d.; that is, the development is valid for any process. If the $\{X_n\}$ are i.i.d., then the conditional pmf's are simply the marginal pmf's since each X_n is independent of past $X_k; k < n$. Thus we have that

$$p_{Y_n | Y_{n-1},\dots,Y_1}(y_n | y_{n-1},\dots,y_1) = p_X(y_n - y_{n-1}) , \qquad (9.14)$$

and hence the vector pmf is

$$p_{Y^n}(y^n) = \prod_{i=1}^{n} p_X(y_i - y_{i-1}) , \qquad (9.15)$$

providing the desired specification.

To apply this formula to the special case of the binomial counting process, we need only plug in the binary pmf for p_X. A useful way of writing the binary pmf for this purpose is

$$p_X(x) = p^x(1-p)^{1-x}; \; x = 0,1 ,$$

which with the given formula yields the desired specification of the binomial counting process:

$$p_{Y^n}(y^n) = \prod_{i=1}^{n} p^{(y_i - y_{i-1})}(1-p)^{1-(y_i - y_{i-1})} ,$$

$$\text{where } y_i - y_{i-1} = 0 \text{ or } 1, i = 1,2,\dots,n; \; y_0 = 0 . \qquad (9.16)$$

The specification of the discrete random walk process follows in a like manner from (9.15).

The complete specification just given permits us to point out an interesting property of the binomial counting process. We could perform a derivation almost the same as the foregoing to evaluate the conditional pmf for Y_n given only its immediate predecessor, as follows:

$$p_{Y_n | Y_{n-1}}(y_n | y_{n-1}) = \Pr(Y_n = y_n | Y_{n-1} = y_{n-1})$$

$$= \Pr(X_n = y_n - y_{n-1} | Y_{n-1} = y_{n-1}) .$$

The conditioning event, however, depends only on values of X_k for $k < n$, and X_n is independent of its past; hence

$$p_{Y_n | Y_{n-1}}(y_n | y_{n-1}) = p_X(y_n - y_{n-1}) . \qquad (9.17)$$

The same conclusion can be reached by the longer route of using the joint pmf for Y^n previously computed to find the joint pmf for Y_n and Y_{n-1}, which in turn can be used to find the conditional pmf. Comparison with (9.14) reveals that discrete time independent increment processes with stationary increments (such as the binomial counting process and the discrete random walk) have the property that

$$p_{Y_n | Y_{n-1},\dots,Y_1}(y_n | y_{n-1},\dots,y_1) = p_{Y_n | Y_{n-1}}(y_n | y_{n-1}) \qquad (9.18)$$

or, equivalently,

$$\Pr(Y_n = y_n \mid Y_i = y_i;\ i = 1,\ldots,n-1) = \Pr(Y_n = y_n \mid Y_{n-1} = y_{n-1})\ . \quad (9.19)$$

Roughly speaking, given the most recent past sample (or the current sample), the remainder of the past does not affect the probability of what happens next. The same conclusion can be reached from the representation of (9.6). Because the input process is i.i.d., X_n is independent of Y_{n-1}. Therefore, from (9.6), the conditional probability of Y_n given Y_{n-1} is dependent only on the pmf of X_n. Alternatively stated, given the present, the future is independent of the past. A discrete time discrete alphabet random process with this property is called a *Markov process*. (Some authors call such processes a *Markov chain*.) Thus we have shown that all discrete alphabet independent increment processes (and the binomial counting process and the discrete random walk in particular) are Markov processes (or Markov chains).

CONDITIONAL PDF'S AND SPECIFICATION

We next focus on a similar specification in the continuous alphabet case. As in the discrete alphabet case, this is most easily done by focusing on conditional probabilities. Formally we can mimic the relation (9.11) of conditional and joint pmf's to write

$$f_{Y_1,\ldots,Y_n}(y_1,\ldots,y_n) =$$

$$\frac{f_{Y_1,\ldots,Y_n}(y_1,\ldots,y_n)}{f_{Y_1,\ldots,Y_{n-1}}(y_1,\ldots,y_{n-1})} f_{Y_1,\ldots,Y_{n-1}}(y_1,\ldots,y_{n-1})$$

$$\cdot$$
$$\cdot$$
$$\cdot$$

$$f_{Y_1}(y_1) \prod_{i=2}^{n} \frac{f_{Y_1,\ldots,Y_i}(y_1,\ldots,y_i)}{f_{Y_1,\ldots,Y_{i-1}}(y_1,\ldots,y_{i-1})}\ .$$

This suggests the following definition:

Define the *conditional pdf* of Y_n given Y_i; $i = 1,\ldots,n-1$ by

$$f_{Y_n \mid Y_{n-1},\ldots,Y_1}(y_n \mid y_{n-1},\ldots,y_1) = \frac{f_{Y_1,\ldots,Y_n}(y_1,\ldots,y_n)}{f_{Y_1,\ldots,Y_{n-1}}(y_1,\ldots,y_{n-1})}\ .$$

With this definition the previous formula becomes a pdf parallel to (9.11):

$$f_{Y_1,\ldots,Y_n}(y_1,\ldots,y_n) =$$

$$f_{Y_1}(y_1)\prod_{i=2}^{n}f_{Y_i|Y_{i-1},...,Y_1}(y_i|y_{i-1},...,y_1) \ . \qquad (9.20)$$

As in the pmf case, this formula can be viewed as a generalization of the i.i.d. example since if the random variables are independent, the conditional densities are simply the marginal densities.

Although (9.20) provides an analogy to (9.11), it is not clear what a conditional pdf means or how it relates to conditional probability. Thus before using the formula to compute specific examples of specifications, we consider the general issue of conditional probability for continuous random variables or nonelementary conditional probability. This will in turn provide the required properties of conditional pdf's.

CONDITIONAL PROBABILITY

We defined and studied elementary conditional probability in chapter 4 and we have seen in this chapter that it plays a fundamental role in describing the relation between random variables and, in particular, in describing the relation among samples of a random process. Perhaps the most important such probabilities have the general form of $\Pr(X \in F | Y = y)$, that is, the probability that one random variable or vector X takes on a value in a specified event given that another random variable or vector Y assumes a specific value. We consider a general X event F instead of an event of the form $X = x$ since, as we have seen, the latter is of little use in studying continuous random variables and vectors. We permit X and Y to be random variables or random vectors to simplify notation. In the case of discrete random variables or vectors, this conditional probability is well defined as an elementary conditional probability since the conditioning event has nonzero probability. This led to the development of conditional pmf's.

Conditional pmf's are seen to be ordinary elementary conditional probabilities. That is, given discrete random variables or vectors X and Y that occur with nonzero probability, we can define $p_{X|Y}(x|y)$ as the elementary conditional probability $\Pr(X = x | Y = y)$. The elementary definition makes sense here because the conditioning event $\{Y = y\}$ has nonzero probability. Then, given the conditional pmf, we can compute the conditional probability $P(X \in F | Y = y)$ from the formula

$$P(X \in F | Y = y) = \sum_{x \in F} p_{X|Y}(x|y) \ . \qquad (9.21)$$

Equation (9.21) has a simple and intuitive interpretation: A conditional pmf gives the mass of conditional probability; that is, just as we sum an ordinary pmf over points in F to find the ordinary probability of F, we sum the conditional pmf over points in F to find the conditional probability of F. The properties of pmf's give us an additional intuitive property of the conditional probability $P(X\epsilon F| Y=y)$: If we multiply both sides of (9.21) by the pmf for y and sum over all y in some set G, we find that

$$\sum_{y\epsilon G} P(X\epsilon F| Y=y)p_Y(y) = \sum_{y\epsilon G}\sum_{x\epsilon F} p_{X|Y}(x|y)p_Y(y)$$

$$= \sum_{y\epsilon G}\sum_{x\epsilon F} p_{X,Y}(x,y) ,$$

so

$$\sum_{y\epsilon G} P(X\epsilon F| Y=y)p_Y(y) = P(X\epsilon F,Y\epsilon G)$$

$$= P_{X,Y}(F\times G) . \qquad (9.22)$$

Thus (9.22) provides a relation in the discrete case between a conditional probability and a joint probability—the conditional probability of an X event F averaged with respect to the pmf on the Y event G gives the joint probability of F and G.

In the continuous case, the conditioning event has zero probability, so care must be used in defining (nonelementary) probability. Nonetheless, as we will see, the correct extension is obvious. A natural means of extending the idea of conditional probability $P(X\epsilon F| Y=y)$ to the continuous case is to mimic the relations with pdf's instead of pmf's, that is, to guess that a conditional probability should be an integral over a conditional pdf just as an ordinary probability is an integral over an ordinary pdf. In other words, it seems reasonable to suspect that a conditional pdf is a density of conditional probability. Say that we have two random variables X and Y and suppose that $f_Y(y)$ is nonzero for some fixed y so that the conditional pdf

$$f_{X|Y}(x|y) = \frac{f_{X,Y}(x,y)}{f_Y(y)}$$

is well defined. First observe that for each fixed y, the conditional pdf $f_{X|Y}(x|y)$ is indeed a pdf considered as a function of x because it satisfies the defining properties of a pdf. It is nonnegative and, if we integrate over x, we have

$$\int f_{X|Y}(x|y)dx = \int \frac{f_{X,Y}(x,y)}{f_Y(y)}dx$$

$$= \frac{1}{f_Y(y)}\int f_{X,Y}(x,y)dx$$

$$= \frac{f_Y(y)}{f_Y(y)} = 1 .$$

Thus we can tentatively define the conditional probability as the integral analog to (9.21):

$$P(X \in F | Y = y) = \int_F f_{X|Y}(x|y)dx \, , \qquad (9.23)$$

that is, we define a conditional pdf as a density of conditional probability. Is this a useful definition? The real test is whether or not it plays the same role that it did in the discrete case; that is, whether or not we can average over conditional probabilities to find joint probabilities. To see if the integral analog of (9.22) holds, we multiply both sides of (9.23) by the marginal pdf for Y and integrate over an event G to obtain

$$\int_G dy P(X \in F | Y = y) f_Y(y) = \int_G dy \int_F dx f_{X|Y}(x|y) f_Y(y)$$

$$= \int_G dy \int_F dx f_{X,Y}(x,y)$$

$$= P(X \in F, Y \in G) \, , \qquad (9.24)$$

as desired. Thus the definition of conditional probability of (9.23) behaves in the manner that one would like. Using the Stieltjes notation we can combine (9.22) and (9.24) into the equation

$$\int_G P(X \in F | Y = y) dF_Y(y) =$$

$$P(X \in F, Y \in G) = P_{X,Y}(F \times G); \text{ all events } G, \qquad (9.25)$$

which is valid in both the discrete case and in the continuous case when one has a conditional pdf. In advanced probability, (9.25) is taken as the *definition* for the general (nonelementary) conditional probability $P(X \in F | Y = y)$; that is, the conditional probability is defined as any function of y and F that satisfies (9.25) for all events G. This reduces to the given constructive definitions of (9.21) in the discrete case and (9.23) in the continuous case when one has a conditional pdf. It also provides a useful theory for continuous cases when the conditional pdf is not well defined. Like the standard theory for impulse or Dirac delta functions, (9.25) defines a function by its behavior under an integral rather than by a constructive definition. In this book we will always encounter conditional probabilities that have a constructive definition using pmf's or pdf's, but it is useful to remember that (9.25) is really the common defining characteristic of such conditional probabilities.

Before leaving the abstract discussion and returning to concrete considerations, observe that if G is an event with nonzero probability $P_Y(G) > 0$, then we can find the elementary conditional probability $P(X \in F | Y \in G)$ using (9.25) as

$$P(X \epsilon F | Y \epsilon G) = \frac{P_{X,Y}(F \times G)}{P_Y(G)} =$$

$$\frac{1}{P_Y(G)} \int_G P(X \epsilon F | Y = y) dF_Y(y) , \qquad (9.26)$$

that is, one can view nonelementary conditional probability as being defined by the requirement that when averaged and suitably scaled, it gives the correct formula for elementary conditional probability.

SPECIFICATION OF THE DISCRETE TIME WIENER PROCESS

For the discrete time Wiener process we use the fact that Y_n is the sum of an i.i.d. sequence of random variables in exactly the same way as we did for the discrete amplitude independent increment random process. The only difference is that we now consider nonelementary conditional probabilities and densities of conditional probability instead of elementary conditional probability and conditional pmf's. Using the autoregressive representation (9.6), we have for $n > 0$ that Y_n is the sum of two independent random variables, Y_{n-1} and X_n. The first quantity is treated as if it were a constant in calculating conditional probabilities for Y_n given all Y_i for $i < n$, and hence the *conditional* probability of an event for Y_n can be specified in terms of the *marginal* probability of an easily determined event for X_n. Specifically, the conditional cdf for Y_n is

$$P(Y_n \leq y_n | y_{n-1}, y_{n-2}, \ldots) = P(X_n \leq y_n - y_{n-1}) \qquad (9.27)$$

$$= F_X(y_n - y_{n-1}) ,$$

where $P(Y_n \leq y_n | y_{n-1}, y_{n-2}, \ldots)$ is shorthand for the conditional probability $P(Y_n \leq y_n | Y_{n-1} = y_{n-1}, Y_{n-2} = y_{n-2}, \ldots)$. Differentiating both sides with respect to y_n we find that the conditional pdf's satisfy

$$f_{Y_n | Y_{n-1}, \ldots, Y_1}(y_n | y_{n-1}, \ldots, y_1) = f_X(y_n - y_{n-1}) , \qquad (9.28)$$

and hence

$$f_{Y^n}(y^n) = \prod_{i=1}^{n} f_X(y_i - y_{i-1}) , \qquad (9.29)$$

the pdf equivalent to (9.14) and (9.15). This provides the desired specification. Plugging in the Gaussian density for f_X yields

$$f_{Y^n}(y^n) = \frac{e^{-\frac{y_1^2}{2\sigma^2}}}{\sqrt{2\pi\sigma^2}} \prod_{i=2}^{n} \frac{e^{-\frac{(y_i - y_{i-1})^2}{2\sigma^2}}}{\sqrt{2\pi\sigma^2}}$$

$$= (2\pi\sigma^2)^{-n/2} e^{-\frac{1}{2\sigma^2}(\sum_{i=2}^{n}(y_i - y_{i-1})^2 + y_1^2)}$$

This formula shows that the discrete time Wiener process is indeed a Gaussian process since it has the required form with mean function $m(t) = 0$ and covariance function $K_X(t,s) = \sigma^2 \min(t,s)$. The reader is invited to verify that the inverse of the covariance matrix has the form required to yield the given pdf.

As in the discrete alphabet case, a similar argument can be used to show that

$$f_{Y_n \mid Y_{n-1}}(y_n \mid y_{n-1}) = f_X(y_n - y_{n-1})$$

and hence from (9.28) that

$$f_{Y_n \mid Y_{n-1}, \ldots, Y_1}(y_n \mid y_{n-1}, \ldots, y_1) = f_{Y_n \mid Y_{n-1}}(y_n \mid y_{n-1}) .$$

As in the discrete alphabet case, a process with this property is called a *Markov* process. We can combine these two cases with the following definition:

A discrete time random process $\{Y_n\}$ is said to be a Markov process if the conditional cdf's satisfy the relation

$$\Pr(Y_n \le y_n \mid y_{n-1}, y_{n-2}, \ldots) = \Pr(Y_n \le y_n \mid y_{n-1})$$

for all n, $y_n, y_{n-1}, y_{n-2}, \ldots$. More specifically, $\{Y_n\}$ is frequently called a *first − order* Markov process because it depends on only the most recent past value. An extended definition to n^{th}-order Markov processes can be made in the obvious fashion.

*SECOND-ORDER MOMENTS OF CONTINUOUS TIME I.S.I. PROCESSES

In this section we show that several important properties of the discrete time independent increment processes hold for the continuous time case. In the next section we generalize the specification techniques and give two examples of such processes—the continuous time Wiener process and the Poisson counting process. This section is devoted to the proof that (9.5) holds for continuous time processes with independent and stationary increments. The proof is primarily algebraic and can easily be skipped.

We now consider a continuous time random process $\{Y_t; t \in \mathbf{I}\}$, where $\mathbf{I} = [0, \infty)$, having independent stationary increments and initial condition $Y_0 = 0$. The techniques used in this section can also be used for an alternative derivation of the discrete time results.

First observe that given any time t and any positive delay or lag $\tau > 0$, we have that

$$Y_{t+\tau} = (Y_{t+\tau} - Y_t) + Y_t , \tag{9.30}$$

and hence, by the linearity of expectation,

$$EY_{t+\tau} = E[Y_{t+\tau} - Y_t] + EY_t .$$

Since the increments are stationary, however, the increment $Y_{t+\tau} - Y_t$ has the same distribution, and hence the same expectation as the increment $Y_\tau - Y_0 = Y_\tau$, and hence

$$EY_{t+\tau} = EY_\tau + EY_t .$$

This equation has the general form

$$g(t + \tau) = g(\tau) + g(t) . \tag{9.31}$$

An equation of this form is called a *linear functional equation* and has a unique solution of the form $g(t) = ct$, where c is a constant that is determined by some boundary condition. Thus, in particular, the solution to (9.31) is

$$g(t) = g(1)t . \tag{9.32}$$

Thus we have that the mean of a continuous time independent increment process with stationary increments is given by

$$EY_t = tm, t \in \boldsymbol{I} , \tag{9.33}$$

where the constant m is determined by the boundary condition

$$m = EY_1 .$$

Thus (9.5a) extends to the continuous time case.

Since $Y_0 = 0$, we can rewrite (9.30) as

$$Y_{t+\tau} = (Y_{t+\tau} - Y_t) + (Y_t - Y_0) , \tag{9.34}$$

that is, we can express $Y_{t+\tau}$ as the sum of two independent increments. The variance of the sum of two independent random variables, however, is just the sum of the two variances. In addition, the variance of the increment $Y_{t+\tau} - Y_t$ is the same as the variance of $Y_\tau - Y_0 = Y_\tau$ since the increments are stationary. Thus (9.34) implies that

$$\sigma_{Y_{t+\tau}}^2 = \sigma_{Y_\tau}^2 + \sigma_{Y_t}^2 ,$$

which is again a linear functional equation and hence has the solution

$$\sigma_{Y_t}^2 = t\sigma^2 \tag{9.35}$$

where the appropriate boundary condition is

$$\sigma^2 = \sigma_{Y_1}^2 .$$

Knowing the variance immediately yields the second moment:

$$E(Y_t^2) = \sigma_{Y_t}^2 + (EY_t)^2 = t\sigma^2 + (tm)^2 \ . \tag{9.36}$$

Consider next the autocorrelation function $R_Y(t,s)$. Choose $t > s$ and write Y_t as the sum of two increments as

$$Y_t = (Y_t - Y_s) + Y_s,$$

and hence

$$R_Y(t,s) = E[Y_t Y_s] = E[(Y_t - Y_s)Y_s] + E[Y_s^2]$$

using the linearity of expectation. The left term on the right is, however, the expectation of the product of two independent random variables since the increments $Y_t - Y_s$ and $Y_s - Y_0$ are independent. Thus from theorem 6.1 the expectation of the product is the product of the expectations. Furthermore, the expectation of the increment $Y_t - Y_s$ is the same as the expectation of the increment $Y_{t-s} - Y_0 = Y_{t-s}$ since the increments are stationary. Thus we have from this, (9.33), and (9.36) that

$$R_Y(t,s) = (t-s)msm + s\sigma^2 + (sm)^2 = s\sigma^2 + (tm)(sm) \ .$$

Repeating the development for the case $t \leq s$ then yields

$$R_Y(t,s) = \sigma^2 \min(t,s) + (tm)(sm) \ , \tag{9.37}$$

which yields the covariance

$$K_Y(t,s) = \sigma^2 \min(t,s); \ t,s \in \mathbf{I} \ , \tag{9.38}$$

which extends (9.5b) to the continuous time case.

SPECIFICATION OF CONTINUOUS TIME I.S.I. PROCESSES

The specification of processes with independent and stationary increments is almost the same in continuous time as it is in discrete time, the only real difference being that in continuous time we must consider more general collections of sample times. In discrete time the specification was constructed using the marginal probability function of the underlying i.i.d. process, which implies the pmf of the increments. In continuous time we have no underlying i.i.d. process so we instead assume that we are given a formula for the cdf (pdf or pmf) of the increments; that is, for any $t > s$ we have a cdf

$$F_{Y_t - Y_s}(y) = F_{Y_{|t-s|} - Y_0}(y) = F_{Y_{|t-s|}}(y) \tag{9.39}$$

or, equivalently, the corresponding pmf $p_{Y_t - Y_s}(y)$ for a discrete amplitude process or the corresponding pdf $f_{Y_t - Y_s}(y)$ for a continuous amplitude process.

To specify a continuous time process we need a formula for the joint probability functions for all n and all ordered sample times $t_1, t_2, \ldots, t_n$ (that is, $t_i < t_j$ if $i < j$). As in the discrete time case, we consider conditional probability functions. To allow both discrete and continuous alphabet, we first focus on conditional cdf's and find the conditional cdf $P(Y_{t_n} \leq y_n \mid Y_{t_{n-1}} = y_{n-1}, Y_{t_{n-2}} = y_{n-2}, \ldots)$. Then, using (9.7) we can apply the techniques used in discrete time by simply replacing the sample times i by t_i for $i = 0, 1, \ldots, n$. That is, we define the random variables $\{X_n\}$ by

$$X_n = Y_{t_n} - Y_{t_{n-1}} . \tag{9.40}$$

Then the $\{X_n\}$ are independent (but not identically distributed unless the times between adjacent samples are all equal), and

$$Y_{t_n} = \sum_{1=1}^{n} X_i \tag{9.41}$$

and

$$P(Y_{t_n} \leq y_n \mid Y_{t_{n-1}} = y_{n-1}, Y_{t_{n-2}} = y_{n-2}, \ldots) = F_{X_n}(y_n - y_{n-1})$$

$$= F_{Y_{t_n} - Y_{t_{n-1}}}(y_n - y_{n-1}) . \tag{9.42}$$

This conditional cdf can then be used to evaluate the conditional pmf or pdf as

$$p_{Y_{t_n} \mid Y_{t_{n-1}}, \ldots, Y_{t_1}}(y_n \mid y_{n-1}, \ldots, y_1) = p_{X_n}(y_n - y_{n-1})$$

$$= p_{Y_{t_n} - Y_{t_{n-1}}}(y_n - y_{n-1}) \tag{9.43}$$

or

$$f_{Y_{t_n} \mid Y_{t_{n-1}}, \ldots, Y_{t_1}}(y_n \mid y_{n-1}, \ldots, y_1) = f_{X_n}(y_n - y_{n-1})$$

$$= f_{Y_{t_n} - Y_{t_{n-1}}}(y_n - y_{n-1}) , \tag{9.44}$$

respectively. These can then be used to find the joint pmf's or pdf's as before as

$$f_{Y_{t_1}, \ldots, Y_{t_n}}(y_1, \ldots, y_n) = \prod_{i=1}^{n} f_{Y_{t_i} - Y_{t_{i-1}}}(y_i - y_{i-1})$$

or

$$p_{Y_{t_1}, \ldots, Y_{t_n}}(y_1, \ldots, y_n) = \prod_{i=1}^{n} p_{Y_{t_i} - Y_{t_{i-1}}}(y_i - y_{i-1}) ,$$

respectively. Since we can thus write the joint probability functions for any finite collection of sample times in terms of the given probability function for the increments, the process is completely specified.

The most important point of these relations is that if we are told that a process has independent and stationary increments and we are given a cdf or pmf or pdf for $Y_t = Y_t - Y_0$, then the process is completely defined via the specification just given! Knowing the probabilistic description of the jumps and that the jumps are independent and stationary completely describes the process.

As in discrete time, a continuous time random process $\{Y_t\}$ is called a Markov process if and only if for all n and all sample times $t_1, t_2, \ldots, t_n$ we have for all $y_n, y_{n-1}, \ldots$ that

$$P(Y_{t_n} \le y_n \mid Y_{t_{n-1}} = y_{n-1}, Y_{t_{n-2}} = y_{n-2}, \ldots) =$$

$$P(Y_{t_n} \le y_n \mid Y_{t_{n-1}} = y_{n-1}) \tag{9.45}$$

or, equivalently,

$$f_{Y_{t_n} \mid Y_{t_{n-1}}, \ldots, Y_{t_1}}(y_n \mid y_{n-1}, \ldots, y_1) = f_{Y_{t_n} \mid Y_{t_{n-1}}}(y_n \mid y_{n-1})$$

for continuous alphabet processes and

$$p_{Y_{t_n} \mid Y_{t_{n-1}}, \ldots, Y_{t_1}}(y_n \mid y_{n-1}, \ldots, y_1) = p_{Y_{t_n} \mid Y_{t_{n-1}}}(y_n \mid y_{n-1})$$

for discrete alphabet processes. Analogous to the discrete time case, continuous time independent increment processes are Markov processes.

We close this section with the most famous examples of continuous time independent increment processes.

[9.1]   *The Wiener process*

The Wiener process is a continuous time independent increment process with stationary increments such that the increment densities are Gaussian with zero mean; that is, for $t > 0$,

$$f_{Y_t}(y) = \frac{e^{-\frac{y^2}{2t\sigma^2}}}{(2\pi t \sigma^2)^{1/2}}; \; y \in \mathbf{R} \; .$$

The form of the variance follows necessarily from the previously derived form for all independent increment processes with stationary increments. The specification for this process and the Gaussian form of the increment pdf's imply that the Wiener process is a Gaussian process.

[9.2]   *The Poisson counting process*

The Poisson counting process is a continuous time discrete alphabet independent increment process with stationary increments such that the increments have a Poisson distribution; that is, for $t > 0$,

$$p_{Y_t}(k) = \frac{(\lambda t)^k}{k!} e^{-\lambda t}; \; k = 0, 1, 2, \ldots \; .$$

In chapter 11 we will provide a derivation of the Poisson counting process from underlying physical assumptions, which leads to the same result.

MOVING-AVERAGE AND AUTOREGRESSIVE PROCESSES

We have seen in the preceding sections that for discrete time random processes the moving-average representation can be used to yield the second-order moments and also can be used to find the marginal probability function of independent increment processes. The general specification for independent increment processes, however, was found using the autoregressive representation. In this section we consider results for more general processes using virtually the same methods.

First assume that we have a moving-average process representation described by (9.1). We can use characteristic function techniques to find a simple form for the marginal characteristic function of the output process. In particular, assuming convergence conditions are satisfied where needed and observing that Y_n is a weighted sum of independent random variables, the characteristic function of the output random process marginal distribution is calculated as the product of the transforms

$$M_{Y_n}(ju) = \prod_k M_{h_k X_{n-k}}(ju) .$$

The individual transforms are easily shown to be

$$M_{h_k X_{n-k}}(ju) = E[e^{j u h_k X_{n-k}}] = M_{X_{n-k}}(juh_k) = M_X(juh_k) .$$

Thus

$$M_{Y_n}(ju) = \prod_k M_X(juh_k) , \qquad (9.46)$$

where the product is, as usual, dependent on the index sets on which $\{X_n\}$ and $\{h_k\}$ are defined.

Equation (9.46) can be inverted in some cases to yield the output cdf and pdf or pmf. Unfortunately, however, in general this is about as far as one can go in this direction, even for an i.i.d. input process. Attempts to find joint or conditional distributions of the output process by this or other techniques will generally be frustrated by the complexity of the calculations required.

Part of the difficulty in finding conditional distributions lies in the moving-average representation. The techniques used successfully for the independent increment processes relied on an autoregressive representation of the output process. We will now show that the methods used work for more general autoregressive process representations. We will consider specifically causal autoregressive processes represented as in (9.2) with

$a_0 > 0$ and $a_k = 0$ for $k < 0$. Then in (9.2) we can solve for Y_n as a function of X_n and the past values (for causality) of Y as

$$Y_n = \frac{1}{a_0}[X_n - \sum_{k>0} a_k Y_{n-k}] .$$

By the causality condition, the $\{Y_{n-k}\}$ in the sum are independent of X_n. Hence we have a representation for Y_n as the sum of two independent random variables, X_n and the weighted sum of the Y's. The latter quantity is treated as if it were a constant in calculating conditional probabilities for Y_n. Thus the *conditional* probability of an event for Y_n can be specified in terms of the *marginal* probability of an easily determined event for X_n. Specifically, the conditional cdf for Y_n is

$$\Pr[Y_n \le y_n | y_{n-1}, y_{n-2}, \dots] = \Pr[X_n \le \sum_{k \ge 0} a_k y_{n-k}]$$

$$= F_X(\sum_{k \ge 0} a_k y_{n-k}) . \qquad (9.47)$$

The conditional pmf or pdf can now be found. For example, if the input random process is continuous alphabet, the conditional output pdf is found by differentiation to be

$$f_{Y_n | Y_{n-1}, Y_{n-2}, \dots}(y_n | y_{n-1}, y_{n-2}, \dots) = a_0 f_X(\sum_k a_k y_{n-k}) . \qquad (9.48)$$

Finally, the complete specification can be obtained by a product of pmf's or pdf's as in (9.15) or (9.29). The discrete time independent increment result is obviously a special case of this equation. For more general processes, we need only require that the sum converge in (9.48) and that the corresponding conditional pdf's be appropriately defined (using the general conditional probability approach). We next consider an important example of the ideas of this section.

THE DISCRETE TIME GAUSS-MARKOV PROCESS

As an example of the development of the preceding section, consider the filter given in example [8.1]. Let $\{X_n\}$ be an i.i.d. Gaussian process with mean m and variance σ^2. The moving-average representation is

$$Y_n = \sum_{k=0}^{\infty} X_{n-k} r^k , \qquad (9.49)$$

from which (9.46) can be applied to find that

$$M_{Y_n}(ju) = \prod_k e^{j(ur^k)m - 1/2(ur^k)^2 \sigma_X^2}$$

$$= e^{jum(\sum_{k} r^k) - 1/2 u^2 \sigma_X^2 (\sum_{k} r^{2k})},$$

that is, a Gaussian random variable with mean $m_Y = m \sum_{k} r^k = m/(1-r)$ and variance $\sigma_Y^2 = \sigma_X^2 \sum_{k} r^{2k} = \sigma_X^2/(1-r^2)$, the moments found by the second-order theory in example [8.1].

To find a complete specification for this process, we now turn to an autoregressive model. From (9.49) it follows that Y_n must satisfy the difference equation

$$Y_n = X_n + r Y_{n-1}. \tag{9.50}$$

Hence $\{Y_n\}$ is a first-order autoregressive source with $a_0 = 1$ and $a_1 = -r$. Note that as with the Wiener process, this process can be represented as a first-order autoregressive process or as an infinite-order moving-average process. In fact, the Wiener process is the one-sided version of this process with $r = 1$.

Application of (9.48) yields

$$f_{Y^n}(y^n) =$$

$$f_{Y_n}(y_n | y_{n-1}, y_{n-2}, \ldots) f_{Y_{n-1}}(y_{n-1} | y_{n-2}, y_{n-3}, \ldots) \cdots f_{Y_1}(y_1) =$$

$$f_{Y_1}(y_1) \prod_{i=2}^{n} f_X(y_i - r y_{i-1}) = \frac{e^{-\frac{y_1^2}{2\sigma_Y^2}}}{\sqrt{2\pi\sigma_Y^2}} \prod_{i=2}^{n} \frac{e^{-\frac{(y_i - r y_{i-1})^2}{2\sigma^2}}}{\sqrt{2\pi\sigma^2}}. \tag{9.51}$$

GAUSSIAN RANDOM PROCESSES

We have seen how to calculate the mean, covariance function, or spectral density of the output process of a linear filter driven by an input random process whose mean, covariance function, or spectral density is known. In general, however, it is not possible to derive a complete specification of the output process. We have seen one exception: The output random process of an autoregressive filter driven by an i.i.d. input random process can be specified through the conditional pmf's or pdf's, as in equation (9.48). In this section we develop another important exception by showing that the output process of a linear filter driven by a Gaussian random process—not necessarily i.i.d.—is also Gaussian. Thus simply knowing the output mean and autocorrelation or covariance functions—the only parameters of a Gaussian distribution—provides a complete specification.

The invariance of the Gaussian distribution property under linear operations is directly shown with multidimensional characteristic functions.

Toward this end, the multidimensional characteristic function of a distribution is defined as follows:

Given a random vector $\mathbf{X} = (X_0,\ldots,X_{n-1})$ and a vector parameter $\mathbf{u} = (u_0,\ldots,u_{n-1})$, the n-dimensional characteristic function $M_{\mathbf{X}}(j\mathbf{u})$ is defined by

$$M_{\mathbf{X}}(j\mathbf{u}) = M_{X_0,\ldots,X_{n-1}}(ju_0,\ldots,ju_{n-1})$$

$$= E(e^{j\mathbf{u}'\mathbf{X}}) = E\left[e^{j\sum_{k=0}^{n-1} u_k X_k}\right] .$$

It can be shown (exercise 9.14) that a Gaussian random vector with mean vector $\mathbf{m}$ and covariance matrix Λ has characteristic function

$$e^{j\mathbf{u}'\mathbf{m} - 1/2\,\mathbf{u}'\Lambda\mathbf{u}} = \exp[j\sum_{k=0}^{n-1} u_k m_k - 1/2 \sum_{k=0}^{n-1}\sum_{m=0}^{n-1} u_k \Lambda(k,m) u_m] . \qquad (9.52)$$

Observe that the Gaussian characteristic function has the same form as the Gaussian pdf—an exponential quadratic in its argument. However, unlike the pdf, the characteristic function depends on the covariance matrix directly, whereas the pdf contains the inverse of the covariance matrix. Thus the Gaussian characteristic function is in some sense simpler than the Gaussian pdf. As a further consequence of the direct dependence on the covariance matrix, it is interesting to note that, unlike the Gaussian pdf, the characteristic function is well-defined even if Λ is only nonnegative definite and not strictly positive definite. Previously we gave a definition of a Gaussian random vector in terms of its pdf. Now we can give an alternate, more general (in the sense that a strictly positive definite covariance matrix is not required) definition of a Gaussian random vector (and hence also of a Gaussian random process):

A random vector is Gaussian if and only if it has a characteristic function of the form of (9.52).

We are now equipped to prove the basic result of this section, which we formally state as a theorem:

Theorem 9.1.   Let $\mathbf{X}$ be a k-dimensional Gaussian random vector with mean vector $\mathbf{m}$ and covariance matrix Λ. Let $\mathbf{Y}$ be the new random vector formed by a linear operation on $\mathbf{X}$:

$$\mathbf{Y} = \mathbf{HX} + \mathbf{b}, \qquad (9.53)$$

where $\mathbf{H}$ is a $n \times k$ matrix and $\mathbf{b}$ is an n-dimensional vector. Then $\mathbf{Y}$ is a Gaussian random vector of dimension n with mean $\mathbf{Hm} + \mathbf{b}$ and covariance matrix $\mathbf{H}\Lambda\mathbf{H}'$.

Proof.   The characteristic function of $\mathbf{Y}$ is found by direct substitution of the expression for $\mathbf{Y}$ in terms of $\mathbf{X}$ into the definition, a little matrix algebra, and (9.52):

$$M_Y(j\mathbf{u}) = E[e^{j\mathbf{u}'\mathbf{Y}}] = E[e^{j\mathbf{u}'(\mathbf{HX}+\mathbf{b})}]$$

$$= e^{j\mathbf{u}'\mathbf{b}}E[e^{j(\mathbf{H}'\mathbf{u})'\mathbf{X}}] = e^{j\mathbf{u}'\mathbf{b}}M_X(j\mathbf{H}'\mathbf{u})$$

$$= e^{j\mathbf{u}'\mathbf{b}}e^{j(\mathbf{H}'\mathbf{u})'\mathbf{m})-\frac{1}{2}(\mathbf{H}'\mathbf{u})'\Lambda(\mathbf{H}'\mathbf{u})} = e^{j\mathbf{u}'(\mathbf{Hm}+\mathbf{b})-\frac{1}{2}\mathbf{u}'(\mathbf{H}\Lambda\mathbf{H}')\mathbf{u}}.$$

It can be seen by reference to (9.52) that the resulting characteristic function is the transform of a Gaussian random vector pdf with mean vector $\mathbf{Hm}+\mathbf{b}$ and covariance matrix $\mathbf{H}\Lambda\mathbf{H}'$. (Note in passing that this vector mean and matrix covariance are effectively vector and matrix versions of the linear system second-moment I/O relations (8.3) and (8.5).) This completes the proof.

The output of a discrete time FIR linear filter can be expressed as a linear operation on the input as in (9.53), that is, a finite dimensional matrix times an input vector plus a constant. Therefore we can immediately extend theorem 9.1 to FIR filtering. It is also possible to extend theorem 9.1 to include more general impulse responses and to continuous time by using appropriate limiting arguments. We will not prove such extensions. Instead we will merely state the result as a corollary:

Corollary 9.1. If a Gaussian random process $\{X_t\}$ is passed through a linear filter, then the output is also a Gaussian random process with mean and covariance given by (8.3) and (8.5).

*EXPONENTIAL MODULATION

We close this chapter with an example of a class of processes generated by a nonlinear operation on another process. While linear techniques rarely work for nonlinear systems, the systems that we shall consider form an important exception where one can find second-order moments and sometimes even complete specifications. The primary examples of processes generated in this way are phase-modulated (PM) and frequency-modulated (FM) Gaussian random processes and the Poisson random telegraph wave.

Let $\{X(t)\}$ be a random process and define a new random process

$$Y(t) = a_0 e^{j(a_1 t + a_2 X(t) + \Theta)}, \qquad (9.54)$$

where a_0, a_1, and a_2 are fixed real constants and where Θ is a uniformly distributed random phase angle on $[0,2\pi]$. The process $\{Y(t)\}$ is called an exponential modulation of $\{X(t)\}$. Observe that it is a nonlinear function of the input process. Note further that, as defined, the process is a complex-valued random process, and hence we must modify some of our techniques. In some, but not all, of the interesting examples of exponentially modulated random processes we will wish to focus on the real part of the modulated process, which we will call

$$U(t) = Re(Y(t)) = {}^1/_2 Y(t) + {}^1/_2 Y(t)^*$$

$$= a_0\cos(a_1 t + a_2 X(t) + \Theta) . \qquad (9.55)$$

In this form, exponential modulation is called *phase modulation* (PM) of a carrier of angular frequency a_1 by the input process $\{X(t)\}$. If the input process is itself formed by integrating another random process, say $\{W(t)\}$, then the U process is called the *frequency modulation* (FM) of the carrier by the W process. Phase and frequency modulation are extremely important in practice and are the most important examples of complex exponential modulation.

A classic example of a random process arising in communications that can be put in the same form is obtained by setting $\Theta = 0$, choosing $a_1 = 0$, $a_2 = \pi$, and letting the input process be the Poisson counting process $\{N(t)\}$, that is, to consider the random process

$$V(t) = a_0(-1)^{N(t)}. \qquad (9.56)$$

This is a real-valued binary random process that changes value with every jump in the Poisson counting process. Because of the properties of the Poisson counting process, this process is such that jumps in nonoverlapping time windows are independent, the probability of a change of value in a differentially small interval is proportional to the length of the interval, and the probability of more than one change is negligible in comparison. It is usually convenient to consider a slight change in this process, which makes it somewhat better behaved. Let Z be a binary random variable, independent of $N(t)$ and taking values of $+1$ or -1 with equal probability. Then the random process $Y(t) = ZV(t)$ is called the *random telegraph wave* and has long served as a fundamental example in the teaching of second-order random process theory. The purpose of the random variable Z is to remove an obvious nonstationarity at the origin and make the resulting process equally likely to have either of its two values at time zero. This has the obvious effect of making the process zero-mean. In the form given, it can be treated as simply a special case of exponential modulation.

We develop the second-order moments of exponentially modulated random processes and then apply the results to the preceding examples. We modify our definitions slightly to apply to the complex-valued process. This is done as follows: The expectation of a complex-valued random variable is defined as the vector consisting of the expectations of the real and imaginary parts; that is, if $X = Re(X) + jIm(X)$, with $Re(X)$ and $Im(X)$ the real and imaginary parts of X, respectively, then

$$EX = (ERe(X), EIm(X)) .$$

In other words, the expectation of a vector is defined to be the vector of ordinary scalar expectations of the components. The autocorrelation function of a complex random process is defined somewhat differently as

$$R_Y(t,s) = E[Y(t)Y(s)^*] \, ,$$

which reduces to the usual definition if the process is real valued. The autocorrelation in this more general situation is not in general symmetric, but it is Hermitian in the sense that

$$R_Y(s,t) = R_Y(t,s)^*.$$

Being Hermitian is, in fact, the appropriate generalization of symmetry for developing a useful transform theory, and it is for this reason that the autocorrelation function includes the complex conjugate of the second term.

It is an easy exercise to show that for the general exponentially modulated random process of (9.54) we have that

$$EY(t) = 0 \, .$$

This can be accomplished by separately considering the real and imaginary parts and using (9.52), exactly as was done in the AM case of chapter 8. The use of the auxiliary random variable Z in the random telegraph wave definition means that both examples have zero mean. Note that it is *not* true that $Ee^{j(a_1 t + a_2 X(t) + \Theta)}$ equals $e^{j(a_1 t + a_2 EX(t) + E\Theta)}$; that is, expectation does not in general commute with nonlinear operations.

To find the autocorrelation of the exponentially modulated process, observe that

$$
\begin{aligned}
E[Y(t)Y(s)^*] &= a_0^2 E[e^{j(a_1(t-s) + a_2(X(t) - X(s)))}] \\
&= a_0^2 e^{ja_1(t-s)} E[e^{ja_2(X(t) - X(s))}] \, ,
\end{aligned}
$$

and hence

$$R_Y(t,s) = a_0^2 e^{ja_1(t-s)} M_{X(t)-X(s)}(ja_2) \, . \tag{9.57}$$

Thus the autocorrelation of the nonlinearly modulated process is given simply in terms of the characteristic function of the increment between the two sample times! This is often a computable quantity, and when it is, we can find the second-order properties of such processes without approximation or linearization. This is a simple result of the fact that the autocorrelation of an exponentially modulated process is given by an expectation of the exponential of the difference of two samples and hence by the characteristic function of the difference.

There are two examples in which the computation of the characteristic function of the difference of two samples of a random process is particularly easy: a Gaussian input process and an independent increment input process.

If the input process $\{X(t)\}$ is Gaussian with zero mean (for convenience) and autocorrelation function $R_X(t,s)$, then the random variable $X(t)-X(s)$ is also Gaussian (being a linear combination of Gaussian random variables) with mean zero and variance

$$\sigma^2_{X(t)-X(s)} = E[(X(t)-X(s))^2] = R_X(t,t)+R_X(s,s)-2R_X(t,s) .$$

Thus we have shown that if $\{X(t)\}$ is a zero-mean Gaussian random process with autocorrelation function R_X and if $\{Y(t)\}$ is obtained by exponentially modulating $\{X(t)\}$ as in (9.54), then

$$R_Y(t,s) = a_0^2 e^{ja_1(t-s)} M_{X(t)-X(s)}(ja_2) =$$
$$a_0^2 e^{ja_1(t-s)} e^{-1/2 a_2^2(R_X(t,t)+R_X(s,s)-2R_X(t,s))} . \qquad (9.58)$$

Observe that this autocorrelation is not symmetric, but it is Hermitian.

Thus, for example, if the input process is stationary, then so is the modulated process, and

$$R_Y(\tau) = a_0^2 e^{ja_1\tau} e^{-a_2^2(R_X(0)-R_X(\tau))} . \qquad (9.59)$$

We emphatically note that the modulated process is not Gaussian.

We can use this result to obtain the second-order properties for phase modulation as follows:

$$R_U(t,s) = E[U(t)U(s)^*] = E[\frac{Y(t)+Y(t)^*}{2}(\frac{Y(s)+Y(s)^*}{2})^*] =$$
$$\frac{1}{4}\left[E[Y(t)Y(s)^*]+E[Y(t)Y(s)]+E[Y(t)^*Y(s)^*]+E[Y(t)^*Y(s)]\right] .$$

Note that both of the middle terms on the right have the form

$$a_0^2 e^{\pm ja_1(t+s)} E[e^{\pm j(a_2(X(t)+X(s))+2\Theta)}] ,$$

which evaluates to 0 because of the uniform phase angle. The remaining terms are $R_Y(t,s)$ and $R_Y(t,s)^*$ from the previous development, and hence

$$R_U(t,s) = 1/2 a_0^2 \cos(a_1(t-s)) e^{-1/2 a_2^2(R_X(t,t)+R_X(s,s)-2R_X(t,s))} , \qquad (9.60)$$

and hence, in the stationary case,

$$R_U(\tau) = 1/2 a_0^2 \cos(a_1\tau) e^{-a_2(R_X(0)-R_X(\tau))} . \qquad (9.61)$$

As expected, this autocorrelation is symmetric.

Returning to the exponential modulation case, we consider the second example of exponential modulation of independent increment processes. Observe that this overlaps the preceding example in the case of the Wiener process. We also note that phase modulation by independent increment processes is of additional interest because in some examples independent increment processes can be modeled as the integral of another process. For example, the Poisson counting process is the integral of a random telegraph wave with alphabet 0 and 1 instead of -1 and $+1$. (This is accomplished by forming the process $1/2(X(t)+1)$ with $X(t)$ the ± 1 random telegraph wave.) In this case the real part of the output process is the FM modulation of the process being integrated.

If $\{X(t)\}$ is a random process with independent and stationary increments, then the characteristic function of $X(t)-X(s)$ with $t>s$ is equal to that of $X(t-s)$. Thus we have from (9.57) that for $t>s$ and $\tau=t-s$,

$$R_Y(\tau) = a_0^2 e^{ja_1\tau} M_{X(\tau)}(ja_2) .$$

We can repeat this development for the case of negative lag to obtain

$$R_Y(\tau) = a_0^2 e^{ja_1\tau} M_{X(|\tau|)}(ja_2) . \tag{9.62}$$

Observe that this autocorrelation is not symmetric; that is, it is not true that $R_Y(-\tau)=R_Y(\tau)$ (unless $a_1=0$). It is, however, Hermitian.

Equation (9.62) provides an interesting oddity: Even though the original input process is not weakly stationary (since it is an independent increment process), the exponentially modulated output *is* weakly stationary! For example, if $\{X(t)\}$ is a Poisson counting process with parameter λ, then the characteristic function is

$$M_{X(\tau)}(ju) = \sum_{k=0}^{\infty} \frac{(\lambda\tau)^k e^{-\lambda\tau}}{k!} e^{juk} =$$

$$e^{-\lambda\tau} \sum_{k=0}^{\infty} \frac{(\lambda\tau e^{ju})^k}{k!} = e^{\lambda\tau(e^{ju}-1)}; \; \tau \geq 0 .$$

Thus if we choose $a_1=0$ and $a_2=\pi$, then the modulated output process is the random telegraph wave with alphabet $\pm a_0$ and hence is a real process. Equation (9.62) becomes

Note that the autocorrelation (and hence also the covariance) decays exponentially with the delay.

A complete specification of the random telegraph wave is possible, but as this chapter is already too long, we leave it as an exercise.

EXERCISES

9.1 Given a random variable X with characteristic function $M_X(ju)$, define the random variable $Y=aX+b$, for a,b fixed real constants. Find $M_Y(ju)$ in terms of $M_X(ju)$.

9.2 Find the characteristic functions for the Poisson, geometric, and exponential distributions.

9.3 Let $\{X_n\}$ be an i.i.d. process with a Poisson marginal pmf with parameter λ. Let $\{Y_n\}$ denote the induced sum process as in equation (9.4). Find the pmf for Y_n and find $\sigma_{Y_n}^2$, EY_n, and $K_Y(t,s)$.

9.4 Let $\{X_n\}$ be an i.i.d. process. Define a new process $\{U_n\}$ by

$$U_0 = 0$$

$$U_n = X_n - X_{n-1} \; ; \; n = 1,2,\ldots \; .$$

Find the characteristic function and the pmf for U_n. Find $R_U(t,s)$. Is $\{U_n\}$ an independent increment process?

9.5 Let $\{X_n\}$ be a ternary i.i.d. process with $p_{X_n}(+1) = p_{X_n}(-1) = \epsilon/2$ and $p_{X_n}(0) = 1 - \epsilon$. Fix an integer N and define the "sliding average"

$$Y_n = \frac{1}{N} \sum_{i=0}^{N-1} X_{n-i} \; .$$

(a) Find EX_n, $\sigma_{X_n}^2$, $M_{X_n}(ju)$, and $K_X(t,s)$.

(b) Find EY_n, $\sigma_{Y_n}^2$, and $M_{Y_n}(ju)$.

(c) Find the *cross-correlation* $R_{X,Y}(t,s) \equiv E[X_t Y_s]$.

(d) Given $\delta > 0$ find a simple upper bound to $\Pr(|Y_n| > \delta)$ in terms of N and ϵ.

9.6 Find the characteristic function $M_{U_n}(ju)$ for the $\{U_n\}$ process of exercise 8.2.

9.7 Find a complete specification of the binary autoregressive process of exercise 8.9. Prove that the process is Markov. (One name for this process is the binary symmetric Markov source.)

9.8 A stationary continuous time random process $\{X(t)\}$ switches randomly between the values of 0 and 1. We have that

$$\Pr(X(t) = 1) = \Pr(X(t) = 0) = \frac{1}{2},$$

and if N_t is the number of changes of output during $(0,t]$, then

$$p_{N_t}(k) = \frac{1}{1 + \alpha t} \left[\frac{\alpha t}{1 + \alpha t} \right]^k ; k = 0,1,2,\ldots ,$$

where $\alpha > 0$ is a fixed parameter. (This is called the Bose-Einstein distribution.)

(a) Find $M_{N_t}(ju)$, EN_t, and $\sigma_{N_t}^2$.

(b) Find $EX(t)$ and $R_X(t,s)$.

9.9 Given two random processes $\{X_t\}$, called the signal process, and $\{N_t\}$, called the noise process, define the process $\{Y_t\}$ by

$$Y_t = X_t + N_t \; .$$

The $\{Y_t\}$ process can be considered as the output of a *channel with additive noise* where the $\{X_t\}$ process is the input. This is a common model for dealing with noisy linear communication systems; e.g., the noise may be due to atmospheric effects or to front-end noise in a receiver. Assume that the signal and noise processes are independent; that is, any vector of samples of the X process is independent of any vector of samples of the N process. Find the characteristic function, mean, and variance of Y_t in terms of those for X_t and N_t. Find the covariance of the output process in terms of the covariances of the input and noise process.

9.10 Prove the following facts about characteristic functions:

$$|M_X(ju)| \leq 1 \tag{i}$$

$$M_X(0) = 1 \tag{ii}$$

$$|M_X(ju)| \leq M_X(0) = 1 \tag{iii}$$

$$\frac{dM_X(ju)}{du}\Big|_{u=0} = jEX \tag{iv}$$

If a random variable X has a characteristic function $M_X(ju)$, if c is a fixed constant, and if a random variable Y is defined by $Y = X + c$, then

$$M_Y(ju) = e^{juc} M_X(ju) \; . \tag{v}$$

Property (iv) is called the "moment-generating" property of transforms. Property (v) is just a variation of the "shift theorem" of Fourier transforms.

9.11 Given an i.i.d. random process $\{X_n\}$, find the characteristic function of the sample average

$$S_n = \frac{1}{n} \sum_{i=0}^{n-1} X_i$$

in terms of $M_X(ju)$.

9.12 Let $\{X_n\}$ and $\{W_n\}$ be two mutually independent Bernoulli processes with parameters p and ϵ, respectively. The first process will be called a source random process and the second will be called a noise random process. Form a received process $Y_n = X_n \oplus W_n$. This is called a binary additive noise channel analogous to the ordinary (real arithmetic) additive noise channel previously encountered. In particular, it is called the binary symmetric memoryless channel (BSMC) since the noise process is i.i.d. and the probability of a 0 being turned into a 1 is the same as that of a 1 being turned into a 0.

(a)   Find a simple expression for $p_{Y^n|X^n}(y^n|x^n)$.

(b)   Find the marginal pmf for the output process and show that the output process is i.i.d.

(c)   Find the so-called *a posteriori* probability $p_{X_n|Y_n}(x|y)$. If $Y_n = 0$, what is the most likely value of the corresponding X_n? (This is an example of maximum *a posteriori* or MAP detection.)

(d)   Compute the probability $\Pr(X_n \neq Y_n)$.

9.13 Find the inverse of the covariance matrix of the discrete time Wiener process, that is, the inverse of the matrix $\{\min(k,j); \ k=1,2,...,n, \ j=1,2,...,n\}$.

9.14 Find the multidimensional Gaussian characteristic function of equation (9.52) by completing the square in the exponent of the defining multidimensional integral.

9.15 Let $\{X(t)\}$ be a Gaussian random process with zero mean and autocorrelation function

$$R_X(\tau) = \frac{N_0}{2}e^{-|\tau|} .$$

Is the process Markov? Find its power spectral density. Let $Y(t)$ be the process formed by DSB-SC modulation of $X(t)$ as in (8.13) with $a_0=0$. If the phase angle Θ is assumed to be 0, is the resulting modulated process Gaussian? Letting Θ be uniformly distributed, sketch the power spectral density of the modulated process. Find $M_{Y(0)}(ju)$.

9.16 Let $\{X(t)\}$ and $\{Y(t)\}$ be the two continuous time random processes of exercise 8.12 and let

$$W(t) = X(t)\cos(2\pi f_0 t) + Y(t)\sin(2\pi f_0 t) ,$$

as in that exercise. Find the marginal probability density function $f_{W(t)}(w)$ and the joint pdf $f_{W(t),W(s)}(u,v)$. Is $\{W(t)\}$ a Gaussian process? Is it strictly stationary?

9.17 Let $\{N_k\}$ be the binomial counting process and define the discrete time random process $\{Y_n\}$ by

$$Y_n = (-1)^{N_n} .$$

(This is the discrete time analog to the random telegraph wave.) Find the autocorrelation, mean, and power spectral density of the given process. Is the process Markov?

9.18 Find the power spectral density of the random telegraph wave. Is this process a Markov process? Sketch the spectrum of an amplitude modulated random telegraph wave.

9.19 Suppose that (U,W) is a Gaussian random vector with $EU=EW=0$, $E(U^2)=E(W^2)=\sigma^2$, and $E(UW)=\rho\sigma^2$. (The parameter ρ has magnitude less than or equal to 1 and is called the *correlation coefficient*.) Define the new random variables

$$S = U+W$$
$$D = U-W .$$

(a) Find the marginal pdf's for S and D.

(b) Find the joint pdf $f_{S,D}(\alpha,\beta)$ or the joint characteristic function $M_{S,D}(ju,jv)$. Are S and D independent?

Suppose that we have a sequence of i.i.d. pairs $\{ U_n,W_n \}$ with each (U_n,W_n) having the same distribution as (U,W) above and with each pair (U_n,W_n) being independent of (U_k,W_k) for $k \neq n$, that is, the current pair (U_n,W_n) is independent of the past and future pairs. Form the new discrete time random processes

$$S_n = U_n + W_n$$
$$D_n = U_n - W_n .$$

(c) Find the autocorrelations and power spectral densities of the processes S_n and D_n. Find the crosscorrelation function defined by

$$R_{S,D}(k,j) = E(S_k D_j) .$$

Are the random variables S_k and D_j independent? (Specify any requirements on k and j for your answer to be valid.)

9.20 Suppose that K is a random variable with a Poisson distribution, that is, for a fixed parameter λ

$$Pr(K=k) = p_K(k) = \frac{\lambda^k e^{-\lambda}}{k!} \; ; k=0,1,2,...$$

(a) Define a new random variable N by $N = K+1$. Find the characteristic function $M_N(ju)$, the expectation EN, and the pmf $p_N(n)$ for the random variable N.

We define a one-sided discrete time random process $\{Y_n\; ; n=1,2,...\}$ as follows: Y_n has a binary alphabet $\{ -1,1\}$. Y_0 is equally likely to be -1 or $+1$. Given Y_0 has some value, it will stay at that value for a total of T_1 time units, where T_1 has the same distribution as N, and then it will change sign. It will stay at the new sign for a total of T_2 time units, where T_2 has the same distribution as N and is independent of T_1, and then change sign again. It will continue in this way, that is, it will change sign for the kth time at time

$$S_k = \sum_{i=1}^{k} T_i ,$$

where the T_i form an i.i.d. sequence with the marginal distribution found in part (a).

(b) Find the characteristic function $M_{S_k}(ju)$ and the pmf $p_{S_k}(m)$ for the random variable S_k. Is $\{S_k\}$ an independent increment process?

9.21 Suppose that $\{Z_n\}$ is a two-sided Bernoulli process, that is, an i.i.d. sequence of binary $\{0,1\}$ random variables with $\Pr(Z_n=1) = \Pr(Z_n=0)$. Define the new processes

$$X_n = (-1)^{Z_n} ,$$

$$Y_n = \sum_{i=0}^{n} 2^{-i} X_i \; ; \; n=0,1,2,\dots ,$$

and

$$V_n = \sum_{i=0}^{\infty} 2^{-i} X_{n-i} \; ; \; n \in \mathbf{Z} .$$

(a) Find the means and autocorrelation functions of the $\{X_n\}$ process and the $\{V_n\}$ process. If possible, find the power spectral densities.

(b) Find the characteristic functions for both Y_n and V_n.

(c) Is Y_n an autoregressive process? a moving average process? Is it weakly stationary? Is V_n an autoregressive process? a moving average process? Is it weakly stationary? (*Note*: answers to parts (a) and (b) are sufficient to answer the stationarity questions, no further computations are necessary.)

(d) Find the conditional pmf

$$p_{V_n|V_{n-1},V_{n-2},\dots,V_0}(v_n|v_{n-1},\dots,v_0)$$

Is $\{V_n\}$ a Markov process?

10

COMPOUND PROCESSES
AND
CONDITIONAL EXPECTATION

In this chapter we develop a class of processes that strongly resembles the independent increment processes of chapter 9 but has a fundamental difference. The difference is most easily exemplified in the discrete time case: As in chapter 9, we form a process by summing the outputs of an i.i.d. process, but instead of adding one i.i.d. random variable to the sum with each time step, at each time step a random number of i.i.d. random variables are added. Thus the sum consists of the sum of a random number of elements at any given time. In continuous time, a similar sum is formed, but the increment to the sum that occurs in any time *interval* is a random number whose distribution depends on the length of the time interval. As the time interval shrinks to zero, the probability of a nonzero number of incremental summands shrinks to zero. This class of processes—called compound processes—models situations such as the total time occupied by a random number of phone calls, each having a random duration and random starting time. As we shall see in chapter 11 (exercise 11.4), compound processes also model physical phenomena such as shot noise. This class of processes provides a guinea pig for the introduction of the second topic of this chapter, conditional expectation—expectation with respect to a conditional probability distribution. With the addition of the tool of conditional expectation, the techniques of chapter 9 easily extend to the more general class of processes.

DISCRETE TIME COMPOUND PROCESSES

As introduced in chapter 9, let $\{N_k; k = 0,1,2,...\}$ be a discrete time counting process such as the binomial counting process and let $\{X_k; k = 1,2,3,...\}$ be an i.i.d. random process with either discrete or continuous alphabet. Assume that the two processes are mutually independent of one another; that is, all collections of outputs of the N_k process are independent of collections of outputs of the X_k process. Define the random process $\{Y_k; k = 0,1,2,...\}$ by

$$Y_0 = 0$$

$$Y_k = \sum_{i=1}^{N_k} X_i, k = 1,2,... , \tag{10.1}$$

where we take the sum to be zero if $N_k = 0$.

As we mentioned in the chapter introduction the $\{Y_k\}$ process can be used to model many applications including telephone usage. In terms of the telephone call example, say that at most one new call originates at time k, that is, $N_k - N_{k-1} = 0$ or 1 and that the ith call in the sequence lasts X_i seconds altogether. Then Y_k equals the total time used by all calls that originate up through time k. Such a process is called a discrete time *compound process*. The processes are also referred to as being "doubly stochastic" because of the double source of randomness: the random process being summed and the random number of elements in the sum. We turn to the analysis of compound processes. Note that if N_k is known, say $N_k = n$, then Y_k is simply the sum of i.i.d. random variables, and the techniques of chapter 9 immediately apply. Since N_k is a random variable, however, it is not clear how to compute any probabilities or expectations such as the mean

$$EY_k = E[\sum_{i=1}^{N_k} X_i] ,$$

since the number of terms in the sum is random. We obviously cannot just interchange the sum and the expectation as we did in earlier work. However, the fact that we can interchange the sum and expectation if we know $N_k = n$ suggests a trick: Rewrite the expectation in terms of conditional probability functions. For simplicity we focus on the discrete alphabet case:

$$EY_k = \sum_y y p_{Y_k}(y) = \sum_y [\sum_n p_{Y_k,N_k}(y,n)]y$$

$$= \sum_n p_{N_k}(n)[\sum_y p_{Y_k|N_k}(y|n)y] .$$

The final bracketed term is simply an expectation with respect to a conditional pmf. Such an expectation is no different from an expectation with respect to an unconditional pmf except that it is indexed by (i.e., the expectation has a functional dependence on) the conditioning random variable. We call such an expectation a conditional expectation and denote it by $E(Y_k|n)$, or, if the conditioning random variable is not clear from the context, we make it explicit as $E(Y_k|N_k = n)$. With this definition in hand, the formula becomes

$$EY_k = \sum_n p_{N_k}(n)E(Y_k|n) , \qquad (10.2)$$

that is, first we find the conditional expectation wherein N_k is fixed at n, then we average over the distribution for N_k. For $N_k = n$, Y_k is simply the sum of n i.i.d. random variables, *all independent of the value n.* Therefore, we can now interchange expectation and summation to yield

$$E(Y_k|n) = E\left(\sum_{i=1}^{n} X_i \right)$$

$$\qquad (10.3)$$

$$= \sum_{i=1}^{n} EX_i = nEX ,$$

where EX is the mean of the i.i.d. process $\{X_n\}$. Thus, substituting (10.3) in (10.2) yields

$$EY_k = \sum_n p_{N_k}(n)nEX = (EN_k)(EX) , \qquad (10.4)$$

a reasonable and intuitively satisfying result—the expected value of Y_k is directly proportional to both the average number of summands and the average value of the summands. The technique we have described of doing an expectation computation in two steps—first doing a conditional expectation with one of the random variables pinned down and then averaging over that random variable—is called *iterated expectation* (or *nested expectation*). The technique generalizes to a form that can be used in many applications. Before we give the general form of this relation, observe that (10.2) can be abbreviated to

$$EY_k = E[E(Y_k|N_k)] . \qquad (10.5)$$

A function of a random variable is a random variable. In (10.5) $E(Y_k|N_k)$ is a function of N_k and therefore is a random variable. Thus (10.5) states that the expected value of Y_k is the expectation of the random variable $E(Y_k|N_k)$ formed by replacing n in the function of n, $E(Y_k|n)$, by the random variable N_k.

CONDITIONAL EXPECTATION

We summarize the developments of the preceding section in a formal definition of conditional expectation and a lemma on the iterated expectation property. The result is given for both discrete and continuous alphabet random processes.

Given a random variable U and a random variable or vector V, the *conditional expectation of U given V*, $E(U|v) = E(U|V = v)$, is defined as the expectation of U with respect to the conditional probability function of U given $V = v$. Thus if U is discrete,

$$E(U|v) = \sum_u u p_{U|V}(u|v) .$$

If U is continuous, then

$$E(U|v) = \int u f_{U|V}(u|v) du .$$

Lemma 10.1 (Iterated Expectation).   Given a random variable U and a random variable or vector V,

$$EU = E[E(U|V))] .$$

Thus, for example,

$$EU = \begin{cases} \sum_v p_V(v) E(U|v) & \text{if } V \text{ is discrete} \\[2mm] \int f_V(v) E(U|v) dv & \text{if } V \text{ is continuous} . \end{cases}$$

The proof of the iterated expectation property follows from the development of the preceding section with integrals substituted for sums for continuous random variables.

As a specific example of the computations for the expectation of the mean of Y_k in (10.1), consider the case of a binomial counting process $\{N_k\}$ with parameter p and a Bernoulli process $\{X_k\}$ with parameter ϵ. We then have

$$E(Y_k|n) = nEX = n\epsilon ,$$

$$E(Y_k|N_k) = N_k\epsilon ,$$

$$E(Y_k) = E[E(Y_k|N_k)] = \epsilon E[N_k] = \epsilon kp .$$

Functions of random variables are also random variables. Thus the iterated expectation property holds for functions of random variables as well. For example, consider the characteristic function of the compound process $\{Y_k\}$ of (10.1). Using iterated expectation,

$$M_{Y_k}(ju) = E(e^{juY_k}) = E[E(e^{juY_k}|N_k)] .$$

As before, given $N_k = n$, we know that Y_k is a sum of n i.i.d. random variables. Thus

$$E(e^{juY_k}|n) = M_X(ju)^n ,$$

where $M_X(ju)$ is the characteristic function of the i.i.d. process $\{X_n\}$. Thus we have for a general discrete time compound process that

$$M_{Y_k}(ju) = \sum_{n=0}^{k} p_{N_k}(n)M_X(ju)^n . \tag{10.6}$$

Equation (10.6) can be expressed in terms of the characteristic function of the random variable N_k (exercise 10.6). Consider the special case of a binomial counting process with parameter p for N_k and a Bernoulli i.i.d. process with parameter ϵ for X. N_k is a binomial random variable as shown in chapter 9. From (10.6) (using the binomial theorem),

$$M_{Y_k}(ju) = \sum_{n=0}^{k} \binom{k}{n} p^n(1-p)^{k-n}M_X(ju)^n$$

$$= (pM_X(ju)+(1-p))^k$$

$$= (p(1-\epsilon+\epsilon e^{ju})+1-p))^k . \tag{10.7}$$

$\{Y_k\}$ is a fairly complex process. The characteristic function for the process is a complicated computation if the iterated expectation technique is not used. The calculation, however, was rendered simple by using iterated expectation to break the computation up into two easy steps.

The same approach can be used to find second order moments of compound processes (exercise 10.1).

We next present a more general version of iterated expectation. Just as one can pull constants out of ordinary expectation, the following shows that one can pull functions of the conditioning random variable out of the conditional expectation.

Lemma 10.2. For any random variable or vector U, any random variable or random vector V, and any measurable functions g,h,

$$E(g(V)h(U,V)) = E[g(V)E(h(U,V)|V)] . \tag{10.8}$$

Proof. As usual, we prove the result only for the discrete case:

$$E(g(V)h(U,V)) = \sum_v \sum_u g(v)h(u,v)p_{U,V}(u,v)$$

$$= \sum_v p_V(v)g(v) \sum_u h(u,v)p_{U|V}(u|v)$$

$$= \sum_v p_V(v)g(v)E(h(U,v)|v)$$

$$= E[g(V)E(h(U,V)|V)] \ .$$

Analogous to the abstract definition (9.25) of general conditional probability, equation (10.8) can be taken as a defining property of general conditional expectation. That is, conditional expectation can be defined when conditional pmf's or pdf's do not exist even though we used their existence in our constructive definition.

Note that (10.8) of lemma 10.2 reduces to lemma 10.1 if we choose g to be the identity function and set $h = 1$. Note also that in taking a conditional expectation, functions of the conditioning random variable are treated as a "constant"; i.e.,

$$E[g(V)|V = v] = g(v) \ . \tag{10.9}$$

In the next section we consider a simple but powerful application of equation (10.8).

*AN APPLICATION TO ESTIMATION THEORY

Say that we observe the outcome of a random variable or vector V—that is, we know that $V = v$—and we wish to estimate another random variable U using some function of v, say $\hat{U}(v)$. Unlike the study in chapter 8, however, we do not require that the estimate be a linear function of the observation. What is the best possible such estimate? Before this question can be answered, we must decide to quantify what is meant by "best." As in the linear case in chapter 8, a common way of quantifying the notion of the "best" estimate is to find the estimate minimizing the mean squared error of the estimate. The mean squared error is chosen as a measure of performance for two reasons: primarily because it is a measurement of average power to which many systems are most sensitive, and secondarily because the resulting problem is not too difficult to analyze and gives results that correspond to intuition. Thus we define the minimum mean squared estimate of U given $V = v$ as the function $\hat{U}(v)$ yielding the minimum possible value of

$$\epsilon^2(\hat{U}) = E[(U - \hat{U}(V))^2] \ . \tag{10.10}$$

In this section we prove that the minimum mean squared estimate of U given $V = v$ is simply the conditional expectation of U given $V = v$! To establish this result, define the supposed optimum estimate $\hat{U}(v) = E(U|v)$ and let $\bar{U}(v)$ denote any other possible estimate. We shall show that always $\epsilon^2(\hat{U}) \le \epsilon^2(\bar{U})$, proving that $\hat{U}$ yields the desired minimum mean squared error. Expanding (10.10) algebraically,

$$\epsilon^2(\bar{U}) = E[(U - \bar{U}(V))^2] = E[(U - \hat{U}(V) + \hat{U}(V) - \bar{U}(V))^2]$$

$$= E[(U - \hat{U}(V))^2] + 2E[(U - \hat{U}(V))(\hat{U}(V) - \bar{U}(V))] + E[(\hat{U}(V) - \bar{U}(V))^2]$$

$$\ge \epsilon^2(\hat{U}) + 2E[(U - \hat{U}(V))(\hat{U}(V) - \bar{U}(V))] .$$

Using lemma 10.2 with $g(V) = \hat{U}(V) - \bar{U}(V)$ and with $h(U,V) = U - \hat{U}(V)$, the cross-term can be expressed as

$$+ E[(\hat{U}(V) - \bar{U}(V))E(U - \hat{U}(V)|V)] . \tag{10.11}$$

We have, however, from the definition of $\hat{U}$ and the linearity of expectation that for all v,

$$E(U - \hat{U}(v)|v) = E(U|v) - E(E(U|v)|v)$$

$$= E(U|v) - E(U|v) = 0 .$$

Thus the cross-term in (10.11) is the expected value of something times 0 and hence is itself zero. This completes the proof.

CONTINUOUS TIME COMPOUND PROCESSES

The extension of the idea of a compound process to continuous time is surprisingly easy, and the techniques adapt virtually without change. We simply replace the discrete time counting process $\{N_k; k = 0,1,...\}$ by a continuous time i.i.d. counting process $\{N_t; t \ge 0\}$ such as the Poisson counting process of chapter 9. Then, given an i.i.d. sequence of random variables $\{X_i; i = 1,2,...\}$ and a continuous time counting process $\{N_t; t \ge 0\}$ such that the two processes are mutually independent of each other, we define the process

$$Y(t) = \sum_{i=1}^{N_t} X_i ,$$

where the sum is taken as zero when $N_t = 0$. $\{Y(t)\}$ is called a *continuous time compound process*.

Observe, just as with discrete time, that the compound process can be continuous alphabet or discrete alphabet depending on the alphabet of the underlying i.i.d. process. Note that the compound process will have a constant value between counts in the counting process and that each time the counting process increases by 1, the compound process will change by the addition of a new independent random variable. Thus the process is continuous and differentiable (with derivative zero) except at the "jumps" where a new random variable is added. The development for the discrete time case extends immediately to continuous time by simply replacing the discrete time index with the continuous time index. In particular,

$$EY_t = (EN_t)(EX)$$

$$M_{Y_t}(ju) = E[M_X(ju)^{N_t}] . \qquad (10.12)$$

If the counting process is the Poisson counting process, then

$$EY_t = (\lambda t)EX \qquad (10.13)$$

and

$$M_{Y_t}(ju) = \sum_{k=0}^{\infty} \frac{(\lambda t)^k e^{-\lambda t}}{k!} M_X(ju)^k \qquad (10.14)$$

$$= e^{-\lambda t} \sum_{k=0}^{\infty} \frac{(\lambda t M_X(ju))^k}{k!} = e^{-\lambda t(1 - M_X(ju))},$$

where we have invoked the Taylor series expansion for an exponential.

EXERCISES

10.1 Let $\{N_k\}$ be a binomial counting process and let $\{X_n\}$ be an i.i.d. process with zero mean and variance σ^2. Let $\{Y_k\}$ denote the corresponding compound process. Use iterated expectation to evaluate the autocorrelation function $R_Y(t,s)$.

10.2 Let $\{W_n\}$ be a discrete time Wiener process. What is the minimum mean squared error estimate of W_n given $W_1,...,W_{n-1}$? (This is called the optimum one step predictor.) How does this compare to the LLSE estimate in chapter 8?

10.3 Let $\{N_k\}$ be a binomial counting process. What is the minimum mean squared error estimate of N_k given $N_1,...,N_{k-1}$?

10.4 Let $\{X_n\}$ be an i.i.d. binary process with $Pr(X_n = +1) = Pr(X_n = -1) = 1/2$ and let N_t be the Poisson counting process. A continuous time random walk can be defined by

$$Y(t) = \sum_{k=1}^{N_t} X_k .$$

Find the expectation, covariance, and characteristic function of Y_t.

10.5 Are compound processes independent increment processes?

10.6 Show that the characteristic function for the compound process of equation (10.1) is

$$M_{Y_k}(ju) = M_{N_k}(\ln(M_X(ju))) .$$

10.7 Suppose that X and Y are jointly Gaussian random variables. Find the LLSE $\hat{X} = aY+b$, the minimum mean squared estimate $\tilde{X} = E(X|Y)$, and the resulting expected squared error for both estimates.

10.8 Suppose that many sensors view a common target random variable X with different additive noises: Say the ith sensor produces the random variable $Y_i = X+W_i$ where the W_i are zero mean independent Gaussian random variables with variance σ_i^2 (they are *not* identically distributed!). Suppose that we know, however, that the variances are all bound above by a common value, say $\sigma_i^2 \le \sigma_{max}^2$, all i. All of the sensors deliver their random variable Y_i to a central processor which attempts to guess the value of the target random variable X. For the moment we assume that there are a fixed number n of these sensors.

Suppose that the central processor forms an estimate of the form:

$$\hat{X}_n = \frac{1}{n}\sum_{i=0}^{n-1} Y_i$$

and define the resulting error $\epsilon_n = \hat{X}_n - X$.

(a) Find the probability density function $f_{\epsilon_n}(\alpha)$ and $f_{\hat{X}_n|X}(\alpha|\beta)$.

(b) Find $E\hat{X}_n$ and $E(\epsilon_n^2)$.

(c) Suppose that we model the situation where we have a huge number of sensors by assuming that $n \to \infty$. Does $\hat{X}_n$ converge in some sense to X in this case? (If so, in what sense and why? If not, why not?)

11

MODELING
PHYSICAL PROCESSES

An engineer encounters two types of random processes in practice. The first is the random process whose probability distribution depends largely on design parameters: the type of modulation used, the method of data coding used, etc. The second type of random processes have probability distributions that depend on naturally occurring phenomena over which the engineer has little, if any control: noise in physical devices, speech waveforms, the number of messages in a telephone system as a function of time, etc. This chapter is oriented toward the second type of random processes. Mathematical models of random processes are derived from physical principles that frequently apply in practice. The focus is on two of the basic classes of distributions—the Gaussian distribution class and the Poisson distribution class. Surprisingly enough, as a consequence of the physical properties, these two distribution classes characterize (at least approximately) the majority of naturally occurring random processes. The goal of the chapter is to clarify the physical origins of these distributions and to provide examples of physical processes that yield these distributions. In doing so, the development provides two examples from the more general applications area of finding good mathematical models for real-world processes.

This subject is somewhat removed from the mainstream of the book. Our intent is simply to remove some of the mystery of the functional forms of two important distributions by showing how these apparently complicated distributional forms arise from nature. Therefore, the development presented is somewhat brief, without consideration of all the mathematical details.

THE CENTRAL LIMIT THEOREM

We first consider the origins of the Gaussian or normal distribution. The Gaussian pdf, originally introduced following example [3.14], seems rather arcane and artificial at first introduction. As we will show in this section, the Gaussian distribution is found in practice as a consequence of physical principles—many random processes are the result of the addition of a large number of small, unpredictable contributing forces, i.e., random variables. At the same time, with very mild restrictions, the probability distribution of the summation of a large number of random variables is necessarily Gaussian. The statement of this fact is embodied in the central limit theorem. Just as we found with laws of large numbers and ergodic theorems, there is really no single central limit theorem—there are many versions of central limit theorems. The various central limit theorems differ in the conditions of applicability. However, they have a common conclusion: convergence of a sum in distribution to a random variable. The random variable is almost always (but not always) a Gaussian random variable. We will present only the simplest form of central limit theorem, a central limit theorem for i.i.d. random variables with convergence to a Gaussian random variable.

Suppose that $\{X_n\}$ is an i.i.d. random process with arbitrary pmf or pdf except that it has a finite mean $EX_n = m$ and finite variance $\sigma^2_{X_n} = \sigma^2$. Consider the "standardized" or "normalized" sum

$$R_n = \frac{1}{n^{1/2}} \sum_{k=0}^{n-1} \frac{X_i - m}{\sigma} . \tag{11.1}$$

By subtracting the means and dividing by the square root of the variance (the standard deviation), the resulting random variable is easily seen to have zero mean and unit variance; that is,

$$ER_n = 0, \ \sigma^2_{R_n} = 1 ,$$

hence the description "standardized," or "normalized." Note that unlike the sample average that appears in the law of large numbers, the sum here is normalized by $n^{-1/2}$ and not n^{-1}. This slight change results in remarkably different behavior: Convergence still occurs, but to a random variable, not to a constant.

Using characteristic functions, we have from the independence of the $\{X_i\}$ and lemma 9.1 that

$$M_{R_n}(ju) = M_{\frac{X-m}{\sigma}}(\frac{ju}{n^{1/2}})^n . \tag{11.2}$$

We wish to investigate the asymptotic behavior of the characteristic function of (11.2) as $n \to \infty$. To accomplish this we form a Taylor series expansion of $M_{\frac{X-m}{\sigma}}(\frac{ju}{n^{1/2}})$ about $u=0$ and then study the limiting behavior of the expression. Recall that the Taylor series of a function $f(u)$ about the point $u=0$ has the following form: Define the derivatives

$$f^{(k)}(0) = \frac{d^k}{du^k} f(u)|_{u=0} ;$$

then

$$f(u) = \sum_{k=0}^{\infty} u^k \frac{f^{(k)}(0)}{k!} =$$

$$f(0) + uf^{(1)}(0) + u^2\frac{f^{(2)}(0)}{2} + \text{terms in } u^k ; k \geq 3 , \tag{11.3}$$

assuming, of course, that all of the derivatives exist. As seen in exercises 6.22 and 9.10, however, derivatives of characteristic functions can be related to moments. To see this in general, consider the derivatives of the characteristic function of a discrete random variable Y:

$$\frac{d^k}{du^k} M_Y(ju) = \frac{d^k}{du^k} \sum_y p_Y(y)e^{juy}$$

$$= \sum_y p_Y(y)\frac{d^k}{du^k}e^{juy} = \sum_y p_Y(y)(jy)^k e^{juy} .$$

Thus if we evaluate the derivative at 0 we have that

$$M_Y^{(k)}(0) = \frac{d^k}{du^k} M_Y(ju)|_{u=0} = j^k \sum_y y^k p_Y(y)$$

or

$$E(Y^k) = \frac{M_Y^{(k)}(0)}{j^k} . \tag{11.4}$$

Equation (11.4) is called the moment-generating property of characteristic functions. Note that if we make the substitution $w = ju$ and differentiate with respect to w, instead of u,

$$\frac{d^k}{dw^k} M_Y(w)|_{w=0} = E(Y^k) .$$

Because of this property, characteristic functions with $ju = w$ are called *moment-generating functions.* (However, from the defining integral for characteristic functions in example [6.6], the moment-generating function may not exist, whereas the characteristic function always does exist.) This equation is useful if one has to find several moments of a given random variable since it permits one to perform a single integration or sum to find the transform and then several differentiations to find the moments. Since differentiation is usually easier than integration, this is usually easier than the direct route. The reader should try using this formula to compute the first and second moments of a Gaussian random variable and compare it with the direct computation.

A similar manipulation with integrals shows that (11.4) also holds in the continuous case. That is, because differentiation and expectation are linear operations that generally commute, we can write

$$M_Y^{(k)}(0) = \frac{d^k}{du^k} M_Y(ju)\big|_{u=0} = \frac{d^k}{du^k} E[e^{juY}]\big|_{u=0}$$

$$E[\frac{d^k}{du^k} e^{juY}]\big|_{u=0} = j^k E[Y^k] \ ,$$

provided that the integrals and derivatives exist.

Combining the Taylor series expansion with the moment-generating property we have that

$$M_Y(ju) = \sum_{k=0}^{\infty} u^k \frac{M_Y^{(k)}(0)}{k!}$$

$$= \sum_{k=0}^{\infty} (ju)^k \frac{E(Y^k)}{k!}$$

$$= 1 + juE(Y) - u^2 E(Y^2) + o(u^2)/2 \ . \tag{11.5}$$

This result has an interesting implication: If the Taylor series can be shown to be valid over the entire range of u rather than just in the area around 0, then knowing *all* of the moments of a random variable is sufficient to know the transform. Since the transform in turn implies the distribution, this guarantees that knowing *all* of the moments of a random variable completely describes the distribution. This is true, however, only when the distribution is sufficiently "nice," that is, when the technical conditions ensuring the existence of all of the required derivatives and of the convergence of the Taylor series hold.

Returning to the development of the distribution of the sum of (11.1), let $Y = (X - m)/\sigma$. Y has zero mean and a second moment of 1, and hence in (11.5)

$$M_{\frac{X-m}{\sigma}}(jun^{-1/2}) = 1 - \frac{u^2}{2n} + o(u^2/n) , \qquad (11.6)$$

where the rightmost term goes to zero faster than u^2/n. Observe that as $n \to \infty$ the argument of the transform is increasingly close to zero. Hence the Taylor series expansion need hold only in a small region about the origin and not for all u for the argument to be valid. Combining this result with (11.2), we have that

$$\lim_{n \to \infty} M_{R_n}(ju) = \lim_{n \to \infty} [1 - \frac{u^2}{2n} + o(\frac{u^2}{n})]^n .$$

From elementary analysis, however, this limit is simply

$$\lim_{n \to \infty} M_{R_n}(ju) = e^{-\frac{u^2}{2}} ,$$

the characteristic function of a Gaussian random variable with zero mean and unit variance! Thus, provided that (11.6) holds, a standardized sum of a family of i.i.d. random variables has a transform that converges to the transform of a Gaussian random variable regardless of the actual marginal distribution of the i.i.d. sequence. Equation (11.6) holds provided that the random variables are nice enough for the described development. Obviously the requirement that the transform be infinitely differentiable at $u = 0$ so that the infinite series of (11.5) converges for all u is stronger than necessary. From the formula for a Taylor series with remainder, (11.6) holds whenever M_Y has a continuous second derivative. This is equivalent to requiring only that X possess a finite second order moment.

By taking inverse transforms, the convergence of transforms implies that the cdf's will also converge to a Gaussian cdf. This does *not* imply convergence to a Gaussian *pdf*, however, because, for example, a finite sum of discrete random variables cannot be approximated by a pdf. A slight modification of the above development shows that if $\{X_n\}$ is an i.i.d. sequence with mean m and variance σ^2, then

$$n^{-1/2} \sum_{k=0}^{n-1} (X_i - m)$$

will have a transform and a cdf converging to those of a Gaussian random variable with mean 0 and variance σ^2. This result and generalizations of this result are called the central limit theorem. We summarize the central limit theorem that we have established as follows:

Theorem 11.1 (A Central Limit Theorem).   Let $\{X_n\}$ be an i.i.d. random process with a finite mean m and variance σ^2. Then

$$n^{-1/2}\sum_{k=0}^{n-1}(X_i - m)$$

converges in distribution to a Gaussian random variable with mean m and variance σ^2.

Intuitively the theorem states that if we sum up a large number of independent random variables and normalize by $n^{-1/2}$ so that the variance of the normalized sum stays constant, then the resulting sum will be approximately Gaussian. For example, a current meter across a resistor will measure the effects of the sum of millions of electrons randomly moving and colliding with each other. Regardless of the probabilistic description of these microevents, the global current will appear to be Gaussian. Making this precise yields a model of thermal noise in resistors. We shall provide a more detailed development of this model later in this chapter. Similarly, if dust particles are suspended on a dish of water and subjected to the random collisions of millions of molecules, then the motion of any individual particle in two dimensions will appear to be Gaussian. Making this rigorous yields the classic model for what is called "Brownian motion." A similar development in one dimension yields the Wiener process.

Note that in (11.2), if the Gaussian characteristic function is substituted on the right-hand side, a Gaussian characteristic function appears on the left. Thus the central limit theorem says that if you sum up random variables, you approach a Gaussian distribution. Once you have a Gaussian distribution, you "get stuck" there—adding more random variables of the same type (or Gaussian random variables) to the sum does not change the Gaussian characteristic. The Gaussian distribution is an example of an *infinitely divisible* distribution. The nth root of its characteristic function is a distribution of the same type as seen in (11.2). Equivalently stated, the distribution class is invariant under summations. The Poisson distribution has the same property, as we will see in the next section.

THE POISSON COUNTING PROCESS

Next consider modeling a continuous time counting process $\{N_t; t \geq 0\}$ with the following properties:

1. $N_0 = 0$ (the initial condition).
2. The process has independent and stationary increments. Hence the changes, called jumps, during nonoverlapping time intervals are independent random variables. The jumps in a given time

interval are memoryless, and their amplitude does not depend on what happened before that interval.

3.   In the limit of *very small* time intervals, the probability of an increment of 1, that is, of increasing the total count by 1, is proportional to the length of the time interval. The probability of an increment greater than 1 is negligible in comparison, e.g., is proportional to powers greater than 1 of the length of the time interval.

These properties well describe many physical phenomena such as the emission of electrons and other subatomic particles from irradiated objects (remember vacuum tubes?), the arrival of customers at a store or phone calls at an exchange, and other phenomena where events such as arrivals or discharges occur randomly in time. The properties naturally capture the intuition that such events do not depend on the past and that for a very tiny interval, the probability of such an event is proportional to the length of the interval. For example, if you are waiting for a phone call, the probability of its happening during a period of τ seconds is proportional to τ. The probability of more than two phone calls in a very small period τ is, however, negligible in comparison.

The third property can be quantified as follows: Let λ be the proportionality constant. Then for a small enough time interval Δt,

$$\Pr(N_{t+\Delta t} - N_t = 1) \cong \lambda \Delta t$$

$$\Pr(N_{t+\Delta t} - N_t = 0) \cong 1 - \lambda \Delta t$$

$$\Pr(N_{t+\Delta t} - N_t > 1) \cong 0 . \qquad (11.7)$$

The relations of (11.7) can be stated rigorously by limit statements, but we shall use them in the more intuitive form given.

We now use the properties 1 to 3 to derive the probability mass function $p_{N_t - N_0}(k) = p_{N_t}(k)$ for an increment $N_t - N_0$, from the starting time at time 0 up to time $t > 0$ with $N_0 = 0$. For convenience we temporarily change notation and define

$$p(k,t) = p_{N_t - N_0}(k) ; t > 0 .$$

Let Δt be a differentially small interval as in (11.7), and we have that

$$p(k,t+\Delta t) =$$

$$\sum_{n=0}^{k} \Pr(N_t = n)\Pr(N_{t+\Delta t} - N_t = k - n \mid N_t = n) .$$

Since the increments are independent, the conditioning can be dropped so that, using (11.7),

$$p(k, t+\Delta t) =$$

$$\sum_{n=0}^{k} \mathrm{Pr}(N_t = n)\mathrm{Pr}(N_{t+\Delta t} - N_t = k - n)$$

$$\cong p(k,t)(1 - \lambda \Delta t) + p(k-1,t)\lambda \Delta t \ ,$$

which with some algebra yields

$$\frac{p(k,t+\Delta t) - p(k,t)}{\Delta t} = p(k-1,t)\lambda - p(k,t)\lambda \ .$$

In the limit as $\Delta t \rightarrow 0$ this becomes the differential equation

$$\frac{d}{dt}p(k,t) + \lambda p(k,t) = \lambda p(k-1,t) \ , t > 0 \ .$$

The initial condition for this differential equation follows from the initial condition for the process, $N_0 = 0$; i.e.,

$$p(k,0) = \begin{cases} 0, & k \neq 0 \\ 1, & k = 0, \end{cases}$$

since this corresponds to $\mathrm{Pr}(N_0 = 0) = 1$. The solution to the differential equation with the given initial condition is

$$p_{N_t}(k) = p(k,t) = \frac{(\lambda t)^k e^{-\lambda t}}{k!} \ ; \ k = 0,1,2,\dots \ ; t \geq 0 \ . \tag{11.8}$$

(This is easily verified by direct substitution.)

The pmf of (11.8) is the Poisson pmf, and hence the given properties produce the Poisson counting process. Note that (11.8) can be generalized using the stationarity of the increments to yield the pmf for k jumps in an arbitrary interval $(s,t), t \geq s$ as

$$p_{N_t - N_s}(k) = \frac{(\lambda(t-s))^k e^{-\lambda(t-s)}}{k!} \ ; \ k = 0,1,\dots \ ; t \geq s \ . \tag{11.9}$$

As developed in chapter 9, these pmf's and the given properties yield a complete specification of the Poisson counting process.

Note that sums of Poisson random variables are Poisson. This follows from the development of this section. That is, for any $t > s > r$, all three of the indicated quantities in $(N_t - N_s) + (N_s - N_r) = N_t - N_r$ are Poisson. Thus the Poisson distribution is infinitely divisible in the sense defined at the end of the preceding section. Of course the infinite divisibility of Poisson random variables can also be verified by characteristic functions as in (11.2). Poisson random variables satisfy the requirements of the central limit theorem. Thus, it can be concluded that with appropriate normalization, the Poisson cdf and the Gaussian cdf approach each other asymptotically.

*THERMAL NOISE

Thermal noise is one of the most important sources of noise in communications systems. It is the "front-end" noise in receivers that is caused by the random motion of electrons in a resistance. The resulting noise is then greatly amplified by the amplifiers that magnify the noise along with the possibly tiny signals. Thus the noise is really in the receiver itself and not in the atmosphere, as some might think, and can be comparable in amplitude to the desired signal. In this section we sketch the development of a model of thermal noise. The development provides an interesting example of a process with both Poisson and Gaussian characteristics.

Say we have a uniform conducting cylindrical rod at temperature T. Across this rod we connect an ammeter. The random motion of electrons in the rod will cause a current $I(t)$ to flow through the meter. We wish to develop a random process model for the current based on the underlying physics. The following are the relevant physical parameters:

A = cross-sectional area of the rod

L = length of the rod

q = electron charge

n = number of electrons per cubic centimeter

α = average number of electron collisions with heavier particles per second (about 10^3)

m = mass of an electron

ρ = resistivity of the rod = $\dfrac{m\alpha}{nq^2}$

R = resistance of the rod = $\dfrac{\rho L}{A}$

k = Boltzmann's constant

The current measured will be due to electrons moving in the longitudinal direction of the rod, which we denote x. Let $V_{x,k}(t)$ denote the component of velocity in the x direction of the kth electron at time t. The total current $I(t)$ is then given by the sums of the individual electron currents as

$$I(t) = \sum_{k=1}^{nAL} i_k(t) = \sum_{k=1}^{nAL} \frac{q}{L/V_{x,k}(t)}$$

$$= \sum_{k=1}^{nAL} \frac{q}{L} V_{x,k}(t).$$

We assume that (1) the average velocity, $EV_{x,k}(t) = 0$, all k,t; (2) $V_{x,k}(t)$ and $V_{x,j}(s)$ are independent random variables for all $k \neq j$; and (3) the $V_{x,k}(t)$ have the same distribution for all k.

The autocorrelation function of $I(t)$ is found as

$$R_I(\tau) = E[I(t)I(t+\tau)] = \sum_{k=1}^{nAL} \frac{q^2}{L^2} E[V_{x,k}(t)V_{x,k}(t+\tau)]$$

$$= \frac{nAq^2}{L} E[V_x(t)V_x(s)] , \tag{11.10}$$

where we have dropped the subscript k since by assumption the distribution, and hence the autocorrelation function of the velocity, does not depend on it.

Next assume that, since collisions are almost always with heavier particles, the electron velocities before and after collisions are independent—the velocity after impact depends only on the momentum of the heavy particle that the electron hits. We further assume that the numbers of collisions in disjoint time intervals are independent and satisfy (11.7) with a change of parameter:

$$\text{Pr(no collisions in } \Delta t) \cong (1 - \alpha \Delta t)$$

$$\text{Pr(one collision in } \Delta t) \cong \alpha \Delta t$$

This implies that the number of collisions is Poisson and that from (11.9)

$$\text{Pr(a particle has } k \text{ collisions in } [t, t+\tau)) = e^{-\alpha\tau}\frac{(\alpha\tau)^k}{k!} ; k = 0,1,2,....$$

Thus if $\tau \geq 0$ and $N_{t,\tau}$ is the number of collisions in $[t, t+\tau)$, then, using iterated expectation and the independence with mean zero of electron velocities when one or more collisions have occurred,

$$E[V_x(t)V_x(t+\tau)] = E(E[V_x(t)V_x(t+\tau)|N_{t,\tau}])$$

$$= E(V_x(t)^2)\text{Pr}(N_{t,\tau}=0) + (EV_x(t))^2\text{Pr}(N_{t,\tau}\neq 0)$$

$$= E(V_x(t)^2)e^{-\alpha\tau} . \tag{11.11}$$

It follows from the equipartition theorem for electrons in thermal equilibrium at temperature T that the electron velocity variance is

$$E(V_x(t)^2) = \frac{kT}{m} . \tag{11.12}$$

Therefore, after some algebra, we have from (11.10) through (11.12) that

$$R_I(\tau) = \frac{kT}{R}\alpha e^{-\alpha|\tau|} .$$

Thevinin's theorem can then be applied to model the conductor as a voltage source with voltage $E(t) = R I(t)$. The autocorrelation function of $E(t)$ is

$$R_E(\tau) = kTR\alpha e^{-\alpha|\tau|} ,$$

an autocorrelation function that decreases exponentially with the delay τ. Observe that as $\alpha \to \infty$, $R_E(\tau)$ becomes a taller and narrower pulse with constant area $2kTR$; that is, it looks more and more like a Dirac delta function with area $2kTR$. Since the mean is zero, this implies that the process $E(t)$ is such that samples separated by very small amounts are approximately uncorrelated. Thus thermal noise is approximately white noise. The central limit theorem can be used to show that the finite dimensional distributions of the process are approximately Gaussian. Thus we can conclude that an approximate model for thermal noise is a Gaussian white noise process!

EXERCISES

11.1 Let $\{X_n\}$ be an i.i.d. binary random process with equal probability of $+1$ or -1 occurring at any time n. Show that if Y_n is the standardized sum

$$Y_n = n^{-1/2} \sum_{k=0}^{n-1} X_k ,$$

then

$$M_{Y_n}(ju) = e^{n \log \cos \frac{u}{\sqrt{n}}} .$$

Find the limit of this expression as $n \to \infty$.

11.2 Suppose that a fair coin is flipped 1,000,000 times. Write an exact expression for the probability that between 400,000 and 500,000 heads occur. Next use the central limit theorem to find an approximation to this probability. Use tables to evaluate the resulting integral.

11.3 In the development of the Poisson counting process, we fixed time values and looked at the random variables giving the counts and the increments of counts at the fixed times. In this problem we explore the reverse description: What if we fix the counts and look at the times at which the process achieves these counts? For example, for each strictly positive integer k, let r_k denote the time that the kth count occurs; that is, $r_k = \alpha$ if and only if

$$N_\alpha = k; N_t < k; \text{ all } t < \alpha .$$

Define $r_0 = 0$. For each strictly positive integer k, define the *interarrival times* τ_k by

$$\tau_k = r_k - r_{k-1} ,$$

and hence

$$r_k = \sum_{i=1}^{k} \tau_i .$$

(a)   Find the pdf for r_k for $k = 1, 2, \ldots$.
      Hint: First find the cdf by showing that

$$F_{r_k}(\alpha) = \Pr(k^{\text{th}} \text{ count occurs before or at time } \alpha)$$

$$= \Pr(N_\alpha \geq k) ,$$

and then using the Poisson pmf to write an expression for this sum, differentiate to find the pdf. You may have to do some algebra to reduce the answer to a simple form not involving any sums. This is most easily done by writing a difference of two sums in which all terms but one cancel. The final answer is called the *Erlang* family of pdf's. You should find that the pdf of r_1 is an exponential density.

(b)   Use the basic properties of the Poisson counting process to prove that the interarrival times are i.i.d.
      Hint: Prove that

$$F_{\tau_n \mid \tau_1, \cdots, \tau_{n-1}}(\alpha \mid \beta_1, \cdots, \beta_{n-1}) =$$

$$F_{\tau_n}(\alpha) = 1 - e^{-\lambda\alpha} ; n = 1, 2, \ldots; \alpha \geq 0 .$$

11.4 Let $\{N_t\}$ be a Poisson counting process. Let $i(t)$ be the deterministic waveform defined by

$$i(t) = \begin{cases} 1 & \text{if } t \in [0, \delta] \\ 0 & \text{otherwise} \end{cases}$$

—that is, a flat pulse of duration δ. For $k = 1, 2, \ldots$, let t_k denote the time of the kth jump in the counting process (that is, t_k is the smallest value of t for which $N_t = k$). Define the random process $\{Y(t)\}$ by

$$Y(t) = \sum_{k=1}^{N_t} i(t - t_k).$$

This process is a special case of a class of processes known as filtered Poisson processes. This particular example is a model for shot noise in vacuum tubes.

Draw some sample waveforms of this process. Find $M_{Y(t)}(ju)$ and $p_{Y(t)}(n)$.
Hint: You need not consider any properties of the random variables $\{t_k\}$ to solve this problem.

11.5 Using an expansion of the form of equation (11.6), show directly that the central limit theorem is satisfied for a sequence of i.i.d. random variables with pdf

$$p(x) = \frac{2}{\pi(1+x^2)^2} \ , x \in \mathbf{R} \ .$$

Try to use the same expansion for

$$p(x) = \frac{1}{\pi(1+x^2)} \ , x \in \mathbf{R} \ .$$

Explain your result.

APPENDIX:
SUPPLEMENTARY READING

In this appendix we provide some suggestions for supplementary reading. Our goal is to provide some leads for the reader interested in pursuing the topics treated in more depth. Admittedly we only scratch the surface of the large literature on probability and random processes. The books referred to are selected based on our own tastes—they are books from which we have learned and from which we have drawn useful results, techniques, and ideas for our own research.

A good history of the theory of probability may be found in Maistrov [35], who details the development of probability theory from its gambling origins through its combinatorial and relative frequency theories to the development by Kolmogorov of its rigorous axiomatic foundations. A somewhat less serious historical development of elementary probability is given by Huff and Geis [26]. Several early papers on the application of probability are given in Newman [38]. Of particular interest are the papers by Bernoulli on the law of large numbers and the paper by George Bernard Shaw comparing the vice of gambling and the virtue of insurance.

An excellent general treatment of the theory of probability and random processes may be found in Ash [1], along with treatments of real analysis, functional analysis, and measure and integration theory. Ash is a former engineer turned mathematician, and his book is one of the best available for someone with an engineering background who wishes to pursue the mathematics beyond the level treated in this book. The only subject of this book completely absent in Ash is the second-order theory and linear systems material of chapter 8 and the related examples of chapter 9.

Other good general texts on probability and random processes are those of Breiman [6] and Chung [9]. These books are mathematical treatments that are relatively accessible to engineers. All three books are a useful addition to any library, and most of the mathematical details avoided here can be found in these texts. Wong's book [53] provides a mathematical treatment for engineers with a philosophy similar to ours but with an emphasis on continuous time rather than discrete time random processes.

Another general text of interest is the inexpensive paperback book by Sveshnikov [49], which contains a wealth of problems in most of the topics covered here as well as many others. While the notation and viewpoint often differ, this book is a useful source of applications, formulas, and general tidbits.

The set theory preliminaries of chapter 2 can be found in most any book on probability, elementary or otherwise, or in most any book on elementary real analysis. In addition to the general books mentioned, more detailed treatments can be found in books on mathematical analysis such as those by Rudin [46], Royden [44], and Simmons [47]. These references also contain discussions of functions or mappings. A less mathematical text that treats set theory and provides an excellent introduction to basic applied probability is Drake [12].

The linear systems fundamentals are typical of most electrical engineering linear systems courses. Good developments may be found in Chen [7], Kailath [27], Bose and Stevens [4], and Papoulis [40], among others. A treatment emphasizing discrete time may be found in Steiglitz [48].

Detailed treatments by Fourier and Laplace techniques may be found in Bracewell [5], Papoulis [39], and the early classic of Wiener [51]. This background is useful both for the system theory applications and for the manipulation of characteristic functions or moment-generating functions of probability distributions.

Although the development of probability theory is self-contained and elementary probability is not, strictly speaking, a prerequisite, an introductory text to the subject can be a useful source of intuition, applications, and practice of some of the basic ideas. Two books that admirably fill this function are Drake [12] and the classic introductory text by two of the primary contributors to the early development of probability theory, Gnedenko and Khinchin [19]. The more complete text by Gnedenko [20] also provides a useful backup text. A virtual encyclopedia of basic probability, including a wealth of examples, distributions, and computations, may be found in Feller [15].

The axiomatic foundations of probability theory presented in chapter 3 were developed by Kolmogorov and first published in 1933. (See the English translation [30].) Although not the only theory of probability (see, e.g., Fine [16] for a survey of other approaches), it has become the standard approach to the analysis of random systems. The general references cited previously provide good additional material for the basic development of probability spaces, measures, Lebesgue integration, and expectation. The reader interested in probing more deeply into the mathematics is referred to the classics by Halmos [23] and Loeve [33].

As observed in chapter 6, instead of beginning with axioms of probability and deriving the properties of expectation, one can go the other way and begin with axioms of expectation or integration and derive the properties of probability. Some texts treat measure and integration theory in this order, e.g., Asplund and Bungart [2]. A nice paperback book treating probability and random processes from this viewpoint in a manner accessible for engineers is that by Whittle [50].

A detailed and quite general development of the Kolmogorov extension theorem of chapter 5 may be found in Parthasarathy [41], who treats probability theory for general metric spaces instead of just Euclidean spaces. The mathematical level of this book is high, though, and the going can be rough. It is useful, however, as a reference for very general results of this variety and for detailed statements of the theorem.

Good background reading for chapters 6 and 7 are the book on convergence of random variables by Lukacs [34] and the book on ergodic theory by Billingsley [3]. The Billingsley book is a real gem for engineers interested in learning more about the varieties and proofs of ergodic theorems for discrete time processes. The book also provides nice tutorial reviews on advanced conditional probability and a variety of other topics. Several proofs are given for the mean and pointwise ergodic theorems. Most are accessible given a knowledge of the material of this book plus a knowledge of the projection theorem of Hilbert space theory. The book also provides insight into applications of the general formulation of ergodic theory to areas other than random process theory. Another nice survey of ergodic theory is that of Halmos [24].

As discussed in chapter 7, stationarity and ergodicity are sufficient but not necessary conditions for the ergodic theorem to hold, that is, for sample averages to converge. A natural question, then, is what conditions are both necessary and sufficient. The answer is known for discrete time processes in the following sense: A process is said to be *asymptotically mean stationary* or a.m.s. if its process distribution, say m, is such that the limits

$$\lim_{n \to \infty} \frac{1}{n} \sum_{i=0}^{n-1} m(T^{-i}F)$$

exist for all process events F, where T is the left-shift operation. The limits trivially exist if the process is stationary. They also exist when the die out with time and in a variety of other cases. It is known that a process will have an ergodic theorem in the sense of having all sample averages of bounded measurements converge if *and only if* the process is a.m.s [21]. The sample averages of an a.m.s. process will converge to constants with probability one if and only if the process is also ergodic.

Second-order theory of random processes and its applications to filtering and estimation form a bread-and-butter topic for engineering applications and are the subject of numerous good books such as Grenander and Rosenblatt [22], Cramér and Leadbetter [10], Rozanov [45], Yaglom [54], and Liptser and Shiryayev [32].

It is emphasized in our book that the focus is on discrete time random processes because of their simplicity. While many of the basic ideas generalize, the details can become far more complicated, and much additional mathematical power becomes required. For example, the simple product sigma fields used here to generate process events are not sufficiently large to be useful. A simple integral of the process over a finite time window will not be measurable with respect to the resulting event spaces. Most of the added difficulties are technical—that is, the natural analogs to the discrete time results may hold, but the technical details of their proof can be far more complicated. Many excellent texts emphasizing continuous time random processes are available, but most require a solid foundation in functional analysis and in measure and integration theory. Perhaps the most famous and complete treatment is that of Doob [11]. Several of the references for second-order theory focus on continuous time random processes, as do Gikhman and Skorokhod [18], Hida [25], and McKean [36]. Lamperti [31] presents a clear summary of many facets of continuous time and discrete time random processes, including second-order theory, ergodic theorems, and prediction theory.

In chapter 8 we briefly sketched some basic ideas of Wiener and Kalman filters as an application of second-order theory. A detailed general development of the fundamentals and recent results in this area may be found in Kailath [28] and the references listed therein. In particular, the classic development of Wiener [52] is an excellent treatment of the fundamentals of Wiener filtering.

Of the menagerie of processes considered in chapters 9 and 10, most may be found in the various references already mentioned. The communication modulation examples may also be found in Gagliardi [17], among others. Compound Poisson processes are treated in detail in Parzen [42]. There is an extensive literature on Markov processes and their applications, as examples we cite Kemeny and Snell [29], Chung [8], Rosenblatt [43], and Dynkin [14].

Perhaps the most notable beast absent from our menagerie of processes is the class of Martingales. Had the book and the target class length been longer, Martingales would have been the next topic to be added. They were not included simply because we felt the current content already filled a semester, and we did not want to expand the book past that goal. An excellent mathematical treatment for the discrete time case may be found in Neveu [37], and a readable description of the applications of Martingale theory to gambling may be found in the classic by Dubins and Savage [13].

REFERENCES

1. R.B. Ash, *Real Analysis and Probability,* Academic Press, New York, 1972.

2. E. Asplund and L. Bungart, *A First Course in Integration,* Holt, Rinehart and Winston, New York, 1966.

3. P. Billingsley, *Ergodic Theory and Information,* Wiley, New York, 1965.

4. A.G. Bose and K.N. Stevens, *Introductory Network Theory,* Harper & Row, New York, 1965.

5. R. Bracewell, *The Fourier Transform and Its Applications,* McGraw-Hill, New York, 1965.

6. L. Breiman, *Probability,* Addison-Wesley, Menlo Park, Calif., 1968.

7. C.T. Chen, *Introduction to Linear System Theory,* Holt, Rinehart and Winston, New York, 1970.

8. K.L. Chung, *Markov Chains with Stationary Transition Probabilities,* Springer-Verlag, New York, 1967.

9. K.L. Chung, *A Course in Probability Theory,* Academic Press, New York, 1974.

10. H. Cramér and M.R. Leadbetter, *Stationary and Related Stochastic Processes,* Wiley, New York, 1967.

11. J.L. Doob, *Stochastic Processes,* Wiley, New York, 1953.

12. A.W. Drake, *Fundamentals of Applied Probability Theory,* McGraw-Hill, San Francisco, 1967.

13. L.E. Dubins and L.J. Savage, *Inequalities for Stochastic Processes: How to Gamble If You Must,* Dover, New York, 1976.

14. E.B. Dynkin, *Markov Processes,* Springer-Verlag, New York, 1965.

15. W. Feller, *An Introduction to Probability Theory and Its Applications,* 3rd ed., Wiley, New York, 1960.

16. T.L. Fine, *Theories of Probability,* Academic Press, New York, 1973.

17. R. Gagliardi, *Introduction to Communications Engineering,* Wiley, New York, 1978.

18. I.I. Gikhman and A.V. Skorokhod, *Introduction to the Theory of Random Processes,* Saunders, Philadelphia, 1965.

19. B.V. Gnedenko and A.Ya. Khinchin, *An Elementary Introduction to the Theory of Probability,* Dover, New York, 1962. Translated from the 5th Russian edition by L.F. Boron.

20. B.V. Gnedenko, *The Theory of Probability,* Chelsea, New York, 1963. Translated from the Russian by B.D. Seckler.

21. R.M. Gray and J.C. Kieffer, Asymptotically mean stationary measures, *Annals of Probability*, vol. 8, pp. 962–973, Chelsea, New York, Oct. 1980.

22. U. Grenander and M. Rosenblatt, *Statistical Analysis of Stationary Time Series,* 8, pp. 962–973, Wiley, New York, 1957.

23. P.R. Halmos, *Measure Theory,* 8, pp. 962–973, Van Nostrand Reinhold, New York, 1950.

24. P.R. Halmos, *Lectures on Ergodic Theory,* 8, pp. 962–973, Chelsea, New York, 1956.

25. T. Hida, *Stationary Stochastic Processes,* 8, pp. 962–973, Princeton University Press, Princeton, N.J., 1970.

26. D. Huff and I. Geis, *How to Take a Chance,* 8, pp. 962–973, W.W. Norton, New York, 1959.

27. T. Kailath, *Linear Systems,* 8, pp. 962–973, Prentice-Hall, Englewood Cliffs, N.J., 1980.

28. T. Kailath, *Lectures on Wiener and Kalman Filtering,* 8, pp. 962–973, Springer-Verlag, New York, 1981. CISM Courses and Lectures No. 140.

29. J.G. Kemeny and J.L. Snell, *Finite Markov Chains,* 8, pp. 962–973, D. Van Nostrand, Princeton, N.J., 1960.

30. A.N. Kolmogorov, *Foundations of the Theory of Probability,* 8, pp. 962–973, Chelsea, New York, 1950.

31. J. Lamperti, *Stochastic Processes: A Survey of the Mathematical Theory,* 8, pp. 962–973, Springer-Verlag, New York, 1977.

32. R.S. Liptser and A.N. Shiryayev, *Statistics of Random Processes,* 8, pp. 962–973, Springer-Verlag, New York, 1977. Translated by A.B. Aries.

33. M. Loeve, *Probability Theory,* 8, pp. 962–973, D. Van Nostrand, Princeton, N.J., 1963.

34. E. Lukacs, *Stochastic Convergence,* 8, pp. 962–973, Heath, Lexington, Mass., 1968.

35. L.E. Maistrov, *Probability Theory: A Historical Sketch,* 8, pp. 962–973, Academic Press, New York, 1974. Translated by S. Kotz.

36. H.P. McKean, Jr., *Stochastic Integrals,* 8, pp. 962–973, Academic Press, New York, 1969.

37. J. Neveu, *Discrete-Parameter Martingales,* 8, pp. 962 – 973, North-Holland, New York, 1975. Translated by T.P. Speed.

38. J.R. Newman, *The World of Mathematics,* 3, pp. 962 – 973, Simon & Schuster, New York, 1956.

39. A. Papoulis, *The Fourier Integral and Its Applications,* 3, pp. 962 – 973, McGraw-Hill, New York, 1962.

40. A. Papoulis, *Signal Analysis,* 3, pp. 962 – 973, McGraw-Hill, New York, 1977.

41. K.R. Parthasarathy, *Probability Measures on Metric Spaces,* 3, pp. 962 – 973, Academic Press, New York, 1967.

42. E. Parzen, *Stochastic Processes,* 3, pp. 962 – 973, Holden Day, San Francisco, 1962.

43. M. Rosenblatt, *Markov Processes: Structure and Asymptotic Behavior,* 3, pp. 962 – 973, Springer-Verlag, New York, 1971.

44. H.L. Royden, *Real Analysis,* 3, pp. 962 – 973, Macmillan, London, 1968.

45. Yu.A. Rozanov, *Stationary Random Processes,* 3, pp. 962 – 973, Holden Day, San Francisco, 1967. Translated by A. Feinstein

46. W. Rudin, *Principles of Mathematical Analysis,* 3, pp. 962 – 973, McGraw-Hill, New York, 1964.

47. G.F. Simmons, *Introduction to Topology and Modern Analysis,* 3, pp. 962 – 973, McGraw-Hill, New York, 1963.

48. K. Steiglitz, *An Introduction to Discrete Systems,* 3, pp. 962 – 973, Wiley, New York, 1974.

49. A.A. Sveshnikov, *Problems in Probability Theory, Mathematical Statistics, and Theory of Random Functions,* 3, pp. 962 – 973, Dover, New York, 1968.

50. P. Whittle, *Probability,* 3, pp. 962 – 973, Penguin Books, Middlesex, England, 1970.

51. N. Wiener, *The Fourier Integral and Certain of Its Applications,* 3, pp. 962 – 973, Cambridge University Press, New York, 1933.

52. N. Wiener, *Time Series: Extrapolation, Interpolation, and Smoothing of Stationary Time Series with Engineering Applications,* 3, pp. 962 – 973, M.I.T. Press, Cambridge, Mass., 1966.

53. E. Wong, *Introduction to Random Processes,* 3, pp. 962 – 973, Springer-Verlag, New York, 1983.

54. A.M. Yaglom, *An Introduction to the Theory of Stationary Random Functions,* 3, pp. 962 – 973, Prentice-Hall, Englewood Cliffs, N.J., 1962. Translated by R.A. Silverman.

INDEX

A

Additivity:
  countable, 47–50
  finite, 47
  uncountable, impossibility of, 48
Affine operation, 191
Alphabet, 76, 80, 99
ASCII, 9
Average, arithmetic, Cesaro mean,
    sample, or time, 67–69 128–29,
    143–45
  weighted, 69

B

Binary expansion, 103
Binomial theorem, 238, 272
Birkhoff-Khinchin Theorem, 177–78
Boolean algebra:
  analogy to set algebra, 15
Brownian motion (*See* Random objects)

C

Cartesian product (*See* Sets; Spaces)
Cauchy-Schwartz inequality, 154
Central limit theorem, 152, 278–82
Cesaro mean (*See* Average)
Channel, 264–65
Chernoff bound, 159
Compound processes (*See* Random
    objects)
Convergence (*See also* Markov,
    Tchebyshev inequality; Central
    limit theorem)
  in distribution, 151–52
  in mean square or in quadratic mean,
    145–49, 151–52
    defined, 145–46
  in probability, 145, 148–50, 152
    defined, 147
  with probability one or almost
    everywhere, 149–52
    defined, 149–50
  relation between types, 152. 157–59

S